Natural Rubber-Based Composites and Nanocomposites

Scrivener Publishing
100 Cummings Center, Suite 541J
Beverly, MA 01915-6106

Publishers at Scrivener
Martin Scrivener (martin@scrivenerpublishing.com)
Phillip Carmical (pcarmical@scrivenerpublishing.com)

Natural Rubber-Based Composites and Nanocomposites

Edited by

Visakh. P.M.
Natural Bioactive Materials Laboratory, Department of Bioengineering Ege University, Bornova/Izmir, Turkey

and

Shiv Kumari Panda
Udayanath Autonomous College of Science & Technology Adaspur, Cuttack, Odisha, India

Scrivener Publishing

WILEY

This edition first published 2025 by John Wiley & Sons, Inc., 111 River Street, Hoboken, NJ 07030, USA and Scrivener Publishing LLC, 100 Cummings Center, Suite 541J, Beverly, MA 01915, USA

For more information about Scrivener publications please visit www.scrivenerpublishing.com.

Wiley Global Headquarters
111 River Street, Hoboken, NJ 07030, USA

For details of our global editorial offices, customer services, and more information about Wiley products visit us at www.wiley.com.

The manufacturer's authorized representative according to the EU General Product Safety Regulation is Wiley-VCH GmbH, Boschstr. 12, 69469 Weinheim, Germany, e-mail: Product_Safety@wiley.com.

Library of Congress Cataloging-in-Publication Data

ISBN 978-1-394-21259-0

Front cover images courtesy of ChatGPT and Adobe Firefly
Cover design by Russell Richardson

Set in size of 11pt and Minion Pro by Manila Typesetting Company, Makati, Philippines

Printed in the USA

10 9 8 7 6 5 4 3 2 1

Contents

Preface

This volume summarizes recent research accomplishments in the area of natural rubber nanocomposites. It covers a wide range of topics, including the state-of-the-art in natural rubber composites and nanocomposites, emerging challenges and opportunities, natural rubber–bamboo composites for noise control, natural rubber-based bio-composites and bio-nanocomposites, mechanical, dynamic mechanical, electrical and thermal properties, biodegradation, radiative protective qualities, advances in recycling, and detailed descriptions of various natural rubber-based composites and nanocomposites.

This book will serve as a valuable reference for academics, researchers, scientists, rubber technologists, polymer scientists, postdoctoral fellows, and graduate students engaged in R&D. It is application-oriented and will help readers find solutions to both fundamental and applied problems. Chapters are authored by prominent researchers from industry, academia, and government or private laboratories worldwide, providing an up-to-date record of major findings and observations in this growing field.

Chapter 1 introduces the scope, state-of-the-art, preparation methods, challenges, and opportunities of natural rubber nanocomposites. Chapter 2 examines natural rubber–bamboo composites for noise control, covering preparation, characterization, and specific noise-reduction applications. Chapter 3 focuses on natural rubber-based bio-composites and bio-nanocomposites, including NR–biopolymer nanocomposites such as NR–chitin, NR–lignin, and NR–cellulose systems. Chapter 4 explores the mechanical, dynamic mechanical, electrical, and thermal properties of NR–carbon nanotube nanocomposites, discussing fabrication techniques, dispersion methods, effects of CNT and hybrid fillers, and applications.

Chapter 5 addresses the biodegradation of natural rubber composites and nanocomposites, exploring synthesis methods, chemical frameworks, mechanical and thermal stability, and factors influencing biodegradability. It also reviews novel nano-reinforced NR composites for engineering applications and proposes new analytical methods for biodegradation studies.

Chapter 6 investigates the radiative protective qualities of NR-based polymer composites, with sections on EMI shielding mechanisms and materials, as well as X- and gamma-ray shielding.

Chapter 7 discusses the use of green additives in eco-friendly NR compounds, including oils, lubrication mechanisms, plasticizers, fillers, and other additives. Chapter 8 examines advances, limitations, and applications in NR recycling, with special focus on devulcanization processes and the effects of processing parameters on final material properties. The final chapter revisits natural rubber-based composites and nanocomposites, emphasizing biodegradation processes for both conventional and nano-reinforced systems.

The completion of this volume was made possible through the commitment and expertise of its contributors, whose work has been essential to its success. We thank the reviewers for their critical input, as well as Martin Scrivener and the team at Scrivener Publishing for recognizing the growing importance of natural rubber-based composites and nanocomposites and for supporting this project in a field few publishers have undertaken.

Dr. Visakh. P. M.
August 2025

1

Natural Rubber Composites and Nanocomposites: State-of-the-Art, New Challenges, and Opportunities

Visakh. P.M.[1]* and Shiv Kumari Panda[2]

[1]*Natural Bioactive Materials Laboratory, Department of Bioengineering Ege University, Bornova/Izmir, Turkey*
[2]*Udayanath Autonomous College of Science & Technology Adaspur, Cuttack, Odisha, India*

Abstract
In this chapter, the we are providing a short versions of all the chapters related to different topics such as natural rubber-bamboo composites for noise control applications, natural rubber-based bio-composites and bio-nanocomposites, mechanical, dynamical/mechanical, electrical, and thermal properties of natural rubber-carbon nanotube nanocomposites and their applications, biodegradation of natural rubber and their composites and nanocomposites, radiative protective qualities of natural rubber-based polymer composites, the use of green additives in eco-friendly compounds based on natural rubber, natural rubber recycling: advances, limitations and applications, elastocaloric effect of natural rubber, and natural-rubber based composites and nanocomposites.

Keywords: Natural-rubber, composites, nanocomposites, elastocaloric, rubber-bamboo, eco-friendly compounds

1.1 Natural Rubber–Bamboo Composites for Noise Control Applications

Rubber can be produced in two ways: naturally from the latex of the *Hevea brasiliensis* tree, which is mostly grown in the Far East, or artificially by

**Corresponding author*: visagam143@gmail.com

Visakh. P.M. and Shiv Kumari Panda (eds.) Natural Rubber-Based Composites and Nanocomposites, (1–28) © 2025 Scrivener Publishing LLC

utilizing several processes on petroleum. The *H. brasiliensis* tree is cultivated in tropical parts of Southeast Asia, mostly in Malaysia and Indonesia. It produces latex, which is used to make natural rubber (NR), which is economically marketed [1]. The lengthy entwined coils that make up the natural rubber's polymer chains are always in motion at room temperature. Carbon black has been the most popular reinforcing filler in the rubber industry, particularly tires, for many years. Crude oil is the source of carbon black, which is a non-renewable resource that can eventually run out. As a result, a natural alternative source is required for carbon black in the rubber industry as a reinforcing agent without significantly altering the final products' mechanical and physical qualities. For damping applications, Praveen *et al.* [2] created a styrene butadiene rubber (SBR) compound that is loaded with carbon black (CB) and organo-modified nanoclay.

They discovered that when the filler was loaded, the mechanical characteristics and storage modulus increased significantly, leading to a low damping ratio. Yurkin *et al.* [3] investigated the vibration damping of three viscoelastic composites. The findings indicate that butyl rubber, EPDM, and EVA had damping ratios of 1.38, 0.6, and 0.42, respectively, suggesting that they could be used as viable options for damping applications. A filler is required to improve the mechanical, chemical, and physical qualities of elastomers when they are used in vibration damping for vehicles, railroads, and aircraft. Fillers have a significant impact on how rubber composites behave in a viscoelastic manner. NR completely satisfies the requirements for cut resistance, flex-fatigue resistance, and tensile qualities of final goods [4]. Except for rubber, all materials are specified as part per hundred rubber (PHR) or the quantity needed for 100 rubbers. A tire rubber compound should offer the best possible performance. For the best grip, the largest coefficient of friction must exist between the wheel and pavement. Simultaneously, the rolling resistance can be decreased to enhance fuel efficiency [5]. Materials with extremely small particle sizes are added to the RB mixture in powdered form as fillers or reinforcement additives [6]. Fillers are used to strengthen rubber, make it easier to process, and lower the cost and color of the material. Fillers can be either organic or inorganic and can be dry powders. Fillers are typically utilized to provide elastomers with the desired characteristics in commercial applications. The primary motivation for this is to lower costs and enhance compositional qualities. Fillers can be divided into two categories based on their color and impact. They can be separated and categorized into three types of fillers: semi-active, inactive, and active (reinforcing). Carbon blacks are known as black fillers, including silica, talc, calcium carbonate, talc, and zinc oxide, are known as white fillers [7].

In 2014, 32 million tons of wood and textile waste and 68 million tons of paper waste were produced in the United States, accounting for approximately 39% of the country's annual municipal solid waste burden [8]. The Central Pollution Control Board (CPCB) in New Delhi conducted research that indicated that the proportion of trash generated in India comprises components of several materials ranging between 3%–5% and 1%–5%. Overall, the unfavorable noise resulting from improvements in mechanical tools has proven to be a more complicated issue. Following this principle, waste products from the processing of fabrics and corn can be mixed with waste products from the processing of paper to produce a substance that absorbs sound and reduces noise pollution [9]. With time and living progress, noise contamination and mechanical oscillations have become inevitable problems in human existence. The need for a better living environment is growing, and people can no longer be exposed to mechanical vibrations and the noise they produce. Furthermore, mechanical vibration shortens machine life and reduces precision [10]. Global bamboo production is estimated to be 20 million tons (MTs) per year, with 10 MTs originating from China, Japan, and India [11]. Bamboo finds its use in a wide range of industries, including apparel, food, furniture, flooring, and interior design. Over the past few decades, there has been increased interest among researchers in the thermochemical conversion of waste bamboo into products rich in carbon and biochar. Researchers have developed a spectrum of damping materials using various methods. The two most common modification strategies are the addition of functional reinforcing fillers to the IIR matrix or their mixing with other fillers to generate composite damping materials. Nanofillers have a large specific surface area that allows them to penetrate deeply into the rubber matrix. Carbon nanotubes, talc, and carbon black are joint-strengthening fillers. To improve the mix of filler and rubber from an environmental perspective, Tang *et al.* [12] substituted waste recycled red brick powder with the conventional filler carbon black and silica in styrene-butadiene rubber (SBR). The production costs were reduced, while the tensile characteristics were improved.

Lignin is a possible reinforcing material for polymeric materials because of its high stiffness, which is attributed to its aromatic portion and intramolecular solid/intermolecular contacts [13]. The butyl rubber's qualities can be enhanced by curing IIR with phenolic resin modified by lignin. Butyl rubber is a hydrophilic substance that absorbs moisture during use because it contains polar functional groups. Owing to their low thermal stability, classic damping materials typically have a limited-service life. Therefore, the development of damping materials with good moisture resistance and thermal stability is crucial. Rubber is a common material found in various products. Several common rubber products that we encounter daily are rubber bands, shoes, and gloves. Because rubber products may return to their original shapes after being stretched or twisted, they are classified as elastomers. Rubber is an elastic substance that can be produced both naturally and artificially. Given that NR is an organic material, its characteristics are dictated by the biochemical mechanism by which the plant produces resin. Unlike synthetic rubber, NR polymerization cannot be changed. Natural rubber cannot be altered after it is removed from the plant. The modified forms of natural rubber that are considered significant are epoxidized natural rubber, poly (methyl methacrylate)-grafted natural rubber, hydro-halogenated natural rubber, cyclized natural rubber, depolymerized liquid natural rubber, resin-modified natural rubber, and poly (methyl methacrylate)-grafted natural rubber.

Mechanical oscillations and noise pollution have become unavoidable issues in human life as time and life have evolved. People are increasingly in need of improved living conditions, and they are unable to tolerate the noise and vibrations caused by machinery. Moreover, mechanical vibrations decrease the precision and shorten the life of the equipment. Sound-absorbing materials exhibit excellent dampening properties and a wide range of possible applications. Rubber-like materials offer excellent viscoelasticity, friction, and damping capabilities. These rubbers usually have a high relative molecular mass and strong molecular chain interactions, including polar, hydrogen, and ionic interactions. These RBs are used as dampers in various industries, including trains, aircraft, and cars. This study analyzed and highlighted natural bamboo rubber composite materials for successful noise and vibration control.

1.2 Natural Rubber-Based Bio-Composites and Bio-Nanocomposites

Sustainable materials, alternatively known as "green" or eco-friendly materials, are currently receiving a lot of attention and are characterized as inexpensive renewable resource materials that are locally produced without causing any kind of harm to living beings [14]. Bio-composites and bio-nanocomposites are not regarded as sustainable if one or more of their parent materials are regarded as eco-unfriendly or if their production process poses serious harm to the environment and human health [15]. Examples include composites that contain synthetic fillers such as carbon black (CB), silica, and carbon fiber, and/or synthetic polymers such as polyethylenes, polybutadienes, polypropylenes, polystyrenes, polyesters, butyls, nitriles, nylons, epoxies, vinyls, and acrylated, chlorinated, fluorinated, and sulfonated polymers [16]. In contrast to bio-composites, which are composed of natural fiber particles that are not on the nanoscale, bio-nanocomposites consist of particles with one, two, or three dimensions of less than 100 nm [17, 18]. Furthermore, composites are generally divided into different classes, with *class 1* comprising composites reinforced with a specific filler and *class 2* encompassing structural composites [19].

Bio-composites and bio-nanocomposites can also be prepared using natural rubber (NR), which is mainly harvested as latex from rubber trees (*H. brasiliensis*), as their base polymer. There are two types of natural rubber: *cis-1,4-polyisoprene* and *trans-1,4-polyisoprene*. The former is a typical bio-based elastomeric matrix for bio-composites and bio-nanocomposites. Moreover, although they are currently in the research and development stage and do not have many applications except in the automotive sector, these composites are expected to play a significant role in the future, particularly in building and construction, sports equipment, agriculture, aerospace, marine, and several other sectors that manufacture goods related to engineering [20–22]. The primary goal is to provide an overview of natural rubber-based bio-composites and bio-nanocomposites, with a particular emphasis on their preparation, properties, as well as long-established and possible applications. Despite being a type of natural fiber (particularly mineral fiber), asbestos is also associated with health concerns owing to its carcinogenic properties [23]. Nevertheless, a variety of plant and animal fibers as well as minerals (such as chalk, limestone, talc, silica (silicon dioxide), and montmorillonite) are still easily accessible. In contrast to the production of avian fibers and silk fibers by birds and chickens, as well as the domestic silk moth, other mammals' hair and fur are sources of cashmere, shearling, Persian, shahtoosh, angora, mohair, and bison wool [24, 25]. Among these, cashmere wool is the most popular owing to its comparatively low density and extremely plush and sumptuous texture [26]. Eggshells, seashells, and cuttlebones have recently been proposed as possible sustainable sources of animal fibers. Therefore, it is possible that the exploration of animal fibers for composites would considerably depend on the presence of protein-based constituents.

In contrast, polysaccharides, particularly crystalline cellulose, semicrystalline chitin, amorphous hemicellulose, and lignin, are the primary chemical components of plant fibers for composite use [27]. The literature indicates that the presence of cellulose, chitin, hemicellulose, or lignin in plant fibers is the key reason why they are substantially beneficial for composite application. Natural rubber (NR)-based bio-composites are an important example of rubber composites that have recently attracted considerable interest in both academia and industry, particularly in terms of their properties, performance, applications, and sustainability [28–31]. These composites, in which NR is mainly *cis-1,4-polyisoprene*, are generally referred to as "green" or eco-friendly materials, where a mechanically and/or chemically transformed natural fiber, serving as a reinforcing filler, is dispersed efficiently in a solid or latex NR matrix [32]. Both NR and natural fibers, which are derived from natural systems, contribute to the outstanding properties and high performance of NR-based bio-composites [33, 34]. Moreover, as various natural fibers and rubbers usually have varied properties, it is expected that the properties (and performances) of NR-based bio-composites will vary for different NR-based bio-composites. This is because the way the matrix transfers stress to the employed fiber determines the mechanical properties of the NR-based bio-composites. NR-based bio-composites comprising plant fibers have been investigated more frequently than those comprising animal fibers, minerals, or mineral fibers. In addition, solid NR has been investigated more for NR-based bio-composites than NR latex. As a result, additional research on the reinforcement of NR-based bio-composites using mineral and animal fiber-sourced fillers, as well as related studies with NR latex, is necessary. Currently, the second most researched bio-filler sources for NR-based bio-composites are palm oil and chitosan. Sugarcane bagasse (also known as megass or bagasse) is another source of key fibers that are widely used as bio-fillers worldwide, with nearly 23,000 and 48.84 lakh registered sugarcane growers in South Africa and India, respectively. The long-established and potential applications of NR-based bio-composites include soles, belts, straps, gloves, hoses, furniture, caskets, sports equipment, brush bristles, architectural elements, and point-of-sale displays, as well as various storage, packaging, and gadget materials [35].

Nonetheless, studies published thus far makes it abundantly evident that the automotive sector is comparatively the one that has made the greatest use of NR-based bio-composites [36]. For instance, the utilization of natural fibers has increased by 73% at Mercedes-Benz, in addition to approximately 10,000 tons of bio-composites that Bayerische Motoren Werke (BMW) utilized in 2004. The C-, S-, E-, and A-class cars of the 164/251 2004 model from Mercedes-Benz had their dashboard support, as well as the surface of the seats and backrests made of NR/coconut fiber bio-composites. Moreover, Shahril *et al.* [37] suggested the use of NR/Kenaf fiber bio-composites to create rubber mounts for automotive engines, whereas Patil *et al.* [38] suggested the use of NR/Coir bio-composites to manufacture insulations, mattresses, toys, and automobile tires. NR latex foam/chitin bio-composites, according to Zhang *et al.* [39], offer strong practical value for future applications and may be broadly utilized in home products, including pillows, beds, and cushions.

The development of natural-based bio-nanocomposites, incorporating nanometer-scale reinforcing bio-fillers such as nanocelluloses, shows their potential to impart improved properties at lower volume fractions compared to macro- or micro-sized fillers [40, 41]. It provides an excellent matrix for studying filler reinforcement effects, allowing the tailoring of properties through the addition of reinforcing fillers with diverse surface chemistries and aggregate size/aspect ratios [42, 43]. Bio-based polymer nanocomposites represent a novel class of sustainable materials derived from bio-renewable sources. There is a growing demand for reinforcing fillers sourced from renewable materials in the research field, owing to their biodegradability, superior mechanical properties, cost-effectiveness, environmental friendliness, biocompatibility, ease of processability, and low density. To develop sustainable polymer nanocomposites, bio-fillers such as chitin, cellulose, and lignin have garnered significant attention among scientists. As highlighted earlier, NR is a biopolymer, and the incorporation of bio-derived nanofillers into NR holds promise for producing environmentally friendly polymer nanocomposites with exceptional properties.

1.3 Mechanical, Dynamical/Mechanical, Electrical, and Thermal Properties of Natural Rubber-Carbon Nanotube Nanocomposites and their Applications

NR has attracted significant scientific and industrial interest owing to its excellent elasticity, flexibility, high tear, general fatigue resistance, low heat build-up, high resilience, retention of strength at elevated temperatures, and superior mechanical and dynamic properties [44]. Owing to its elasticity, resilience, and toughness, NR is used in transportation (e.g., tires), industrial (sealants), consumer (rubber boots, sports materials), hygienic, and medical sectors, and several others. NR was subjected to different modifications to form maleated natural rubber (MNR) and epoxidized natural rubber (ENR). NR in the vulcanization process in the presence of sulfur, an accelerator, and other compounding ingredients, and exhibits excellent mechanical properties owing to the strain-induced crystallization phenomenon [45]. The exceptionally high tensile strength compared to other carbon materials makes them suitable for developing high-performance composite materials for their applications as efficient vibration and shock absorbers [46], biomedical applications [47, 48], tires for cars, buses, and even airplanes [49]; automotive parts, such as bearings, seals, o-rings, oil filter gaskets, manifold gasket brake pads, airbags, drive couplings, springs and window seals, and toys [50]. Mensah *et al.* [51] observed the important role of morphology, dispersion, and functionalization of CNTs in the preparation of NR/CNT nanocomposites in achieving the desired mechanical properties. Thomas *et al.* [52] studied the effect of 1-octadecanol functionalization on acid-treated carbon nanotubes. Subsequently, NR/CNT composites were prepared by adding sonicated CNT (in toluene) to rubber already dissolved in toluene, followed by compounding in a two-roll mill. These findings showed better dispersion of functionalized CNT in the natural rubber matrix compared to the non-functionalized CNT in the NR matrix, as evident from the TEM images. Conductive nanocomposites of NR/CNT by solvent-cast suspensions followed by vulcanization [53] and solution mixing using an ultrasonicator [54] also produced a homogeneous dispersion of CNT.

Investigations have also been conducted to study the effects of *in situ* functionalization of MWCNT with bis(triethoxysilylpropyl) tetrasulfide (TESPT) and 3-aminopropyltriethoxysilane (APTES) as silane coupling agents on the properties of the corresponding ENR/CNT composites. Anand and his co-workers [55] used SDS to enhance the dispersion of SWCNTs in water and this stabilized aqueous solution of SWCNTs was used in the preparation of NR nanocomposites through the latex compounding method. In addition, it resulted in a 4% enhancement in the 300% modulus and a 5% decrease in the Akron abrasion loss compared with those of the NR/CNT composites. The observed enhancement in the mechanical properties could be attributed to the dispersion of CNT in NR and the bonding between the WPRP and the NR matrix. The tensile strength and tensile modulus of the NR/SWNT (2 phr) nanocomposite prepared by latex compounding through surfactant-assisted sonication were significantly enhanced compared to those of NR. Guo *et al.* [56] improved the dispersion of CNT in NR by using waterjet-produced rubber powder (WPRPP) as a carrier. The resulting composite comprising 5 phr of WPRP exhibited an approximately 3% oncrease in the tensile strength. Mohamed *et al.* [57] measured the mechanical properties of MWCNTs in NR latex nanocomposites following the enhanced dispersion of surfactants bearing phenyl groups. Rubber nanocomposites were prepared by incorporating CNT in addition to carbon black into the NR, which significantly improved the tensile strength, hardness, and tear strength of the NR system up to a threshold CNT content of 2 phr [58]. A possible reason for this is the homogeneous dispersion and effective interaction between the CNT and the NR. Wang *et al.* [59] fabricated CNTs@CNC (cellulose nanocrystal)/NR nanocomposites with a 3D hierarchical conductive network following a latex assembly approach.

Jahed *et al.* [60] reported mechanical properties of NR/ethylene propylene diene monomer (EPDM)/ MWCNT composites were prepared in an internal and a two roll-mill mixer in two steps. They observed that the tensile strength and modulus at 300% of NR/EPDM/MWCNT nanocomposites were enhanced by the filler content, functionalized MWCNT, and compatibilizer; however, the elongation at break decreased. Abdullateef *et al.* [61] observed a reduction in the volume resistivity of CNT incorporated in the NR matrix (~5 Ω·m). In contrast, the surface-modified CNT-based NR composite showed an increase in resistivity (~13 Ω·m). In another study, Nakaramontri *et al.* [62] investigated the electrical conductivity of epoxidized NR filled with 1 phr of CNT (1 phr), CNT functionalized with bis(triethoxysilylpropyl) tetrasulfide (CNT-TESPT), and 3-aminopropyltriethoxysilane (CNT-APTES) prepared by *in situ* functionalization. Aguilar-Bolados *et al.* [63] measured the electrical properties of hybrid nanocomposites comprising a combination of few-layer thermally reduced graphite oxide (TRGO) and multiwall carbon nanotubes (MWCNTs) mixed with NR. They observed an increase in the electrical conductivity of the polymer by approximately 13 orders of magnitude for a combination of 2 wt.% TRGOs and 2 wt.% MWCNTs hybrid filler. This is ascribed to the can be attributed to the formation of a highly interconnected percolation network. Nanocomposites have made major contributions in the field of flexible and stretchable electronics such as sensing, aerospace, non-metallic flexible capacitors, and robotic arms. It is anticipated that these carbon-nanotube-reinforced natural rubber nanocomposites may have major contributions in the aerospace sectors, robotic arms, sensors, nonmetallic flexible capacitors, and many other fields. Natarajan *et al.* [64] investigated the sensing performance of natural rubber nanocomposites filled with MWCNT, CB, and their combinations operating from 50% to 150% strain depending on the ratios of the two-filler system. They concluded that these elastic materials could be used in structural health monitoring and sensors in different dynamic elastomeric parts such as tires, valves, gaskets, and engine mounts. In another study, piezoresistive CNT-filled rubber-based composites were used to determine the strain sensing application and main mechanisms underlying the conductivity of NR/CNT composites at different strains [65].

1.4 Biodegradation of Natural Rubber and their Composites and Nanocomposites

Aerobic biodegradation uses oxygen in chemical reactions that break down degrading substances [66–69]. The end product of the degradation process is carbon dioxide or methane. Anaerobic digestion, on the other hand, is a series of processes in which microorganisms break down biodegradable materials in the absence of oxygen. Anaerobic biodegradation is widely used to treat wastewater sludge and biodegradable waste because it provides volume and mass reduction of the input material [70–73]. Recent research has focused on improving the properties of biopolymer composites, including starch-based materials, cellulose-derived polymers, bacterial polyesters, and natural rubber composites [74]. The blending process, which involves the addition of biofillers (such as agricultural waste, cellulose, and ash), can be used to modify the properties of biopolymer composites. Biodegradable polymers have a wide range of applications in electronics, including in dielectric material, electronic packaging, insulators, and (semi) conductors [75]. Diverse biodegradation environments, such as soil, aquatic settings, landfills, and compost, present unique conditions, and host distinct microorganisms. For instance, fungi assume a pivotal role in soil, predominantly facilitating the degradation of organic matter, including polymers. In aquatic environments, two types of bacteria, surface and sediment bacteria, dominate, and their concentration diminishes as depth increases [76].

Research on bio-nanofillers made from biomass has accelerated because of the need for fillers other than petroleum-based fillers. Bio-based nanocomposites occasionally exhibit special benefits over conventional inorganic nanoparticles [77]. NR are frequently used in cellulose whiskers, chitosan, starch nanocrystals, and other bio-nanofillers. The goal of these filler investigations is to produce them at a production cost that is comparable with other petroleum-based fillers while maintaining equivalent qualities [78]. Furthermore, it is anticipated that the bio-based fillers' biocompatibility and biodegradability will be preserved after their dispersion in the NR matrix [79]. While cellulose nanocrystals are 100% crystalline, starch nanocrystals contain approximately 45% crystalline content. To produce starch nanocrystals from waxy maize starch, starch granules were mixed with a known concentration of H_2SO_4 solution and stirred at a specific speed with orbital shaking. Following various durations of hydrolysis, the resulting suspension was washed with distilled water until neutral pH was achieved [80]. Angellier *et al.* [81] synthesized nanocrystal platelets from waxy maize starch, characterized by dimensions of 6–8 nm thickness, 40–60 nm length, and 15–30 nm width. These starch nanocrystals, present in an aggregated form, exhibit an average size of 4.4 mm. This study also examined the properties of unvulcanized natural rubber (NR) reinforced with these nanocrystals. The resulting films were conditioned at 0% relative humidity (RH).

The elongation at break of the NR filled with starch nanocrystals decreased with increasing nanocrystal content, dropping from 1960% (unfilled) to 920% (30 wt.%). Conversely, the tensile modulus experiences almost exponential growth with starch content, rising from 0.64 MPa for the unfilled NR matrix to 77.8 MPa for composite films filled with 30 wt.% of starch nanocrystals. However, at higher filler contents (e.g., 20 wt.%), there is a reduction in tensile modulus when exposed to 43% RH compared to the same loading at 0% RH. Furthermore, the elongation at break decreases to 1,500% for filler loadings of 5–20 wt.% at 43% RH. This deterioration in the properties of starch nanocrystal-filled NR at a high moisture content or RH is attributed to the hydrophilic nature of the starch nanocrystals. According to Ismail *et al.* [82], who studied the incorporation of chitosan as a new type of filler in rubber compounds, one of the most important requirements for chitosan to be successfully employed as a filler is ensuring good interaction with rubber at a reasonable loading level, with the loading level being the amount or concentration of chitosan added to the composite. The hydrophobicity of rubber and the hydrophilicity of chitosan were cited as reasons why the rubber–filler interaction weakened with the addition of chitosan and further weakened with higher chitosan loading.

1.5 Radiative Protective Qualities of Natural Rubber-Based Polymer Composites

Natural Rubber (NR) is known for its flexibility and good resistance to extreme temperatures. As such, NR is not recommended as EMI shielding because of its poor conductivity and SE value. However, composites of natural rubber with some metals as fillers can enhance the conductivity and SE of the NR composite materials. Many researchers worldwide are conducting experiments to develop better EMI shielding materials using NR as part of their composite matrix. Zhan *et al.* [83] attempted to enhance the EMI shielding properties of natural-rubber-based supercritical CO_2 foams. The authors have developed an EMI Shielding material comprising NR/

CNT composites. In addition, the authors concluded that the addition of CNT to the NR matrix enhanced the EMI Shielding effectiveness and electrical conductivity of the NR/CNT composites. Wang *et al.* [84] developed $Ti_3C_2T_x$ MXene/NR composite to have a better EMI Shielding material. In the study, authors modified the structure of the matrix from brick-mortar to honeycomb structure by the help of vulcanization reaction. It is observed that the SE value of the composite matrix with a honeycomb structure is 63.5 dB when compared to that of 57.1 dB. Jeevakumari *et al.* [85] investigated the role of conductive graphene nano platelet and cobalt nano wire in the natural rubber matrix in enhancing the mechanical property of the composite material.

This proves that, the addition of graphene nano platelet and cobalt nano wire to NR matrix enhances the SE and mechanical properties of shielding material. Rashid *et al.* [86] studied the EMI shielding effect by adding PANI (polyaniline) and NiFe (nickel ferrites nanoparticles) to the natural rubber. The authors attempted to prepare three combinations (NBR/PANI, NBR/NiFe, and NBR/PANI/NiFe) of the composite to understand the effect of additives on the NBR matrix. The electromagnetic interference shielding effectiveness of Natural Rubber (NR) and Ethylene Vinyl Acetate (EVA) with conductive short carbon fibers (SCF) and carbon black was studied by Das *et al.* [87]. In this study, an attempt was made to compare the shielding effectiveness (SE) of two different composites. Kanking *et al.* [88] have developed the NR polymer composite in which bagasse fiber ash (BFA) was used as filler along with carbon black or precipitated silica. The authors attempted to replace carbon black or precipitated silica in polymer composites. However, the study shows that the mechanical properties of both BFA:CB/NR and BFA:Psi/NR composites are reduced by the addition of BFA to the polymer matrix. Ninyong *et al.* [89] have used lead as filler particles in natural rubber (NR) based polymer composites. The study revealed no significant improvements in the curing characteristics upon adding lead to the natural rubber matrix as lead does not interact chemically with the rubber matrix, whereas wood/natural rubber composites were found to be better than natural rubber composites.

1.6 The Use of Green Additives in Eco-Friendly Compounds Based on Natural Rubber

In a study by Dasgupta *et al.* [90], natural rubber (NR) compounds containing various vegetable oils (soybean oil, castor oil, etc.) were compared to compositions of aromatic, naphthenic, and paraffinic oils with low levels of polycyclic aromatic compounds (PCA). The addition of these oils affected the action of the vulcanization system, resulting in low ML, MH, and optimum vulcanization time (t_{90}) in relation to the compounds containing conventional petroleum-derived oils. Da Costa *et al.* [91] conducted a study on castor oil used as an activator in the process of NR vulcanization using compositions containing silica. They observed that neat castor oil did not show a significant improvement in vulcanization characteristics. However, when castor oil was combined with stearic acid, better performance was achieved. This effect was attributed mainly to the castor oil lubricant, which promoted an improvement in the vulcanization characteristics. Gujel *et al.* [92, 93] studied the effect of the incorporation of soybean oil in ethylene-propylene diene monomer rubber (EPDM). The results showed that vulcanization properties were affected, and there was a reduction in vulcanization and, consequently, a reduction in the processing time. Another conclusion was that the incorporation of 10 phr of soybean oil (MD600) reduced the amount of lubricant oil required by approximately 60%. Jayewardhana *et al.* [94] analyzed the reason for a better dispersion of fillers such as carbon black on NR matrix containing vegetable oils.

In this case, the interactions between the long chains of unsaturated fatty acid esters of vegetable oil with the polar groups present on the surface of carbon black (phenol, carboxyls, and lactones) and the non-polar interaction of these oils with the molecules of NR provided such a result. Some additives that are used are unsafe to human health and may cause illnesses. NR latex is widely used as medical items like gloves, catheters, among others. Dithiocarbamate accelerators, commonly used in NR latex, are likely to cause allergies, which can be avoided by changing the sulfur vulcanization by radiation vulcanization [95–97] or by peroxide curing. Concerning the environmental problems related to zinc, water pollution by the zinc contained in rubber formulations mainly occurs indirectly because the disposal of waste in the soil (mainly due to the natural wear of tires) is of great concern to the environment because of the contamination of surface water bodies or groundwater. Human health, aquatic fauna, and flora are polluted by inorganic pollutant residues. Organisms may also be sensitive to toxic actions, such as bioaccumulation, affecting the food chain and endangering the organisms located at the top of the chain [98, 99].

1.7 Natural Rubber Recycling: Advances, Limitations, and Applications

The application of granulated vulcanized elastomer in equestrian soils [100] has been shown to be promising. The granules, incorporated into other materials and sand, provide horse cushioning during the race and impulse in jumps, due to the elasticity and flexibility of the material, so muscle and joint problems can be minimized in the animals. Unlike pure sand, the compound does not produce such an amount of dust during the exercise of the horses, in addition to avoiding excessive wear of horseshoes. The same can also be applied to the comfort of other large animals such as cattle. Similarly, athletes of different sports have demonstrated the advantages of recycled rubber in multi-sport courts because of characteristics such as elasticity, shock absorption, drainage, and high strength [101]. Rattanasom *et al.* [102] studied the vulcanization characteristics and mechanical properties of polymeric blends raw NR/recycled NR (rubber powder). The results indicated that the scorch time and optimum vulcanization decreased with an increase in the incorporated residue content. Regarding the mechanical properties, an increase in the modulus at 100% and 300% elongation was observed because of the increase in elastomeric stiffness resulting from the increase in the residue concentration. In rubbers such as NR, SBR, and EPDM, this capacity is low and can be improved by the incorporation of another material in the formulation with a higher microwave absorption capacity. Carbon black, a filler widely used in rubber goods such as tires, can be used as a conductive filler [103, 104], which induces a phenomenon known as Maxwell–Wagner polarization [105]. According to De *et al.* [106], a large number of chemical reclaiming agents such as diphenyl disulfide, dibenzyl disulfide, diamyl disulfide, bis(alkoxy-aryl) disulfides, butyl mercaptans, thiophenols, xylene thiols, other mercaptans, phenol sulfides and disulfides, iron oxide phenyl hydrazine-based catalysts, and copper chloride-tributyl amine catalyst have been used for the treatment of scrap ground rubber crumbs or powders at elevated temperatures. Desulfurization of ground rubber is performed by bacteria and fungi, which can convert cross-linked sulfur to simple sulfur compounds or inorganic sulfate [107]. This method is considered eco-friendly because it does not use chemicals or energy. On the other hand, some disadvantages are that the process is slow, only superficial devulcanization/desulfurization occurs, and some additives may inhibit the function of bacteria [108] and reduce the culture responsible for the process due to toxicity.

1.8 The Elastocaloric Effect of Natural Rubber

Refrigeration technology is considered as one of the most transformative engineering achievements of the 20th century [109, 110]. Despite its enormous social benefits, its use in various applications incurs significant environmental costs [111]. For this reason, there has been growing interest in developing alternative refrigeration technologies over the last few decades [112]. Natural rubber (NR), a promising elastocaloric material, is a natural polymer comprising isoprene as its main component. NR has a high strength [113, 114] and resistance to crack expansion [115–117]. It is environmentally-friendly, recyclable, low cost and non-toxic [118–120]. Its most notable physical property is its low modulus of elasticity [121, 122]. Moreover, it condenses under normal ambient conditions, whereas it is rarely studied at low temperatures. Sebald *et al.* [123] investigated the effect of fatigue on the thermal effects of NR elasticity. The engineering strain cycles at different amplitudes were investigated experimentally, and it was observed that the number of natural rubber fracture cycles in the experiment was limited (approximately 800 cycles). This indicates an extremely short fatigue life. The experiments mainly investigated the fatigue dependence of the elasto-thermal effect in three small-amplitude strain states with strain amplitudes of 3 for strains 0–3, 2–5, and 4–7.

British physicist Gough [124] first discovered the elastocaloric effect of NR in the early 19th century and proposed related concepts. Gough's theory was confirmed 50 years later by Joule [125].

Gerlach *et al.* [126] proposed that a heat pump or refrigeration system operating with rubber as the elastic thermal mass is divided into four paths: adiabatic stretching, isothermal stretching, adiabatic shrinkage, isothermal shrinkage. Tianying [127] independently constructed a cyclic stretching experimental bench. The elastocaloric effects of different rubber materials were investigated in experiments where the Mullins effect was eliminated. The lab bench consisted of two main parts. The first part was an FLS40 series screw linear module with a controller, drive, and DC power supply. The second part is a self-designed fixture fixed on the module, which is composed of an aluminum plate that is tightened by nuts to fix the rubber. Farris [128] designed a small rubber fiber engine. Confirmation of the high-power potential of these engines when using modern elastane fiber designs. A specific cooling power of 1 W/g was applied at a temperature difference of 30 K between the heat source and heat sink. This demonstrates the high-power potential of engines using elastane fibers. Lyon *et al.* [129] performed thermomechanical cycling experiments on two polyurethane elastomers at different strains and temperature differences. The two elastic fibers used in the experiments were commercially available polycondensate yarns. Zhongjian *et al.* [130] investigated the effect of pre-stretching on the thermal effect of NR elasticity. Compared to deformation without pre-stretching, pre-stretching can skip the low elastic heat coefficient region and move directly to the high elastic heat coefficient region induced by the SIC. In particular, the elastic thermal coefficient reaches its maximum value when pre-stretching is applied before SIC. The elastic thermal coefficient decreased again when pre-stretching was applied to the SIC. In addition, the stress input is maximum when pre-stretching is applied at the beginning of SIC.

Aprea *et al.* [131] studied the application of thermal effects in heat pumps: application of acetoxy silicone rubber and vulcanized natural rubber to active thermal storage heat pump systems. A comparison between electrothermal, piezothermal, and magnetothermal materials was performed. The system consists of a cold heat exchanger (in contact with the outdoor environment), hot heat exchanger (in contact with the indoor environment), parallel-plate heat accumulator made of thermophysical material, and auxiliary fluid (water).

1.9 Natural Rubber-Based Composites and Nanocomposites

Natural Rubber as a polymer matrix for applicational level composites are important because of its wide applications, but it generates vast amounts of waste. At the same time, special consideration is given to NR composites made with natural fillers, such as cellulose and starch. Natural fillers are particularly significant in the realm of environmentally friendly composite applications owing to their widespread availability, low cost, and straightforward manufacturing processes [132]. The abiotic degradation of NR involves hydrolysis of rubber by water. Hydrolysis in rubber refers to the chemical breakdown of the polymer and the resulting physical breakdown or crumbling of rubber. This process was accelerated by warmth and high humidity. NR can undergo hydrolysis in water under certain conditions. One methods for degrading NR by hydrolysis is hydrous pyrolysis [133]. The biodegradation of NR is influenced by several factors, including the properties of the polymer (such as molecular weight, crystallinity, presence of functional groups, substituents, or additives), the specific microorganisms involved, and any pretreatment applied to the polymer [134].

NR degrades slowly in nature because of certain microorganisms, and natural environments contain a few bacteria and fungi that can degrade NR as a carbon and energy source. The degradation of rubber by fungi was initially examined by De Vries through the utilization of several strains of Penicillium and Aspergillus in a liquid medium consisting of 10% (wt/vol) NaCl. The results revealed a 6% increase in biomass and a weight reduction of 15.5% to 30.9% in the rubber material over an incubation period ranging from 19 months to 5 years [135].

However, in 2013, an investigation into the isolation of two fungal species, Penicillium and *Aspergillus niger*, degraded the NR sample by 20% and 25.9% of its weight loss, respectively, over a two-month period [136]. Many researchers have verified the substantial potential of natural fillers as alternative reinforcements, due to their advantages including low cost, low density, eco-environmentally friendly, biodegradability, and large availability in nature, in addition to their ability to improve the mechanical properties of composite materials [137–139].

NR composites consist of an NR matrix and various fillers or reinforcements such as carbon black, silica, clay, cellulose, and starch. The addition of these fillers can improve the mechanical, thermal, and electrical properties of NR, and reduce its costs. However, the presence of fillers can also affect the biodegradability of NR composites, depending on their nature, amount, and interaction with the rubber matrix. A biodegradable NR composite is formed by blending NR with other natural or biodegradable materials, resulting in an environmentally friendly composite that can naturally decay or disintegrate over time by microorganisms, such as bacteria and fungi [140]. Biodegradable fillers can include natural fibers, such as like cellulose, jute, hemp, or sisal, as well as other biodegradable materials, such as starch-based components, chitosan, or other plant-derived materials. The biodegradation of NR composites is an important process that can contribute to sustainable and eco-friendly rubber product development and is a potential solution for reducing the environmental impact of the disposal of rubber waste. Recently, a study produced biodegradable composite sponges of NR and cellulose fibers (NR-C) using the Dunlop process has been reported [141]. Microcellulose fibers from eucalyptus pulp were used as reinforcement fillers, while sodium alginate was utilized as a dispersing agent to improve the compatibility between cellulose and NR. The NR-C sponges exhibited a higher compressive stress, modulus of elasticity, and efficient water absorption capacity. Starch is a natural polymer fiber that is classified as a carbohydrate and is composed of repeating glucose units. Starch is abundant in nearly all plants because of photosynthesis, and serves as the primary energy storage compound. Its abundant supply, renewable nature, affordability, and capacity for decomposition make it a valuable resource to produce bioplastics. Starch contains an integral protein that resides in its granular structure. This protein synthesizes a complex carbohydrate chain consisting of amylose and amylopectin, which is consumed by soil bacteria to generate an enzyme that breaks down the matrix, similar to the NR latex backbone chain.

Another NR/starch-based biodegradation study reported biopolymer sheets comprising a combination of maleated epoxidized natural rubber (MENR) and natural rubber-g-cassava starch (NR-g-CSt) [142]. There is an abundance of biodegradable NR rubber research utilizing natural fibers [68]; for example, sisal/oil palm hybrid fiber that also reinforced NR composites and their biodegradability and aging were studied by Jacon *et al.* [143]. The degradation of the composites was found to be dependent on the fiber content and chemical treatment. Degradation of the composites was discovered owing to the fiber content and chemical treatment.

References

1. Li, C., Zhang, W., Sun, G., Li, Y., Zhang, Y., Research on damping properties and microscopic mechanism of polyurethane cement-based composites. *Constr. Build. Mater.*, *365*, 130137, 2023, https://doi.org/10.1016/j.conbuildmat.2022.130137.
2. Praveen, S., Bahadur, J., Yadav, R., Billa, S., Umasankar Patro, T., Rath, S.K., Ratna, D., Patri, M., Tunable viscoelastic and vibration damping properties of a segmented polyurethane synergistically reinforced with carbon black and anisotropic additives. *Appl. Acoust.*, *170*, 107535, 2020, https://doi.org/10.1016/j.apacoust.2020.107535.
3. Yurkin, Y.V., Mansurova, I.A., Belozerov, V.S., Zlobina, A., Morphological and Dynamic Mechanical Analysis of Vibration Damping Composite Material Based on Different Elastomers. *Mater. Plast.*, *55*, 4, 469–473, 2018, https://doi.org/10.37358/MP.18.4.5055.
4. Sarkar, P. and Bhowmick, A.K., Sustainable rubbers and rubber additives. *J. Appl. Polym. Sci.*, *135*, 24, 45701, 2018, https://doi.org/10.1002/app.45701.
5. Zhao, W., He, J., Yu, P., Jiang, X., Zhang, L., Recent progress in the rubber antioxidants: A review. *Polym. Degrad. Stab.*, *207*, 110223, 2023, https://doi.org/10.1016/j.polymdegradstab.2022.110223.
6. Miyajima, A., Sakemi-Hoshikawa, K., Usami, M., Mitsunaga, K., Irie, T., Ohno, Y., Sunouchi, M., Thyrotoxic rubber antioxidants, 2-mercaptobenzimidazole and its methyl derivatives, cause both inhibition and induction of drug-metabolizing activity in rat liver microsomes after repeated oral

administration. *Biochem. Biophys. Res. Commun.*, *492*, 1, 116–120, 2017, https://doi.org/10.1016/j.bbrc.2017.08.024.
7. Fang, M., Wang, B., Qu, X., Li, S., Huang, J., Li, J., Lu, S., Zhou, N., State-of-the-art of biomass-derived carbon dots: Preparation, properties, and applications. *Chin. Chem. Lett.*, *35*, 1, 108423, 2024, https://doi.org/10.1016/j.cclet.2023.108423.
8. Wang, B., Waterhouse, G.I.N., Lu, S., Carbon dots: mysterious past, vibrant present, and expansive future. *Trends Chem.*, *5*, 1, 76–87, 2023, https://doi.org/10.1016/j.trechm.2022.10.005.
9. Wu, S., Qiu, M., Guo, B., Zhang, L., Lvov, Y., Nanodot-Loaded Clay Nanotubes as Green and Sustained Radical Scavengers for Elastomer. *ACS Sustain. Chem. Eng.*, *5*, 2, 1775–1783, 2017, https://doi.org/10.1021/acssuschemeng.6b02523.
10. Arroyo, M., López-Manchado, M.A., Herrero, B., Organo-montmorillonite as substitute of carbon black in natural rubber compounds. *Polymer*, *44*, 8, 2447–2453, 2003, https://doi.org/10.1016/S0032-3861(03)00090-9.
11. Rathour, R., Kumar, H., Prasad, K., Anerao, P., Kumar, M., Kapley, A., Pandey, A., Kumar Awasthi, M., Singh, L., Multifunctional applications of bamboo crop beyond environmental management: an Indian prospective. *Bioengineered*, *13*, 4, 8893–8914, 2022, https://doi.org/10.1080/21655979.2022.2056689.
12. Tang, D., Zhang, X., Hu, S., Liu, X., Ren, X., Hu, J., Feng, Y., The reuse of red brick powder as a filler in styrene-butadiene rubber. *J. Clean. Prod.*, *261*, 120966, 2020, https://doi.org/10.1016/j.jclepro.2020.120966.
13. Jong, L., Peterson, S.C., Jackson, M.A., Utilization of Porous Carbons Derived from Coconut Shell and Wood in Natural Rubber. *J. Polym. Environ.*, *22*, 3, 289–297, 2014, https://doi.org/10.1007/s10924-013-0637-4.
14. Elfaleh, I., Abbassi, F., Habibi, M., Ahmad, F., Guedri, M., Nasri, M., Garnier, C., A comprehensive review of natural fibers and their composites: An eco-friendly alternative to conventional materials. *Results Eng.*, 19, 101271, 1–31, 2023.
15. Szczepanowska, H.M. and Akhmedov, N.G., Gutta percha, natural rubber and balata–chemical characterization of polyisoprenes in the context of cultural heritage. *Heritage Sci.*, 11, 125, 1–18, 2023.
16. Pongdong, W., Nakason, C., Kummerlöwe, C., Vennemann, N., Influence of filler from a renewable resource and silane coupling agent on the properties of epoxidized natural rubber vulcanizates. *J. Chem.*, 1–15, 2015.
17. Jawaid, M.H.P.S. and Khalil, H.A., Cellulosic/synthetic fibre reinforced polymer hybrid composites: a review. *Carbohydr. Polym.*, 86, 1, 1–18, 2011.
18. Karine, C., Jean-Paul, J., Moussa, G., Christophe, B., Laurent, B., Joël, B., Morphology and mechanical behavior of a natural composite: The flax fiber. *16th International Conference on Composite Materials, Materials Science, Engineering*, Japan, pp. 1–8, 2007.

19. Dénes, T.-O., Istoan, R., Tămaş-Gavrea, D.R., Manea, D.L., Hegyi, A., Popa, F., Vasile, O., Analysis of sheep wool-based composites for building insulation. *Polymers*, 14, 2109, 1–25, 2022.
20. Zhang, L., The application of composite fiber materials in sports equipment. *5th International Conference on Education, Management, Information and Medicine*, Atlantis Press, pp. 450–453, 2015.
21. Brüning, K., Natural rubber, in: *Encyclopedia of polymeric nanomaterials*, S. Kobayashi, and K. Müllen, (Eds.), pp. 1377–1382, Springer, Berlin, Heidelberg, 2015.
22. Cruz, J. and Fangueiro, R., Surface modification of natural fibers: a review. *Procedia Eng.*, 15, 285–288, 2016.
23. Mohanty, A.K., Misra, M., Drzal, L.T., Selke, S.E., Harte, B.R., Hinrichsen, G., Natural fibers, biopolymers, and biocomposites: An introduction, in: *Natural fibers, biopolymers, and biocomposites*, A.K. Mohanty, M. Misra, L.T. Drzal, (Eds.), pp. 1–35, CRC Press, Boca Raton, 2005.
24. Das, P.P., Chaudhary, V., Ahmad, F., Manral, A., Gupta, S., Gupta, P., Acoustic performance of natural fiber reinforced polymer composites: influencing factors, future scope, challenges, and applications. *Polym. Compos.*, 43, 3, 1221–37, 2022.
25. Perrella, F.W. and Gaspari, A.A., Natural rubber latex protein reduction with an emphasis on enzyme treatment. *Methods*, 27, 77–86, 2002.
26. Ibrahim, S., Riahi, O., Said, S.M., Sabri, M.F.M., Rozali, S., Biopolymers from crop plants, in: *Reference Module in Materials Science and Materials Engineering*, pp. 123–5 Elsevier, 2019, https://doi.org/10.1016/B978-0-12-803581-8.
27. Payungwong, N., Sakdapipanich, J., Wu, J., Ho, C., The interplay of protein hydrolysis and ammonia in the stability of *Hevea* rubber latex during storage. *Polymers*, 1, 4636, 1–15, 2023.
28. Maslowski, M., Miedzianowska, J., Czylkowska, A., Strzelec, K., Horsetail (*Equisetum Arvense*) as a functional filler for natural rubber biocomposites. *Materials*, 13, 1–20, 2020.
29. Mustafa, M.S., Syed Ismail, S.N., Mohd Shaipul Amini, S., Abdul Wahab, N.M., Majid, N.A., Sarip, M.N., Pineapple leaf fibre filled natural rubber composite: The effect of filler loading, in: *Charting the sustainable future of ASEAN in science and technology*, N. Alias, and R. Yusof, (Eds.), pp. 447–454, Singapore, Springer, 2020.
30. Osabohien, E., Soya, U., Ojeifo, N.O., Characteristics of natural rubber–carbonized pawpaw seed composites. *Int. J. Biol. Chem. Sci.*, 14, 8, 451–464, 2020.
31. Da Costa, H.M., Visconte, L.L.Y., Nunes, R.C.R., Furtado, C.R.G., Mechanical and dynamic mechanical properties of rice husk ash–filled natural rubber compounds. *J. Appl. Polym. Sci.*, 83, 2331–2346, 2002.
32. Mohamed, W.Z.W., Baharum, A., Ahmad, I., Abdullah, I., Zakaria, N.E., Mengkuang fiber reinforced thermoplastic natural rubber composites:

Influence of rubber content on mechanical properties and morphology. *Malays. J. Anal. Sci.*, 22, 5, 906–913, 2018.
33. Miedzianowska, J., Masłowski, M., Rybínski, P., Strzelec, K., Modified nanoclays/straw fillers as functional additives of natural rubber biocomposites. *Polymers*, 13, 799, 1–30, 2021.
34. Mitsaichon, V., Ruaesrijan, K., Phinyocheep, P., Amornsakchai, T., Junkong, P., Thanawan, S., Preparation of spent coffee grounds-rubber composites using natural rubber latex as binder. *J. Met. Mater. Miner.*, 32, 4, 161–166, 2022.
35. Zhou, Y., Fan, M., Chen, L., Zhuang, J., Lignocellulosic fibre mediated rubber composites: An overview. *Composites, Part B*, 76, 180–191, 2015.
36. Wazeer, A., Das, A., Abeykoon, C., Sinha, A., Karmakar, A., Composites for electric vehicles and automotive sector: A review. *Green Energy Intell. Transp.*, 2, 100043, 1–23, 2023.
37. Shahril, D.A., Azammi, A.M.N., Zulfadli, S.M., Reinforcement of kenaf fiber in natural rubber composite for automotive engine rubber mounting. *Int. J. Appl. Eng. Res.*, 12, 24, 14490–14494, 2017.
38. Patil, Y. and Sharma, S., Effect of coir fiber reinforcement on mechanical properties of vulcanized natural rubber composites. *Sci. Eng. Compos. Mater.*, 25, 3, 517–528, 2018.
39. Zhang, N. and Cao, H., Enhancement of the antibacterial activity of natural rubber latex foam by blending it with chitin. *J. Materials*, 13, 1039, 1–15, 2020.
40. Khalil, H.A., Davoudpour, Y., Islam, M.N., Mustapha, A., Sudesh, K., Dungani, R., Jawaid, M., Production and modification of nanofibrillated cellulose using various mechanical processes: a review. *Carbohydr. Polym.*, 99, 649–65, 2014.
41. Qing, Y., Sabo, R., Zhu, J.Y., Agarwal, U., Cai, Z., Wu, Y., A comparative study of cellulose nanofibrils disintegrated *via* multiple processing approaches. *Carbohydr. Polym.*, 97, 1, 226–34, 2013.
42. Mariano, M., El Kissi, N., Dufresne, A., Cellulose nanocrystal reinforced oxidized natural rubber nanocomposites. *Carbohydr. Polym.*, 137, 174–83, 2016.
43. Rippel, M.M. and Galembeck, F., Nanostructures and adhesion in natural rubber: new era for a classic. *J. Braz. Chem. Soc.*, 20, 1024–30, 2009.
44. Srivastava, S.K., *Natural Rubber Materials, in Composites and Nanocomposites*, vol. 2, S. Thomas, C. Han Chan, L. Pothen, J. Joy, H. Maria, S. Thomas, H. Maria (Eds.), pp. 504–549, The Royal Society of Chemistry, UK, 2013.
45. Candau, N., Laghmach, R., Chazeau, L., Chenal, J.-M., Gauthier, C., Biben, T., Munch, E., Strain induced crystallization and melting of natural rubber during dynamic cycles. *Macromolecules*, 47, 5815–5824, 2014.
46. MMMM, K.L., Ramlow, H., Floriano, J.F., Acosta, E.D., Faita, F.L., Machado, R.A.F., Latex and natural rubber: recent advances for biomedical applications. *Polímeros*, 32, e2022015, 2022.

47. Davies, B., Natural rubber—Its engineering characteristics. *Mater. Des.*, 7, 68–74, 1986.
48. Guerra, N.B., Pegorin, G.S., Boratto, M.H., R de Barros, N., Graeff, C.F.O., Herculano, R.D., Biomedical applications of natural rubber latex from the rubber tree Hevea brasiliensism. *Mater. Sci. Eng., C*, 126, 112126, 2021.
49. Kohjiya, S. and Ikeda, Y., *Chemistry, Manufacture and Applications of Natural Rubber*, 2nd Edition, Elsevier, USA, 2021.
50. Srivastava, S.K. and Bhuyan, B., Rubber Nanocomposites for Tire Tread Application, in: *Rubber Nanocomposites and Nanotextiles in Automobiles*, B. Banerjee (Ed.), pp. 31–74, Rapra, India, 2019.
51. Mensah, B., Kim, H.G., Lee, J.-H., Arepalli, S., Nah, C., Carbon nanotube-reinforced elastomeric nanocomposites. *Int. J. Smart Nano Mater.*, 6, 211–238, 2015.
52. Selvin Thomas, P., Abdullateef, A.A., Al-Harthi, M.A., Atieh, M.A., De, S.K., Rahaman, M., Chaki, T.K., Khastgir, D., Bandyopadhyay, S., Electrical properties of natural rubber nanocomposites: Effect of 1-octadecanol functionalization of carbon nanotubes. *J. Mater. Sci.*, 47, 3344–3349, 2012.
53. Capezza, A., Andersson, R.L., Strom, V., Wu, Q., Sacchi, B., Farris, S., Hedenqvist, M.S., Olsson, R.T., Protein Nanofibrils and Their Hydrogel Formation with Metal Ions. *ACS Omega, 4*, 3458–3468, 2019.
54. Kang, B.-H., Jeong, I.-Y., Park, S.-H., Synthesis and properties of ABA-triblock copolymers from polyester A-blocks and easily degradable polyacetal B-blocksPolymers. *Polymers*, 14, 2551–2562, 2022.
55. Anand, K.S., Jose, T.R., Alex, R.J., Natural rubber-carbon nanotube composites through latex compounding. *Int. J. Polym. Mater.*, 59, 33–44, 2009.
56. Guo, X., Guo, S., Liu, G., Bai, L., Liu, H., Xu, Y., Zhao, J., Chai, H., Jian, X., Guo, L., Liu, F., Improving dispersion of carbon nanotubes in natural rubber by using waterjet-produced rubber powder as a carrier. *Polymers*, 15, 477, 2023.
57. Mohamed, A., Anas, A.K., Bakar, S.A., Ardyani, T., Manshol W. Zin, W., Ibrahim, S., Sagisaka, M., Brown, P., Eastoe, J., Enhanced dispersion of multiwall carbon nanotubes in natural rubber latex nanocomposites by surfactants bearing phenyl groups. *J. Colloid Interface Sci.*, 455, 179–187, 2015.
58. Jawahar, P., Payak, V., Chandradass, J., Prabhu, P., Optimization of mechanical properties of CNT-rubber nanocomposites. *Mater. Today Proc.*, 45, 7183–7189, 2021.
59. Wang, S., Zhang, X., Wu, X., Lu, C., Tailoring percolating conductive networks of natural rubber composites for flexible strain sensors *via* a cellulose nanocrystal templated assembly. *Soft Matter*, 12, 845–852, 2016.
60. Jahed, M., Naderi, G., Ghoreishy, M.H.R., Microstructure, mechanical, and rheological properties of natural rubber/ethylene propylene diene monomer nanocomposites reinforced by multi-wall carbon nanotubes. *Polym. Compos.*, 39, E745–E753, 2018, DOI: 10.1002/pc.24153.

61. Abdullateef, A.A., Thomas, S.P., Al-Harthi, M.A., De, S.K., Bandyopadhyay, S., Basfar, A.A., Atieh, M.A., Natural rubber nanocomposites with functionalized carbon nanotubes: Mechanical, dynamic mechanical, and morphology studies. *J. Appl. Polym. Sci.*, 125, E76–E84, 2012.
62. Nakaramontri, Y., Nakason, C., Kummerlowe, C., Vennemann, N., Effects of *in-situ* functionalization of carbon nanotubes with bis(triethoxysilylpropyl) tetrasulfide (TESPT) and 3-aminopropyltriethoxysilane (APTES) on properties of epoxidized natural rubber–carbon nanotube composites. *Polym. Eng. Sci.*, 55, 2500–2510, 2015, doi.org/10.1002/pen.24140.
63. Aguilar-Bolados, H., Yazdani-Pedram, M., Contreras-Cid, A., Lopez-Manchado, M.A., May-Pat, A., Avilés, F., Influence of the morphology of carbon nanostructures on the piezoresistivity of hybrid natural rubber nanocomposites. *Compos. B Eng.*, 109, 147–154, 2017.
64. Natarajan, T.S., Bhagavatheswaran, E.S., Stöckelhuber, K.W., Wieβner, S., Potschke, P., Heinrich, G., Das, A., Strong Strain Sensing Performance of Natural Rubber Nanocomposites. *ACS Appl. Mater. Interfaces*, 9, 4860–4872, 2017.
65. Selvan, N.T., Eshwaran, S.B., Das, A., Stöckelhuber, K.W., Wießner, S., Pötschke, P., Nando, G.B., Chervanyov, A.I., Heinrich, G., Piezoresistive natural rubber-multiwall carbon nanotube nanocomposite for sensor applications. *Sens. Actuators, A*, 239, 102–113, 2016.
66. Song, M., Yue, X., Wang, X., Huang, M., Ma, M., Pan, W., Qin, Q., Improved high-temperature damping performance of nitrile-butadiene rubber/phenolic resin composites by introducing different hindered amine molecules. *E-Polymers, 20*, 1, 482–490, 2020, https://doi.org/10.1515/epoly-2020-0054.
67. Chen, J., Wang, J., Li, S., Xiang, K., Song, S., Pyridine terminated polyurethane dendrimer/chlorinated butyl rubber nanocomposites with excellent mechanical and damping properties. *Chin. J. Chem. Eng.*, *53*, 211–221, 2023, https://doi.org/10.1016/j.cjche.2022.02.013.
68. Tang, K. and Wang, J., Chlorinated butyl rubber/two-step modified montmorillonite nanocomposites: Mechanical and damping properties. *Chin. J. Chem. Eng.*, *42*, 437–449, 2022, https://doi.org/10.1016/j.cjche.2021.02.001.
69. Park, K.H., Fujiwara, Y., Mazda, T., Kajita, Y., Evaluation of mechanical properties considering hysteresis characteristic of high damping rubber bearing. *J. Phys. Conf. Ser.*, *1687*, 1, 012019, 2020, https://doi.org/10.1088/1742-6596/1687/1/012019.
70. Jiang, C., Bo, J., Xiao, X., Zhang, S., Wang, Z., Yan, G., Wu, Y., Wong, C., He, H., Converting waste lignin into nano-biochar as a renewable substitute of carbon black for reinforcing styrene-butadiene rubber. *Waste Manage. (Oxford)*, *102*, 732–742, 2020, https://doi.org/10.1016/j.wasman.2019.11.019.
71. Qu, C., Li, S., Zhang, Y., Wang, T., Wang, Q., Chen, S., Surface modification of Ti3C2-MXene with polydopamine and amino silane for high performance nitrile butadiene rubber composites. *Tribol. Int.*, *163*, 107150, 2021, https://doi.org/10.1016/j.triboint.2021.107150.

72. Şen, M., Soydaş, O., Aksüt, D., Preparation and characterization of high damping poly(epichlorohydrin) and poly(epichlorohydrin-co-ethylene oxide-co-allyl glycidyl ether) elastomers: I. Effect of curing system and blending on the damping properties. *React. Funct. Polym.*, *184*, 105520, 2023, https://doi.org/10.1016/j.reactfunctpolym.2023.105520.
73. Xiaolin, J., Min, X., Minhui, W., Yuanhao, M., Wencong, Z., Yanan, Z., Haoxiang, R., Xun, L., Preparation and molecular dynamics study of polyurethane damping elastomer containing dynamic disulfide bond and multiple hydrogen bond. *Eur. Polym. J.*, *162*, 110893, 2022, https://doi.org/10.1016/j.eurpolymj.2021.110893.
74. Zhu, J., Zhao, X., Liu, L., Yang, R., Song, M., Wu, S., Thermodynamic analyses of the hydrogen bond dissociation reaction and their effects on damping and compatibility capacities of polar small molecule/nitrile-butadiene rubber systems: Molecular simulation and experimental study. *Polymer*, *155*, 152–167, 2018, https://doi.org/10.1016/j.polymer.2018.09.040.
75. Gu, X., Wang, Y., Lai, C., Qiu, J., Li, S., Hou, Y., Martens, W., Mahmood, N., Zhang, S., Microporous bamboo biochar for lithium-sulfur batteries. *Nano Res.*, *8*, 1, 129–139, 2015, https://doi.org/10.1007/s12274-014-0601-1.
76. Hodzic, A., Re-use, recycling and degradation of composites, in: *Green Composites*, pp. 252–271, Elsevier, UK, 2004, https://doi.org/10.1016/B978-1-85573-739-6.50015-4.
77. Yin, C., Zhao, X., Zhu, J., Hu, H., Song, M., Wu, S., Experimental and molecular dynamics simulation study on the damping mechanism of C5 petroleum resin/chlorinated butyl rubber composites. *J. Mater. Sci.*, *54*, 5, 3960–3974, 2019, https://doi.org/10.1007/s10853-018-3134-2.
78. Lu, X. and Li, X., Broad temperature and frequency range damping materials based on epoxidized natural rubber. *J. Elastomers Plast.*, *46*, 1, 84–95, 2014, https://doi.org/10.1177/0095244312462161.
79. Doddamani, S., Kulkarni, S.M., Joladarashi, S., Gurjar, A.K., Mohan Kumar, T.S., Development of rubber-sand composite for enhanced impact resistance: Implications of vulcanization. *Eng. Sci. Technol. Int. J.*, 51, 101631, 2024, https://doi.org/10.1016/j.jestch.2024.101631.
80. Noguchi, T., Bamba, Y., Odaka, I., Iwamoto, R., Endo, M., Isogai, A., Natural rubber composites with high strength, modulus, water-resistance, and thermal stability, prepared with cellulose nanofibrils and sodium methacrylate. *Composites, Part A*, 173, 107665, 2023, https://doi.org/10.1016/j.compositesa.2023.107665.
81. Angellier, H., Molina-Boisseau, S., Dufresne, A., Mechanical Properties of Waxy Maize Starch Nanocrystal Reinforced Natural Rubber. *Macromolecules*, *38*, 22, 9161–9170, 2005, https://doi.org/10.1021/ma0512399.
82. Ismail, H., Shaari, S.M., Othman, N., The effect of chitosan loading on the curing characteristics, mechanical and morphological properties of chitosan-filled natural rubber (NR), epoxidised natural rubber (ENR) and

styrene butadiene rubber (SBR) compounds. *Polym. Test.*, *30*, 7, 784–790, 2011, https://doi.org/10.1016/j.polymertesting.2011.07.003.

83. Zhan, Y., Oliviero, M., Wang, J., Sorrentino, A., Buonocore, G.G., Sorrentino, L., Lavorgna, M., Xia, H., Iannace, S., Enhancing the EMI shielding of natural rubber-based supercritical CO_2 foams by exploiting their porous morphology and CNT segregated networks. *Nanoscale*, 11, 3, 1011–20, 2019.
84. Wang, Y., Liu, R., Zhang, J., Miao, M., Feng, X., Vulcanization of $Ti_3C_2T_x$ MXene/natural rubber composite films for enhanced electromagnetic interference shielding. *Appl. Surf. Sci.*, 546, 149143, 2021.
85. Amutha Jeevakumari, S.A., Mothilal, T., Ramesh, G., Arun Prakash, V.R., Mechanically toughened EMI shielding natural rubber composite in microwave frequency bands. *Plast. Rubber Compos.*, 50, 5, 241–8, 2021.
86. Rashid, I.A., Hamza, A., Asim, S., Zubair, K., Shakir, H.F., Afzal, A., Zubair, Z., Achieving enhanced electromagnetic shielding by novel flexible rubber based nanocomposite with the incorporation of nickel spinal ferrites and polyaniline. *Synth. Met.*, 295, 117339, 2023.
87. Das, N.C., Khastgir, D., Chaki, T.K., Chakraborty, A., Electromagnetic interference shielding effectiveness of carbon black and carbon fibre filled EVA and NR based composites. *Composites, Part A*, 31, 10, 1069–81, 2000.
88. Kanking, S., Niltui, P., Wimolmala, E., Sombatsompop, N., Use of bagasse fiber ash as secondary filler in silica or carbon black filled natural rubber compound. *Mater. Des.*, 41, 74–82, 2012.
89. Ninyong, K., Wimolmala, E., Sombatsompop, N., Saenboonruang, K., Properties of natural rubber (NR) and wood/NR composites as gamma shielding materials. *IOP Conf. Ser.: Mater. Sci. Eng.*, 526, 1, 012038, 2019 Jun, IOP Publishing.
90. Dasgupta, S., Agrawal, S.L., Bandyopadhyay, R., Mukhopadhyay, S., MalkaniI, R.K., Ameta, S.C., Characterization of eco-friendly processing aids for rubber compounds: Part II. *Polym. Test.*, 27, 277–238, 2008.
91. da Costa, H.M., Ramos, V.D., Furtado, C.R.G., Efeito do óleo de mamona em composições de borracha natural contendo sílica. *Polimeros-Ciencia E Tecnologia*, 14, 46–50, 2004.
92. Gujel, A.A., Bandeira, M., Giovanela, M., Carli, L.N., Brandalise, R.N., Crespo, J.S., Development of bus body rubber profiles with additives from renewable sources: Part II - Chemical, physical-mechanical and aging characterization of elastomeric compositions. *Mater. Des.*, 53, 1119–1123, 2014.
93. Gujel, A.A., Bandeira, M., Veiga, V.D., Giovanela, M., Carli, L.N., Mauler, R.S., *et al.*, Development of bus body rubber profiles with additives from renewable sources: Part I - Additives characterization and processing and cure properties of elastomeric compositions. *Mater. Des.*, 53, 1112–1118, 2014.
94. Jayewardhana, W.G.D., Perera, G.M., Edirisinghe, D.G., Karunanayake, L., Study on natural oils as alternative processing aids and activators in carbon black filled natural rubber. *J. Natl. Sci. Found. Sri Lanka*, 37, 187–193, 2009.

95. Chirinos, H., Yoshii, F., Makuuchi, K., Lugao, A., Radiation vulcanization of natural rubber latex using 250 keV electron beam machine. *Nucl. Instrum. Methods Phys. Res., Sect. B*, 208, 256–259, 2003.
96. Haque, M.E., Dafader, N.C., Akhtar, F., Ahmad, M.U., Radiation dose required for the vulcanization of natural rubber latex. *Radiat. Phys. Chem.*, 48, 505–510, 1996.
97. Hill, D.J.T., O'Donnell, J.H., Perera, M.C.S., Pomery, P.J., Smetsers, P., Mechanism of radiation vulcanization of natural rubber latex sensitized by monoacrylates. *J. Appl. Polym. Sci.*, 57, 1155–1171, 1995.
98. Gualtieri, M., Andrioletti, M., Vismara, C., Milani, M., Camatini, M., Toxicity of tire debris leachates. *Environ. Int.*, 31, 723–730, 2005.
99. Nelson, S.M., Mueller, G., Hemphill, D.C., Identification of tire leachate toxicants and a risk assessment of water quality effects using tire reefs in canals. *Bull. Environ. Contam. Toxicol.*, 52, 574–581, 1994.
100. Aliapur, https://www.aliapur.fr/fr/les-applications/applications/sol-equestre (accessed May 27, 2017).
101. Ecopneus, http://www.ecopneus.it/en/spazio-sensoriale/sensory-space-the-project.html (accessed May 27, 2017).
102. Rattanasom, N., Poonsuk, A., Makmoon, T., Effect of curing system on the mechanical properties and heat aging resistance of natural rubber/tire tread reclaimed rubber blends. *Polym. Test.*, 24, 728–732, 2005.
103. Sun, X., Isayev, A.I., Joshi, T.R., von Meerwall, E., Molecular mobility of unfilled and carbon-black-filled isoprene rubber: Proton NMR transverse relaxation and diffusion. *Rubber Chem. Technol.*, 80, 854–872, 2007.
104. Amari, T., Themelis, N.J., Wernick, I.K., Resource recovery from used rubber tires. *Resour. Policy*, 25, 179–188, 1999.
105. Menendez, J.A., Arenillas, A., Fidalgo, B., Fernandez, Y., Zubizarreta, L., Calvo, E.G., *et al.*, Microwave heating processes involving carbon materials. *Fuel Process. Technol.*, 91, 1–8, 2010.
106. De, D., Maiti, S., Adhikari, B., Reclaiming of rubber by a renewable resource material (RRM). II. Comparative evaluation of reclaiming process of NR vulcanizate by RRM and diallyl disulfide. *J. Appl. Polym. Sci.*, 73, 2951–2958, 1999.
107. Jiang, G.M., Zhao, S.H., Li, W.J., Luo, J.Y., Wang, Y.Q., Zhou, Q.S., *et al.*, Microbial desulfurization of SBR ground rubber by Sphingomonas sp. and its utilization as filler in NR compounds. *Polym. Adv. Technol.*, 22, 2344–2351, 2011.
108. Karger-Kocsis, J., Meszaros, L., Barany, T., Ground tyre rubber (GTR) in thermoplastics, thermosets, and rubbers. *J. Mater. Sci.*, 48, 1–38, 2013.
109. Zhu, L., Fiorino, A., Thompson, D., *et al.*, Near-field photonic cooling through control of the chemical potential of photons. *Nature*, 566, 7743, 239–44, 2019.
110. Mancin, S., Zilio, C., Rossetto, L., *et al.*, Heat transfer performance of aluminum foams. *J. Heat Transfer*, 133, 6, 060904–060913, 2011.

111. McLinden, M.O., Seeton, C.J., Pearson, A., New refrigerants and system configurations for vapor-compression refrigeration. *Science*, 370, 6518, 791–6, 2020.
112. Mischenko, A.S., Zhang, Q., Scott, J.F., *et al.*, Giant electrocaloric effect in thin-film PbZr(0.95)Ti(0.05)O3. *Science*, 311, 5765, 1270–1, 2006.
113. Zhang, H., Niemczura, J., Dennis, G., *et al.*, Toughening effect of strain-induced crystallites in natural rubber. *Phys. Rev. Lett.*, 102, 24, 245503, 2009.
114. Cadwell, S., Merrill, R., Sloman, C., *et al.*, Dynamic fatigue life of rubber. *Rubber Chem. Technol.*, 13, 2, 304–15, 1940.
115. Katz, J., Röntgenspektrographische Untersuchungen am gedehnten Kautschuk und ihre mögliche Bedeutung für das Problem der Dehnungseigenschaften dieser Substanz. *Naturwissenschaften*, 13, 19, 410–6, 1925.
116. Albouy, P.-A., Guillier, G., Petermann, D., *et al.*, A stroboscopic X-ray apparatus for the study of the kinetics of strain-induced crystallization in natural rubber. *Polymer*, 53, 15, 3313–24, 2012.
117. Toki, S., Che, J., Rong, L., *et al.*, Entanglements and networks to strain-induced crystallization and stress–strain relations in natural rubber and synthetic polyisoprene at various temperatures. *Macromolecules*, 46, 13, 5238–48, 2013.
118. Trabelsi, S., Albouy, P.-A., Rault, J., Stress-induced crystallization around a crack tip in natural rubber. *Macromolecules*, 35, 27, 10054–61, 2002.
119. Candau, N., Laghmach, R., Chazeau, L., *et al.*, Strain-induced crystallization of natural rubber and cross-link densities heterogeneities. *Macromolecules*, 47, 16, 5815–24, 2014.
120. Cui, J., Wu, Y., Muehlbauer, J., *et al.*, Demonstration of high efficiency elastocaloric cooling with large ΔT using NiTi wires. *Appl. Phys. Lett.*, 101, 7, 073904, 2012.
121. LAW, J.Y., Franco, V., Moreno-Ramírez, L.M., *et al.*, A quantitative criterion for determining the order of magnetic phase transitions using the magnetocaloric effect. *Nat. Commun.*, 9, 1, 1–9, 2018.
122. Disalvo, F.J., Thermoelectric cooling and power generation. *Science*, 285, 5428, 703–6, 1999.
123. Sebald, G., Xie, Z., Guyomar, D., Fatigue effect of elastocaloric properties in natural rubber. *Philos. Trans. R. Soc. A Math. Phys. Eng. Sci.*, 374, 2074, 20150302, 2016.
124. Gough, J., A description of a property of Caoutchouc, or Indian rubber. *Memories Literacy Philos. Soc. Manchester*, 1, 288–95, 1805.
125. Joule, J.P.V., On some thermo-dynamic properties of solids. *Philos. Trans. R. Soc. London*, 149, 91–131, 1859.
126. Gerlach, D.W., Alvarado, J.L., Mina, E.M., *et al.*, Analysis Of Elastomer Refrigeration Cycles, purdue university.
127. Li, T.Y., Study on the elasto-thermal effect of different rubber materials, Tianjin University of Commerce, 2019, (In China).

128. Farris, R.J., Rubber Heat Engines, Analyses and Theory. *Rubber Chem. Technol.*, 52, 1, 159–72, 1979.
129. Lyon, R.E., Wang, D.X., Farris, R.J., *et al.*, Polyurethane–urea elastomers as working substances in rubber heat engines. *J. Appl. Polym. Sci.*, 29, 9, 2857–72, 1984.
130. Xie, Z., Sebald, G., Guyomar, D., Elastocaloric effect dependence on pre-elongation in natural rubber. *Appl. Phys. Lett.*, 107, 8, 980, 2015.
131. Hoffmeister, K.M., Josefsson, E.C., Isaac, N.A., *et al.*, Glycosylation restores survival of chilled blood platelets. *Science*, 301, 5639, 1531–4, 2003.
132. Thomas, S.K., Parameswaranpillai, J., Krishnasamy, S., Begum, P.M.S., Nandi, D., Siengchin, S., George, J.J., Hameed, N., Salim, N.V., Sienkiewicz, N., A comprehensive review on cellulose, chitin, and starch as fillers in natural rubber biocomposites. *Carbohydr. Polym. Technol. Appl.*, 2, 100095, 2021.
133. Ahmad, N., Abnisa, F., Daud, W.M.A.W., Potential use of natural rubber to produce liquid fuels using hydrous pyrolysis – a review. *RSC Adv.*, 6, 73, 68906–68921, 2016.
134. Ali Shah, A., Hasan, F., Shah, Z., Kanwal, N., Zeb, S., Biodegradation of natural and synthetic rubbers: A review. *Int. Biodeterior. Biodegrad.*, 83, 145–157, 2013.
135. De Vries, O., Zersetzung von Kautschuk-Kohlenwasserstoff durch Pilze. *Zentbl. Bakteriol. Parasitenkd. Infektionskr.*, 74, 22–24, 1928.
136. Nayanashree, G. and Thippeswamy, B., Natural rubber degradation by Aspergillus niger and Penicillium sp. *Int. J. Recent Sci. Res.*, 4, 9, 1337–1341, 2013.
137. Zhang, C., Dan, Y., Peng, J., Turng, L.-S., Sabo, R., Clemons, C., Thermal and Mechanical Properties of Natural Rubber Composites Reinforced with Cellulose Nanocrystals from Southern Pine. *Adv. Polym. Tech.,* 33, S1, 2014.
138. Kumar C M, P., B, A.R., Kumar, M., R, C.P., Natural nano-fillers materials for the Bio-composites: A review. *J. Indian Chem. Soc.*, 99, 10, 100715, 2022.
139. Sienkiewicz, N., Dominic, M., Parameswaranpillai, J., Natural Fillers as Potential Modifying Agents for Epoxy Composition: A Review. *J. Polymers*, 14, 2, 265, 2022.
140. Bosco, F. and Mollea, C., Biodegradation of Natural Rubber: Microcosm Study. *Water Air Soil Pollut.*, 232, 6, 227, 2021.
141. Suppanucroa, N., Nimpaiboon, A., Boonchuay, K., Khamkeaw, A., Phisalaphong, M., Green composite sponge of natural rubber reinforced with cellulose filer using alginate as a dispersing agent. *J. Mater. Res. Technol.*, 27, 3119–3130, 2023.
142. Riyajan, S.-A., Robust and biodegradable polymer of cassava starch and modified natural rubber. *Carbohydr. Polym.*, 134, 267–277, 2015.
143. Jacob, M., Thomas, S., Varughese, K.T., Biodegradability and aging studies of hybrid biofiber reinforced natural rubber biocomposites. *J. Biobased Mater. Bioenergy*, 1, 1, 118–126, 2007.

2

Natural Rubber-Bamboo Composites for Noise Control Applications

Samuel Eshorame Sanni[1*], Denen Ashiekaa Vershima[1], Rotimi Emmanuel Sadiku[2], Babalola Aisosa Oni[3] and Emeka Emmanuel Okoro[4]

[1]Department of Chemical Engineering, Covenant University, Ota, Ogun State, Nigeria
[2]Department of Chemical, Metallurgical and Materials Engineering, Tshwane University of Technology, Pretoria, South Africa
[3]Department of Chemical Engineering, North Dakorta University, Grand Forks, United States
[4]Department of Petroleum and Gas, University of Port Harcourt, Choba, Rivers State, Nigeria

Abstract

Noise pollution and vibrations have become very important. This impacts the working efficiency of industrial workers and is very harmful to them. Several methods have been employed to address these concerns, but limitations such as cost and efficiency have hindered the implementation of these techniques. Rubber is a substance that can be obtained *via* natural processes, and it serves various purposes in the everyday lives of industrialists and workers. Research has been conducted on the utilization of these materials to control and mitigate noise in the workplace. The use of reinforcements such as fillers and waste products has been explored to boost the mechanical, chemical, thermal, and hydrostability of the product material to assist in this process. This study examines various existing techniques for handling and controlling noise concerns in the workplace. This study highlights the various synthetic methods for rubber, the characterization methods applied to analyze the properties of natural rubber, and the application of natural rubber-bamboo composites for noise control applications and concludes with potential improvements in these methods to enhance this process.

**Corresponding author*: adexz3000@yahoo.com; samuel.sanni@covenantuniversity.edu.ng

Visakh. P.M. and Shiv Kumari Panda (eds.) Natural Rubber-Based Composites and Nanocomposites, (29–62) © 2025 Scrivener Publishing LLC

Keywords: Rubber, natural rubber, noise control, natural rubber fillers, hydrogenated dicyclopentadiene, molding

2.1 Introduction

Over the past few years, significant advancements have been made in reducing waves and vibrations in automobiles and manufacturing equipment. Idealy, the application of methods such as viscoelastic vibration damping (VVD) can lessen the vibration caused by noise from structures. VVDM components convert force into thermal energy [1]. It is noteworthy that the best device for damping applications has important characteristics [2]. Moreover, the demand for automated and silent equipment in numerous sectors has resulted in a dramatic increase in the demand for visco elastic damping materials (VVDM). To meet the growing global demand, researchers have developed viscoelastic damping materials with a variety of matrices, fillers, and additives. In addition, the VVDM are divided into three classes according to their matrix materials: (1) water-based (resin) damping materials, (2) asphalt, and (3) rubber-based [3]. When rubber is used as their matrix, Rubber-based (RB) damping components such as graphite and mica are used as fillers. Scientists have used several types of rubber, along with commercial fillers and additives, to create a variety of vibration-dampening materials. Rubber can be produced organically from the milky fluid of numerous crops or chemically by mixing petroleum and alcohol. Rubber is a crucial technical material that has a broad range of applications [4].

The term "rubber" is now used to describe any substance, regardless of chemical makeup, that has qualities similar to those of natural rubber, rather than just the original natural rubber [5–7]. Synthetic polymers that have characteristics akin to rubber are referred to as "elastomers" in more recent times. Rubbers have several applications. It is present in a wide range of everyday products, particularly those manufactured in the automotive industry. The tires for automobiles are manufactured using approximately 70% of the rubber produced. Among the important qualities that tires are supposed to have include resistance to tire fatigue and rips. These qualities were provided by natural rubber. Because natural rubber cannot supply the necessary qualities desired in tires on its own, synthetic rubber polymers are used in combination with it to produce tires. The chemistry and physics of artificial rubber materials influence the rolling resistance, abrasion resistance, and traction of each tire component as well as the tire's overall performance. Additionally, synthetic rubbers increase the internal permeability of tires [8].

Carbon black has been the most famous reinforcing material in rubber manufacturing, especially tires, for many years now. Crude oil is the source of carbon black, which is a nonrenewable resource that could run out eventually. As a result, a natural alternative source is required for carbon composites in rubber manufacturing as a reinforcing agent without significantly altering the final products' mechanical and physical qualities. Lu and Li [9] studied rubber composite damping for various uses using NR filled with carbon black (CB) and a variety of reinforcing chemicals. Phenolic resins exhibit damping characteristics over the greatest operating condition range when compared to alternative curing materials. Yurkin *et al.* [10] investigated the vibration damping of three viscoelastic materials. The findings indicated that butyl rubber, ethylene propylene rubber (EPDM or EPR) and ethylene-vinyl acetate (EVA) had damping ratios of 1.38, 0.6, and 0.42, respectively, suggesting that they may be used as viable options for damping applications. In another study, the structural and vibration damping properties of hybrid SBR materials loaded with fillers were investigated [11]. The authors observed that when the filler content increased, so did the several other properties of the material. However, it has been noted that an increase in filler percentage results in a decrease in essential RB properties [12]. In Figure 2.1, Buaksuntear *et al.* [13] presented the synthesis of the rubber samples.

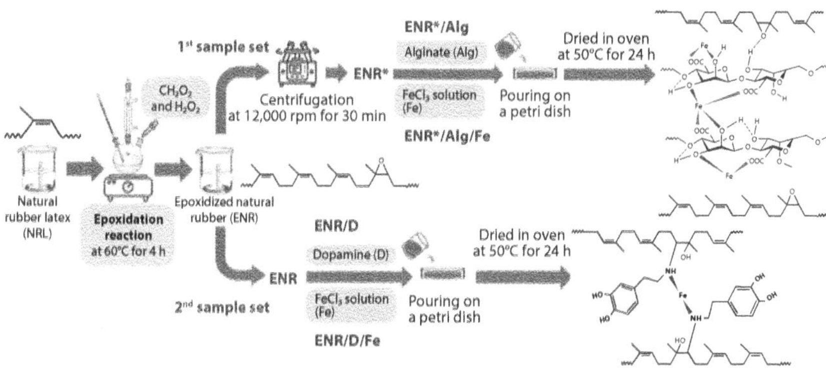

Figure 2.1 Synthesis of the rubber samples (i) Epoxydized natural rubber (ENR)*, (ii) ENR*/Alg, and (iii) ENR*/Alginate (Alg)/Fe, Adapted from Buaksuntear *et al.* [13].

Furthermore, hybrid RB materials loaded with fillers exhibit improved mechanical and structural damping properties [14]. Using modal analysis, the system loss factor values of the composite were determined and were found to be between 0.1 and 0.2, indicating that they may be useful in real-world vibration damping applications. Yoon *et al.* [15] evaluated the dynamic mechanical properties of hybrid SBR reinforced with silica. They found that a higher styrene concentration improved the damping qualities and decreased the physical characteristics of the resulting polymers. Furthermore, hybrid RB packed with N330 CB and short aramid fibers has improved material and structural damping properties. Using modal analysis, the system loss factor values of the composite were determined and found to be between 0.1 and 0.2, indicating that they may be useful in real-world vibration damping applications. Filler is needed to improve the mechanical, chemical, and physical qualities of elastomers when they are used in vibration damping for vehicles, railroads, and aircrafts. Fillers have a significant impact on how rubber composites behave in a viscoelastic manner. The properties of the reinforcements can substantially influence the mechanical characteristics and particle size of rubber composites [16].

In Figure 2.2, Yang *et al.* [17] described the synthesis of some rubber samples.

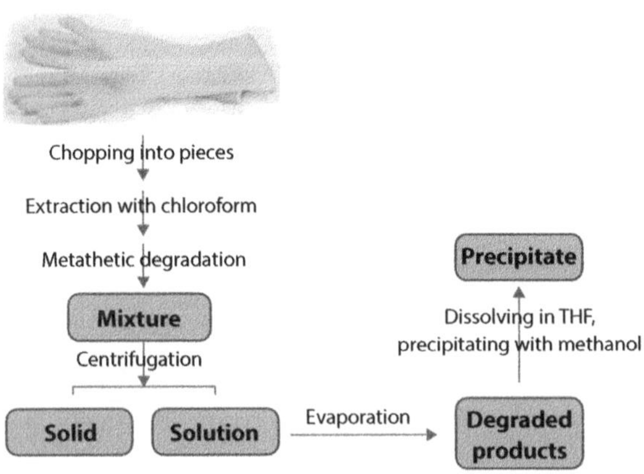

Figure 2.2 Preparation of the rubber samples adapted from Yang *et al.* [17].

2.2 Natural Rubber-Bamboo Composites: Preparation

Rubber is used on regular basis for different purposes. Rubber bands, rubber shoes, and rubber gloves are popular items made of rubber. Rubber products are categorized as elastomers because they can regain their original shape after being stretched or twisted. Rubber is an elastic material that can be artificially or organically produced [18]. Table 2.1 shows some hybrid RB materials, their synthesis, maximum vibration damping ratio, and sound transmission loss.

2.3 Natural Bamboo Composite Characterization

Because NR is an organic substance, its characteristics are evaluated by the biochemical process that the plant uses to generate the resin. NR cannot undergo altered polymerization, in contrast to synthetic rubbers [28]. NR cannot be changed once it is extracted from a plant. Epoxidized natural rubber, hydro-halogenated natural rubber, cyclized natural rubber, depolymerized liquid natural rubber, resin-modified natural rubber, and poly (methyl methacrylate) grafted natural rubber are advanced types of NR that are deemed significant. One form of natural rubber that has undergone physical modification is thermoplastic natural rubber, which is created by combining natural rubber with polypropylene [29].

Figure 2.3 [17] shows the synthesis of some of the aforementioned samples.

2.3.1 Preparation of Natural Rubber

- **Mixing:** Two-roll mills or internal mixers such as Banbury mixers are typically used to combine compounding components with dry natural rubber. For every batch, there should be a specific duration, temperature, and order of ingredient addition for both mixers [30].
- **Rubber tapping:** White latex fluid is to be collected in a cup after a tiny V-shaped cut is made in the rubber tree bark. The recovered latex is cleaned, filtered, and exposed to an acid reaction to coagulate the rubber particles [31].

Table 2.1 Hybrid RB materials, their synthesis, maximum vibration damping ratio, and sound transmission loss.

Material	Synthesis	Max vibration damping ratio (ξ)	Operation (synthesis) conditions		Sound transmission loss (STL)	Ref.
			Time (min/h)	Temp. (°C)		
NR-BB (Natural rubber-bamboo biochar)	• Four steps were used to create NR-BB composites in accordance with ASTM D3182-21a standard. • Stage one: Two-roll mill processing and mastication of raw materials. • Stage two: Mixing compounding ingredients to give the masterbatch. • Stage three: Addition of the remaining ingredients and mixing. • Stage four: Curing and mixing	33.19	5a (stage one), 15a (stage two), 5a (stage three), 10a (stage four).	80 (stage one)	8 % and 11 %	[19]
Natural rubber (SMR20)/ Schizostachyum zollingeri (Buluh Dinding)	• Stage one: Cutting of bamboo sticks and alkalization reaction. • Stage two: Immersion in NaOH and washing with HCl. • Stage three: rinse with distilled water. • Stage four: Grinding and pulverization. • Stage five: Two-roll mill processing. • Stage six: curing.	–	24b (stage two), 24b (stage three)	80 (stage three)	–	[20]

(*Continued*)

Table 2.1 Hybrid RB materials, their synthesis, maximum vibration damping ratio, and sound transmission loss. (*Continued*)

Material	Synthesis	Max vibration damping ratio (ξ)	Operation (synthesis) conditions		Sound transmission loss (STL)	Ref.
			Time (min/h)	Temp. (°C)		
Natural rubber (NR)/ carbon black	• Stage one: Two-roll mixing mill processing of NR. • Stage two: The mixture/integrating of the rubber with the various reinforcement materials. • Stage three: *In-situ* processing/vulcanization of test specimens.	–	13a each (stage three)	65 ± 5 (stage two), 145 (stage three)	–	[21]

(*Continued*)

Table 2.1 Hybrid RB materials, their synthesis, maximum vibration damping ratio, and sound transmission loss. (*Continued*)

Material	Synthesis	Max vibration damping ratio (ξ)	Operation (synthesis) conditions		Sound transmission loss (STL)	Ref.
			Time (min/h)	Temp. (°C)		
NR/cellulosic nanoparticles	• Stage one: Defibering of Coconut husk and bamboo by mechanical beating and retting process. • Stage two: Treatment of the fibers of the various biomass using a mixture of H_2O_2 + NaOH. • Stage three: Drying, washing, rinsing and distillation of the mixture. • Stage four: Cutting and drying of component fibers with scissors and Thomas-Wiley laboratory mill. • Step five: Screening of the ground particles recovered from this process. • Step six: Balling of the milled microparticles recovered from the above process. • Step seven: Preparation of composites of nanoparticles and NR using a two-roll mill (2RM) in accordance. • Step eight: *In-situ* reinforcing and processing. • Step nine: Mixing/compounding of the rubber with nanoparticles and other ingredients using two roll mixing mills.	–	4b (stage two), 48b (stage three), 30b (step six)	75 ± 5 (stage two)	–	[22]

(*Continued*)

Table 2.1 Hybrid RB materials, their synthesis, maximum vibration damping ratio, and sound transmission loss. (*Continued*)

Material	Synthesis	Max vibration damping ratio (ξ)	Operation (synthesis) conditions		Sound transmission loss (STL)	Ref.
			Time (min/h)	Temp. (°C)		
NR/ground tire rubber (GTR)	• Step one: The first step involved devulcanizing GTR in a prototype built in our laboratory. • Step two: The sample is placed in a container. • Step three: Devulcanization process. • Step four: NR/GTR composites were made with a batch mixer called a Brabender Plastograph. • Step five: preliminary mastication of NR. • Combining carbon black and GTR that has not been treated or microwave devulcanized. • Step six: combining the mixture with the sulfur curing apparatus. • Step seven: NR/GTR composites were gotten and molded into modified specimens using a Dr. Collin GmbH laboratory plate press type P 200E.		3a, 5a, and 10a (step three), 8a (step four), 2a (step five), 4a (step five), 2a (step six), 12a (step seven)	70 (step four), 160 (step seven)		[23]

(*Continued*)

Table 2.1 Hybrid RB materials, their synthesis, maximum vibration damping ratio, and sound transmission loss. (*Continued*)

Material	Synthesis	Max vibration damping ratio (ξ)	Operation (synthesis) conditions		Sound transmission loss (STL)	Ref.
			Time (min/h)	Temp. (°C)		
NR/AO-80/SIS	• Step one: An open refiner was used to plasticize natural rubber. • Step two: then, for the NR to properly react, AO-80 and activators were mixed and integrated. • Step three: The product material as well as other rubber materials were mixed uniformly to produce integrated NR. • Step four: The NR/AO-80/SIS was kept cooling to standard conditions and processed.	–	–	–	–	[24]
Epoxidized natural rubber (ENR)	• Step one: The processing activators, accelerators and agents were mixed into the ENR and carbon black mixture by stages during the following mixing in a 2RM. • Step two: after phases of process methods, a piece of modified rubber was generated.	–	–	–	–	[25]

(*Continued*)

Table 2.1 Hybrid RB materials, their synthesis, maximum vibration damping ratio, and sound transmission loss. (*Continued*)

Material	Synthesis	Max vibration damping ratio (ξ)	Operation (synthesis) conditions		Sound transmission loss (STL)	Ref.
			Time (min/h)	Temp. (°C)		
Rubber-sand composites	• Step one: The sulfur + carbon black mixture was weighted accurately following standard procedures. • Step two: The NR underwent a mastication procedure. • Step three: Sand particles were mixed into product material. • Step four: The filler agents were added and mixed. • Step five: This was companied by the addition of NPN. • Step six: Sulfur was added into the mixer and mixed to avoid premature cross-linking. • Step seven: The prepared product material underwent a molding procedure.		1a (step two), 2a (step three), 1a (step four), 1a (step five), 1a (step six)			[26]

(*Continued*)

Table 2.1 Hybrid RB materials, their synthesis, maximum vibration damping ratio, and sound transmission loss. (*Continued*)

Material	Synthesis	Max vibration damping ratio (ξ)	Operation (synthesis) conditions		Sound transmission loss (STL)	Ref.
			Time (min/h)	Temp. (°C)		
Cross-linked natural rubber (NR) nanocomposites	• Step one: Isolation of nanocellulose whiskers. • Step two: The product material was processed with H_2SO_4 and hydrolysis. • Step three: Removal of extra acid was gotten by repeated processing with distilled water. • Step four: The masterbatch processing procedure involves, mixing of NR latex and CNW in a solution. • Step five: The product material was modified. • Step six: This dispersion was then processed using fillers. • Step seven: This parent composite was then powdered and utilized for the next processing step. • Step eight: In this step, the integrating of the master batch with solid NR and vulcanizing chemicals was carried out using a 2RM. • Step nine: Nanocomposite compositions were modified during this procedure. • Step ten: The cure time for the vulcanization was determined. • Step eleven: Curing.		130a (step two), for 12b (step 4), 9a (step eleven)	40 (step two), 150 (step eleven)		[27]

Legend: [a*]minute (min); [b]hour (h).

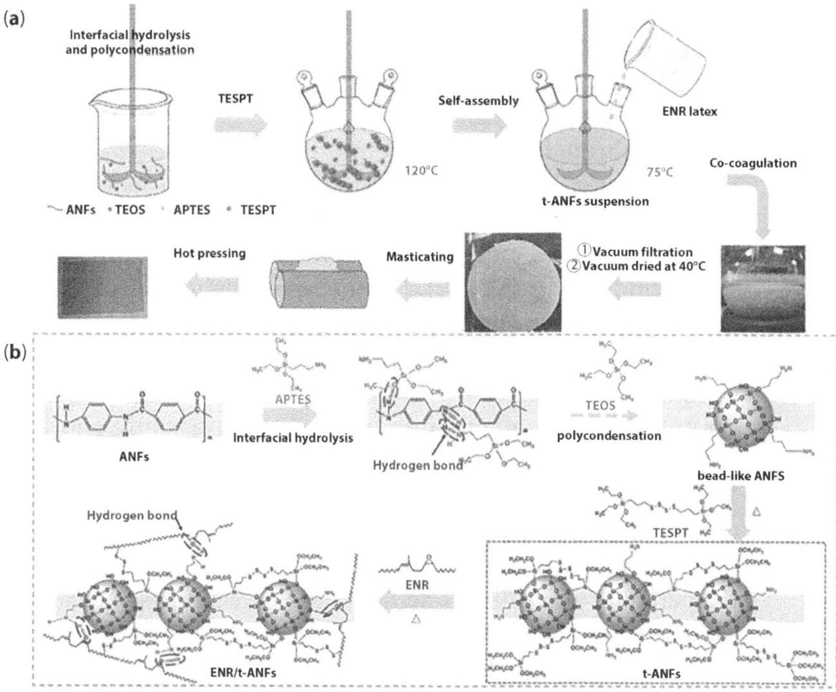

Figure 2.3 (a) Synthesis of ENR/t-ANFs components; (b) Reaction mechanism for the t-ANFs and ENR compounds, Adapted from Zhang *et al.* [68].

- **Mastication:** Rubber (RB) that remains unusable following the tapping process. Its nature is very gluey in warm conditions and quite delicate in cold conditions. After being compressed to make it pliable and softer for usage, the RB is moved through devices to eliminate its strong odor and brittleness. This procedure was iterated based on the characteristics required for the RB. Through this procedure, more chemical components are combined with rubber to enhance its characteristics [32].
- **Calendering:** The primary goal of the calendering process is to use rollers to produce rubber shapes. Subsequently, the completed material is fed through precisely designed perforations in an extrusion machine to create hollow tubes [33].
- **Extrusion:** Rubber is pressed through a die or nozzle by extruders to produce a material strip with a profile. Ram and

screw extruders are the two primary types of extruders. The components of a screw extruder are advanced parts such as a feed hopper. The compound is inserted into the feed hopper and moved through screw flights. The RB component is forced through a barrel and into the top by means of a reduction gear and electric motor as the screw turns [34]. The compound's built-up pressure is then released by allowing it to move *via* an aperture or die, assuming the necessary form. The discharge end of the screw flights is smaller than the feedstock end to produce efficient compression to remove air and maintain a constant head pressure. A heated compound then forced into the ram extruder cylinder using a ram. This type of machine operates inconsistently and requires additional maintenance.

- **Vulcanization:** Despite following all the preceding instructions, rubber will not be sufficiently hard to be utilized in machinery and automobile tires. The RB is heated to a certain temperature, and sulfur is added to improve all these qualities. This process is called vulcanization. Rubber becomes cross-linked and hardens following vulcanization owing to the action of sulfur as an integrating agent [35]. The process of molding and vulcanizing a rubber compound into a suitable-shaped mold using heat and pressure is known as molding. Utilizing appropriately designed and constructed hardened metal molds with appropriate surface finishes is essential, contingent upon the product's desired level of surface quality.

The three primary molding methods are injection, transfer, and compression molding, depending on how the material is introduced into the mold [36].

Compression molding: This process involves inserting a blank RB compound into a two dimensional- or multi-piece mold, sealing the cavity, and introducing pressure within the device's cylinders. The extra material is released from the mold through an appropriate opening in the mold. Typically, pneumatic tires are cured in a modified compression mold, wherein the green tire is forced and held against the mold surface during the vulcanization process using a rubber bladder [37].

In Figure 2.4, Zainal *et al.* [38] show the synthesis of some of the already described rubber samples.

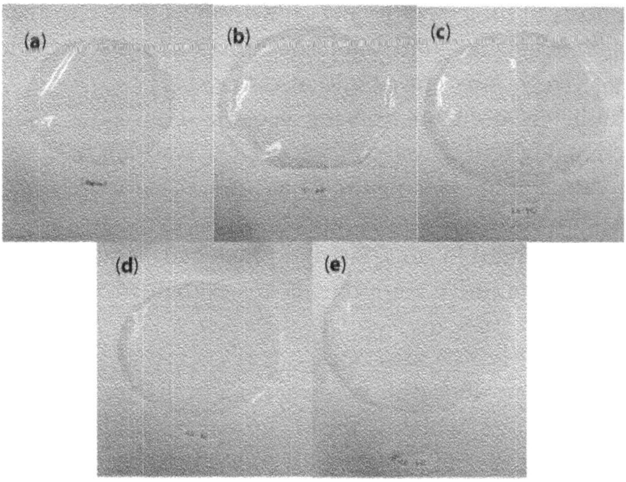

Figure 2.4 Synthesis of the rubber samples (i) ENR*, (ii) ENR*/Alg, and (iii) ENR*/Alg/Fe.

Transfer molding: During this procedure, the unprocessed rubber composites are shifted from the loading chamber of the mold into the cavity. This method works well for fragile inserts that need to be anchored in molds, as well as for molding items with intricate shapes [39].

- **Injection molding:** Injection molding is the most widely used molding technique for thermoplastic materials because of its high degree of reproducibility and speed. In the case of natural rubber, this method has recently gained popularity for large-scale manufacture of certain rubber goods. Using this method, the cycle time required to create moldings can be reduced to a few seconds with careful RB compound composition and process temperature management. The injection molding technology also has the benefit of significant lower finishing costs and rejection rates [24, 40].

2.3.2 Uses of Rubber

Rubber can be utilized for various types of operations/purposes and along a variety of platforms, a few of which are highlighted below.

- Industrial mixers, bins, and chute linings were constructed. Their resilience and water resistances allowed them to be effectively transformed into insulators [41].
- Rubber is used for flooring because it provides padding, reduces fatigue, is waterproof and slip-resistant [42].
- It can also be used in the garment sector for wetsuits and expandable apparel such as cycling and gym shorts.
- Its application in tires, airbags, brake pads, seats, roofs, and other automotive components are evident in the materials [42].

Rubber can be produced in two ways: naturally from the fluid of the *Hevea brasiliensis* tree, which is mostly grown in Southeast Asia, or artificially by utilizing several processes to convert petroleum. *H. brasiliensis* is grown in the tropical parts of Southeast Asia, mostly in Malaysia and Indonesia. It produces latex, which is used to make natural rubber (NR), which is economically marketed [43]. The lengthy, entwined coils that make up natural rubber's polymer chains are always in motion at room temperature. Natural rubber polymers twist and wrap because of the suppression of the 3D order created by the methyl group and hydrogen atom, which are on the same surface as the double bond that binds to carbon. Today, NR provides over 46% of the rubber utilized worldwide. For instance, NR is almost the only material that has been used to date, with no compromise between craftsmanship and characteristics. NR exhibits superior crystallization properties compared to its alternatives. NR completely satisfies the requirements for cut resistance, flex-fatigue resistance, and tensile qualities of the final products it is used to manufacture. Zeng *et al.* [44] presented the synthesis of the rubber samples in Figure 2.5.

The process of choosing and combining the right rubber with other raw materials and compounds to give the final product the required qualities is known as rubber combination. A prescription or formula is a set of goods chosen based on their intended use and proportionate integrity with one another [45]. Rubber serves as the main raw material, and the ratios used in rubber innovation are based on the notion that one unit of rubber is equal to 100 weight units. Except for rubber, all materials are specified as parts per hundred rubber (PPHR), or the quantity needed for 100 rubbers. It is also desired that the tire rubber compound would offer the best possible performance. For the best grip, there must be the largest coefficient of friction between the wheel and the pavement. Simultaneously, the rolling resistance can be decreased to enhance fuel efficiency [46]. The tire rolling resistance performance is mostly influenced by the filler. The wear

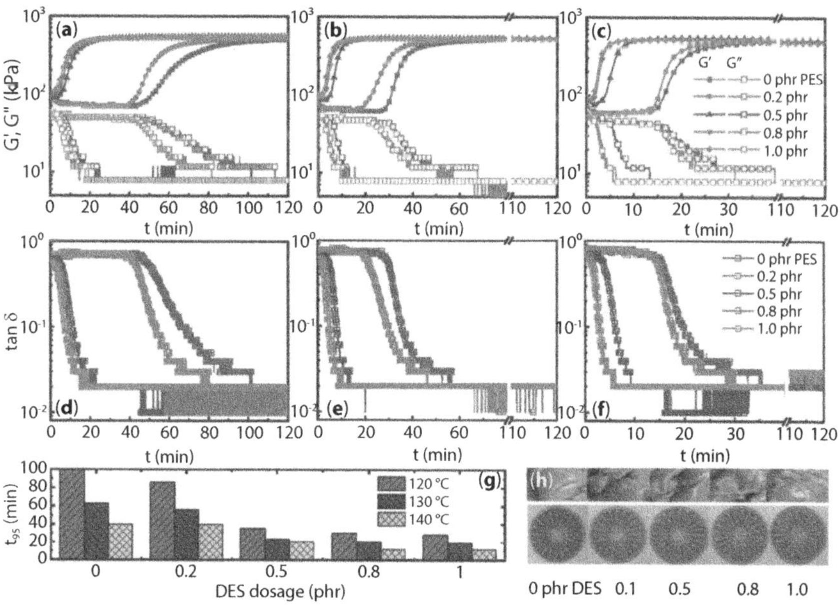

Figure 2.5 Impact of DES on vulcanization kinetics for the gums cured at 120°C (a and d), 130°C (b and e), and 140°C (c and f), t95 as a function of DES dosage (g) and coloration. Adapted from Zeng *et al.* [44].

resistance, the final triangle corner, is influenced by the tire's longevity and the surrounding conditions. Specifically, the tire's outside surface that touches the road should be abrasion-resistant. Because improving one of these characteristics will have to come at the expense of others, the tire compound needs to be specifically made for the rubber's intended use. Materials with extremely small particle sizes are added to the RB mixture in powdered form as fillers or reinforcement additives [47]. Fillers are used to strengthen rubber, make it easier to process, and lower the cost and color of the material. Fillers can be either organic or inorganic, and they can be dry powders. Fillers are typically utilized to give elastomers the desired characteristics in commercial applications. The primary motivation for this is to lower costs and enhance some compositional qualities. Fillers can be classified into two categories based on their color and impact. These can be separated and categorized into three types of fillers: semi-active, inactive, and active (reinforcing). Carbon blacks are known as black fillers, silica, talc, calcium carbonate, and zinc oxide are known as white fillers [48]. Table 2.2 shows the materials, composites, characterization methods, and properties to test for in some natural rubber-based composites.

Table 2.2 The materials, composites, characterization methods, properties to test for and results.

Material	**Composites**	**Tests (characterization)**	**Properties**	**Results**	**Ref.**
NCDG/SBR composites	• SBR (SBR 1502) • natural coal-derived graphite (NCDG)	• X-ray Diffraction (XRD) • Raman Spectroscopy analysis (RSA) • X-ray Photoelectron Spectroscopy (XPS) • Elemental Analysis • Particle Size Analysis (PSA) • Small-angle Neutron Scattering (SANS) • Small-angle X-ray Scattering (SAXS) • Brunauer-EmmettTeller (BET) analysis • Contact Angle Measurements • Transmission Electron Microscopy (TEM) • Scanning Electron Microscopy (SEM)	• Mechanical properties • Electrical conductivity • Particle size • Chemical composition	• The chemical properties of NCDG samples also change with the degree of graphitization. • Substantial alternations in both the dimensions and particle distribution among the NCDG fillers.	[49, 68]

(Continued)

Table 2.2 The materials, composites, characterization methods, properties to test for and results. *(Continued)*

Material	Composites	Tests (characterization)	Properties	Results	Ref.
NR composites/ TEMPO–CNFs/ mAANa	• The TEMPO–CNF/water dispersion • dried NR sheet • The mAANa • Dicumyl peroxide • ZnO • 2-Mercaptobanzimidazole • poly (2,2,4- trimethyl-1,2-dihydroqunoline • CB powder	• Tensile tester • DMA7100 analyzer • SEM	• Tensile properties • Dynamic thermomechanical properties • Tensile facture • Water-uptakes of the composite sheets • Viscosity	• The presence of TEMPO–CNFs in the NR sheet resulted in a higher storage modulus.	[69]
Silica/NR nanocomposites	• Natural rubber • Hydrophilic fumed silica (A200) • Sulfur, N-(1,3-dimethylbutyl)-N′-phenyl-p-phenylenediamine • Stearic acid • N-Cyclohexyl-2-benzothiazolesulfenamide • Zinc chloride • Tin chloride • Magnesium chloride hexahydrate • Ferric chloride hexahydrate • Urea	• Broadband dielectric spectrometer • FTIR analysis • SEM • TEM	• Dielectric loss • Molecular interaction and microstructure of nanocomposites • Rheological behaviors • Payne effect • Mullins effect	• The surface property of silica influences the cure behavior and filler dispersion.	[70]

(Continued)

Table 2.2 The materials, composites, characterization methods, properties to test for and results. (*Continued*)

Material	Composites	Tests (characterization)	Properties	Results	Ref.
Flexible $Ti_3C_2T_x$/ natural rubber composites	• Ti_3AlC_2 MAX phase • Hydrochloric acid • Lithium fluoride • NR latex	• SEM • TEM • AFM • XRD patterns • Zeta potentials of materials • FTIR • XPS • 4-probe tester • Resistivity test system • Rohde Schwarz ZNA43 2 Port vector network analyzer • Mark-10 testing stand	• Microstructures and morphology of $Ti_3C_2T_x$ MXene/NR composites • Electrical and EMI shielding properties of $Ti_3C_2T_x$ /NR composites • Mechanical behavior and electrical stability of MXene/NR composites	• Intact conductive framework. • Outstanding electrical conductivity.	[71]

(*Continued*)

Table 2.2 The materials, composites, characterization methods, properties to test for and results. (*Continued*)

Material	Composites	Tests (characterization)	Properties	Results	Ref.
Cys-GO@GNP/NR (CGG/NR)	• Natural rubber latex • GO aqueous dispersion • GNP • Nhydroxysuccinimide • Ethyl-3-(3-dimethylaminopropyl) carbodiimide hydrochloride • Triton X-100 • Zinc diethyldithiocarbamate • Sulfur	• FSEM • XRD • FT-IR • XPS • Raman spectra • Low-field nuclear magnetic resonance • The Infrared Thermal Imager • Resistivity meter • The digital multimeter • Universal Testing Machine	• Morphology and structure of 3D CGG/NR composites • Filler-matrix interaction of CGG/NR composites • Thermal conductivity of 3D CGG/NR composites • Thermal management capability of 3D CGG/NR • Electromechanical properties of CGG/NR composites	• The NR/GO composites showed excellent thermal conductivity. • It had excellent mechanical characteristics. • It had antistatic characteristics.	[72]

(*Continued*)

Table 2.2 The materials, composites, characterization methods, properties to test for and results. (*Continued*)

Material	Composites	Tests (characterization)	Properties	Results	Ref.
Bamboo fiber reinforced natural rubber composites	• Natural rubber (SMR L) • Sulfur • Zinc oxide • Stearic acid • n-cyclohexylbenthiazyl sulfenamide (CBS) • silica • Flectol H (poly-1,2-dihydro-2,2,4-trimethylquinoline • phenol formaldehyde • hexamethylenetetramine	• SEM	• Morphology	• The presence of bonding agent gave shorter curing time. • Enhanced mechanical properties.	[73]
Carbonized bamboo fiber (CBF) and Carbon black (CB)	• Natural rubber (Standard African Rubber, SAR 3) • carbon black (N330), • Sulphur, • accelerator (MBT), • Zinc oxide, stearic acid, • wax and antioxidant (TMQ) • The bamboo Maleic anhydride-graft-Polyisoprene	• Gravimetrically according to ASTM 1509 standard test method • Iodine adsorption number method	• Decrease in ignition • PH • Bulk density • Surface area • Moisture content and oil absorption	• The lesser particle-sized carbonized bamboo composite vulcanizates showed higher specific gravity. • Specific gravity of the vulcanizates increased with increases in filler loading.	[74]

2.4 Natural Rubber-Bamboo Composites: Noise Control Applications

As an increasing number of environmental awareness campaigns are being promoted around the world, Industries need to process rubber-trash as a material that can be used to create new items. In 2014, 32 million tons of wood and textile waste and 68 million tons of paper waste were produced in the United States, accounting for approximately 39% of the country's annual municipal solid waste burden [49]. The Central Pollution Control Board (CPCB) in New Delhi conducted research that indicated that the proportion of trash generated in India comprises components of several materials ranging between 3–5%, and 1–5%, respectively. Nonetheless, the amount of garbage produced by industry has been steadily rising annually and needs to be recycled or repurposed. It is preferable to reuse these wastes rather than recycle them [50]. The unfavorable noise that has resulted from improvements in mechanical tools has proven to be a more complicated issue overall. By following this principle, waste products from the processing of fabrics and corn can be mixed with waste products from the processing of paper to produce a substance that absorbs sound and reduces noise pollution [51]. The use of recyclable elements to create composite materials has attracted considerable attention. The influence of these materials on production costs is low. The utilization of plant leaves, hardwood fibers, and textile fibers (bamboo and jute) has an impact on noise reduction [52]. As time and life progress, noise contamination and physical oscillations are inevitable concerns in human existence. The need for a better living environment is growing, and people can no longer put up with mechanical vibration and the noise it produces. Furthermore, mechanical vibration shortens machine life and reduces precision [53]. There are many potential applications of sound-absorbing materials, which have superior damping qualities. Materials that resemble rubber exhibit good damping, friction, and excellent viscoelasticity. These rubbers typically exhibit strong molecular chain interactions, such as ionic, hydrogen, and polar groups, and a high relative molecular mass [54]. RBs are employed as dampers in automobiles, trains, airplanes, and other industries. However, several factors restrict their usability in practice. Consequently, it remains important to create RB compounds with outstanding damping; the ideal component should not only be effectively damped but also have relevant characteristics such as physicochemical stability. Owing to its unique qualities, Isobutylene-isopropene rubber (IIR), a polymer composed primarily of isobutylene and a small amount of isoprene, is used extensively in the production of inner tubes, industrial rubber sheets, anti-vibration rubber, and other materials. The macromolecular chains of IIR have a high number of methyl groups, resulting in significant internal friction between them. Macromolecular chains (MC) also convert the damping that occurs across the relaxation procedure into internal consumption. The IIR has a particular damping influence at the glass transition temperature when the MC sections and MC hysteresis are free to move [55]. For the past few decades, conventional fillers such as CB derived from non-renewable resources, including tar, coal, acetylene, and natural gas, have been used as the main filler for rubber composites [56]. However, because of the high cost of producing CB, Dwivedi *et al.* [57] investigated the use of natural fillers as substitutes for CB in rubber composites.

Conversely, because of their large particle size, high structure, low surface area, and nonpolar nature, CBs potential in making RB composites is unsuitable [58]. The use of renewable fillers in the development of composite materials has recently attracted attention. Thus, biochar is a viable substitute for CB because of its high percentage of carbon, low ash, nitrogen, and sulfur components, as well as low cost. Global bamboo production is estimated to be 20 million tons (MTs) per year, with 10 MTs originating from China, Japan, and India [59]. Bamboo finds its use in a wide range of industries, including apparel, food, furniture, flooring, and interior design. Over the past few decades, there has been an increase in interest among researchers in the thermochemical conversion of waste bamboo into products rich in carbon and biochar. Bamboo biochar has been used traditionally to produce recyclable goods, increase water holding ability [60], produce biomaterial [61], and as an electrode in the production of battery cells [62]. For the rubber sector, bamboo charcoal offers another readily available, affordable, and sustainable source of fillers. To the best of the authors knowledge, no research has been published on the development of acoustic and vibration-damping materials using bamboo biochar (BB) as a filler. Thus, to create NR composites, biochar made from bamboo waste can be used as filler ingredient. The primary aim of this study is to discuss NR–BB compounds and different filler materials. Scientists have created a spectrum of damping products using various techniques. The two most common advancement strategies include the addition of efficient and effective reinforcing fillers to IIR composite or mixing them with other materials to generate composite damping products. Nanofillers possess a large specific surface region that allows them to penetrate deeply into the rubber matrix. Carbon nanotubes, talc, and carbon black are some of the joint strengthening fillers. To enhance the mix of filler and RB from an environmental perspective, Tang *et al.* [63] substituted waste recycled red brick powder with carbon black and silica in styrene-butadiene rubber (SBR).

The production costs were reduced, while the tensile characteristics were improved. According to experimental testing, using BB as an NR filler synergistically enhanced the NR matrix's vibration damping capabilities. It is possible to significantly increase the characteristics of butyl rubber compounds by blending the resin with IIR. Song *et al.* [64] added various hindered amine molecules to enhance the butyl rubber/phenolic resin composites' high-temperature damping capabilities. To prepare the desired damping materials, additional relaxation components can be added using IIR as the matrix. By adjusting the MC and polyisobutylene (PIB) concentrations. Xia *et al.* [65] managed the damping qualities of IIR. One major way to increase the damping capabilities of rubber is to use polymerization to introduce a greater volume and a specific number of side sets into the rubber's molecular framework. However, the modification reaction is more intricate and challenging to manage. There are certain practical application restrictions with this approach owing to the low grafting rate of adding side sets to the MC.

The design of damping materials has become more dependent on sustainable development and the green environment because of environmental contamination and the depletion of finite and unsustainable petrochemical resources. Lignin is a frequent material from forests among aromatic compounds and is one of the rare natural resources. It can be used to prepare lignin-based phenolic resins (LPF) by substituting phenol in resins. Lignin is a possible reinforcement material for polymeric materials because of its high stiffness, which is attributed to its aromatic portion and intramolecular solid/intermolecular contacts [66]. Butyl rubber's qualities can be enhanced by curing IIR with phenolic resin that has been modified by lignin. Butyl rubber is a usually hydrophilic substance that absorbs moisture during use because it includes polar functional groups. Because of their low thermal stability, classic damping materials typically have a limited-service life. Therefore, the development of damping materials with good physicochemical characteristics is crucial. In this study, co-blending advancement and NaH activation improvement of the IIR MC framework were combined to enhance the damping qualities of the compounds. To prepare the bio-based phenolic resin, lignin was mixed with the IIR matrix, and then a certain amount of phenol was substituted. Broad Duratek High Rubber hydrophobic butyl rubber compounds were obtained by adding a double bond to the IIR MC framework and activating the IIR with dibenzyl fork acetone (DBA) containing a benzene ring and NaH [67].

2.5 Conclusions

Significant strides have been made over the last several decades to reduce noise and vibration in manufacturing machinery and cars. Importantly, vibrations caused by noise from structures can be reduced using passive approaches, such as viscoelastic vibration damping materials (VVDM). Rubber can be made chemically by combining alcohol and petroleum, or biologically from the milky fluid of certain crops. Rubber is an essential technological substance with many uses. Nowadays, the term "rubber" no longer just refers to the original natural rubber; it also refers to any material with properties comparable to those of NR, independent of its chemical composition. More recently, synthetic polymers with properties similar to rubber have been referred to as "elastomers." Rubber is a material that may be used for many things. CB has long been the most popularly utilized reinforcing material in the RB sector, particularly for tires. Carbon black is derived from crude oil, which is a nonrenewable resource that may ultimately run out. Therefore, the rubber sector needs a natural substitute for carbon black as reinforcing ingredient without materially changing the mechanical and physical properties of the finished goods.

Rubber is a common material found in a wide variety of products. Several common rubber products that are encountered daily are rubber bands, shoes, and gloves. Because rubber products may return to their former shapes after being stretched or twisted, they are classified as elastomers. Rubber is an elastic substance that can be made both naturally and artificially. Given that NR is an organic material, its characteristics are dictated by the biochemical mechanism by which the plant produces the resin. Unlike synthetic rubbers, NR's polymerization process cannot be changed. Natural rubber cannot be altered after it is removed from a plant. Many advanced forms of NR are considered significant and include cyclized NR. The industry must view waste as a resource that can be utilized to develop new products, as global environmental awareness campaigns gain traction. Approximately 39% of the nation's yearly municipal solid trash load comprised 32 million tons of wood and textile waste and 68 million tons of paper garbage produced in the United States in 2014. According to a study conducted by the Central Pollution Control Board (CPCB) in New Delhi, the percentage of garbage produced in India is made up of parts of multiple materials that range from 3–5% to 1–5%. Mechanical oscillations and noise pollution have become unavoidable issues in human life as time and living evolve. People are becoming more and more in need of improved living conditions, and they are unable to tolerate the noise and vibrations caused by machinery. Moreover, mechanical vibration decreases precision and shortens the life of the equipment. Sound-absorbing products exhibit good sound/vibration-damping characteristics and a wide range of possible applications. RB-like materials exhibit good viscoelasticity and friction properties. These RBs usually have high relative molecular masses and excellent molecular chain interactions, including polar, hydrogen, and ionic groups. Hence, they are used as noise-dampers in a variety of industries, including train, aircraft, and automobile industries. This chapter has successfully highlighted natural-bamboo rubber composite materials for successful noise and vibration control.

Acknowledgment

The authors acknowledge and thank Covenant University for their support in completing this study. Access to the School's library resources, publications, and literature is also appreciated.

References

1. Rao, M.D., Fifth International Congress On Sound Automotive Applications Of Viscoelastic Damping Treatment. *Fifth Int. Congr. Sound*, pp. 1–8, 1997.
2. Wang, T.-Y., Liu, S.-C., Tsai, J.-L., Micromechanical stick-slip model for characterizing damping responses of single-walled carbon nanotube nanocomposites. *J. Compos. Mater.*, *50*, 1, 57–73, 2016, https://doi.org/10.1177/0021998315570371.
3. Pang, J., *Noise and Vibration Control in Automotive Bodies*, Wiley, China 2018, https://doi.org/10.1002/9781119515500.
4. Park, K.H., Fujiwara, Y., Mazda, T., Kajita, Y., Evaluation of mechanical properties considering hysteresis characteristic of high damping rubber bearing. *J. Phys. Conf. Ser.*, *1687*, 1, 012019, 2020, https://doi.org/10.1088/1742-6596/1687/1/012019.
5. Qu, C., Li, S., Zhang, Y., Wang, T., Wang, Q., Chen, S., Surface modification of Ti3C2-MXene with polydopamine and amino silane for high performance nitrile butadiene rubber composites. *Tribol. Int.*, *163*, 107150, 2021, https://doi.org/10.1016/j.triboint.2021.107150.
6. Şen, M., Soydaş, O., Aksüt, D., Preparation and characterization of high damping poly(epichlorohydrin) and poly(epichlorohydrin-co-ethylene oxide-co-allyl glycidyl ether) elastomers: I. Effect of curing system and

blending on the damping properties. *React. Funct. Polym.*, *184*, 105520, 2023, https://doi.org/10.1016/j.reactfunctpolym.2023.105520.
7. Xiaolin, J., Min, X., Minhui, W., Yuanhao, M., Wencong, Z., Yanan, Z., Haoxiang, R., Xun, L., Preparation and molecular dynamics study of polyurethane damping elastomer containing dynamic disulfide bond and multiple hydrogen bond. *Eur. Polym. J.*, *162*, 110893, 2022, https://doi.org/10.1016/j.eurpolymj.2021.110893.
8. Azammi, A.M.N., Sapuan, S.M., Ishak, M.R., Sultan, M.T.H., Physical and damping properties of kenaf fibre filled natural rubber/thermoplastic polyurethane composites. *Def. Technol.*, *16*, 1, 29–34, 2020, https://doi.org/10.1016/j.dt.2019.06.004.
9. Lu, X. and Li, X., Broad temperature and frequency range damping materials based on epoxidized natural rubber. *J. Elastomers Plast.*, *46*, 1, 84–95, 2014a, https://doi.org/10.1177/0095244312462161.
10. Yurkin, Y.V., Mansurova, I.A., Belozerov, V.S., Zlobina, A., Morphological and Dynamic Mechanical Analysis of Vibration Damping Composite Material Based on Different Elastomers. *Mater. Plast.*, *55*, 4, 469–473, 2018, https://doi.org/10.37358/MP.18.4.5055.
11. Pöschl, M., Vašina, M., Zádrapa, P., Měřínská, D., Žaludek, M., Study of Carbon Black Types in SBR Rubber: Mechanical and Vibration Damping Properties. *Materials*, *13*, 10, 2394, 2020, https://doi.org/10.3390/ma13102394.
12. Tang, L.-C., Zhao, L., Qiang, F., Wu, Q., Gong, L.-X., Peng, J.-P., Mechanical Properties of Rubber Nanocomposites Containing Carbon Nanofillers, in: *Carbon-Based Nanofillers and Their Rubber Nanocomposites*, pp. 367–423, Elsevier, UK, 2019, https://doi.org/10.1016/B978-0-12-817342-8.00012-3.
13. Buaksuntear, K., Panmanee, K., Wongphul, K., Lim-arun, P., Jansinak, S., Shah, D.U., Smitthipong, W., Enhancing mechanical properties and stabilising the structure of epoxide natural rubber using non-covalent interactions: Metal–ligand coordination and hydrogen bonding. *Polymer*, *291*, 126626, 2024, https://doi.org/10.1016/j.polymer.2023.126626.
14. Praveen, S., Bahadur, J., Yadav, R., Billa, S., Umasankar Patro, T., Rath, S.K., Ratna, D., Patri, M., Tunable viscoelastic and vibration damping properties of a segmented polyurethane synergistically reinforced with carbon black and anisotropic additives. *Appl. Acoust.*, *170*, 107535, 2020, https://doi.org/10.1016/j.apacoust.2020.107535.
15. Yoon, B., Kim, J.Y., Hong, U., Oh, M.K., Kim, M., Han, S.B., Nam, J.-D., Suhr, J., Dynamic viscoelasticity of silica-filled styrene-butadiene rubber/polybutadiene rubber (SBR/BR) elastomer composites. *Composites, Part B*, *187*, 107865, 2020, https://doi.org/10.1016/j.compositesb.2020.107865.
16. Barrera, C.S. and Tardiff, J.L., Static and dynamic properties of eggshell filled natural rubber composites for potential application in automotive vibration isolation and damping. *J. Clean. Prod.*, *353*, 131656, 2022, https://doi.org/10.1016/j.jclepro.2022.131656.

17. Yang, C., Zhang, R., Zhao, H., Xie, P., Pan, C., Liu, P., Traceless metathetic degradation of waste vulcanized natural rubber with styrene as chain transfer agent in closed system. *Polymer, 300*, 127022, 2024, https://doi.org/10.1016/j.polymer.2024.127022.
18. Jantachum, P., Khumpaitool, B., Utara, S., Effect of silane coupling agent and cellulose nanocrystals loading on the properties of acrylonitrile butadiene rubber/natural rubber nanocomposites. *Ind. Crops Prod., 195*, 116407, 2023, https://doi.org/10.1016/j.indcrop.2023.116407.
19. Sunali, Mago, J., Negi, A., Pant, K.K., Fatima, S., Development of natural rubber-bamboo biochar composites for vibration and noise control applications. *J. Clean. Prod., 373*, 133760, 2022, https://doi.org/10.1016/j.jclepro.2022.133760.
20. Zainathul, A.S.A.S., Nurul, A.M.Z., Mohd, M.A., Siti, S.J., Nor, H.Z., The cure characteristics and mechanical behaviour of bamboo fibre filled natural rubber composite. *Adv. Mater. Res., 812*, 53–59, 2013, https://doi.org/10.4028/www.scientific.net/AMR.812.53.
21. Oboh, J.O., Okafor, J.O., Kovo, A.S., Abdulrahman, A.S., Dynamic mechanical properties of crosslinked natural rubber composites reinforced with cellulosic nanoparticles. *Niger. J. Technol., 37*, 3, 668, 2018a, https://doi.org/10.4314/njt.v37i3.16.
22. Oboh, J.O., Okafor, J.O., Kovo, A.S., Abdulrahman, A.S., Investigation of eco-friendly cellulosic nanoparticles potential as reinforcement agent in the production of natural rubber composites. *Niger. J. Technol., 36*, 4, 1078, 2018b, https://doi.org/10.4314/njt.v36i4.13.
23. Colom, X., Marín-Genescà, M., Mujal, R., Formela, K., Cañavate, J., Structural and physico-mechanical properties of natural rubber/GTR composites devulcanized by microwaves: Influence of GTR source and irradiation time. *J. Compos. Mater., 52*, 22, 3099–3108, 2018, https://doi.org/10.1177/0021998318761554.
24. Li, C., Zhang, W., Sun, G., Li, Y., Zhang, Y., Research on damping properties and microscopic mechanism of polyurethane cement-based composites. *Constr. Build. Mater., 365*, 130137, 2023, https://doi.org/10.1016/j.conbuildmat.2022.130137.
25. Li, Q., Liu, B., Hu, Z., Jiang, X., Yang, L., Meng, H., Song, Y., Zheng, Q., Preparation of deep eutectic solvents based on metal ions and their influences on reinforcement and strain softening behaviors of silica filled natural rubber nanocomposites. *Composites, Part A, 181*, 108119, 2024, https://doi.org/10.1016/j.compositesa.2024.108119.
26. Doddamani, S., Kulkarni, S.M., Joladarashi, S., Gurjar, A.K., Mohan Kumar, T.S., Development of rubber-sand composite for enhanced impact resistance: Implications of vulcanization. *Eng. Sci. Technol. Int. J., 51*, 101631, 2024, https://doi.org/10.1016/j.jestch.2024.101631.
27. Visakh, P.M., Thomas, S., Oksman, K., Mathew, A.P., Crosslinked natural rubber nanocomposites reinforced with cellulose whiskers isolated from bamboo

waste: Processing and mechanical/thermal properties. *Composites, Part A*, *43*, 4, 735–741, 2012, https://doi.org/10.1016/j.compositesa.2011.12.015.
28. Zhao, X., Fu, G., Wang, Y., Wu, Y., Shou, T., Yin, D., Li, X., Hu, S., Zhang, L., Bio-based polyurethane/hindered phenol AO-80 composites for room temperature high damping properties. *Composites, Part B*, *243*, 110118, 2022, https://doi.org/10.1016/j.compositesb.2022.110118.
29. Li, C., Cao, D., Guo, W., Wu, C., The Investigation of Miscibility in Blends of ENR/AO-80 by DMA and FT-IR. *J. Macromol. Sci. Part B*, *47*, 1, 87–97, 2007, https://doi.org/10.1080/00222340701746077.
30. Hu, S., Chen, S., Zhao, X., Guo, M., Zhang, L., The Shape-Memory Effect of Hindered Phenol (AO-80)/Acrylic Rubber (ACM) Composites with Tunable Transition Temperature. *Materials*, *11*, 12, 2461, 2018, https://doi.org/10.3390/ma11122461.
31. Eissa, M.M., Botros, S.H., Moustafa, A.F., Effect of triblock copolymers on homogeneity, mechanical properties and swelling behavior of IIR/SBR rubber blends. *Polym. Bull.*, *74*, 2, 393–412, 2017, https://doi.org/10.1007/s00289-016-1720-5.
32. Unwin, A.P., Hine, P.J., Ward, I.M., Guseva, O.A., Schweizer, T., Fujita, M., Tanaka, E., Gusev, A.A., Predicting the visco-elastic properties of polystyrene/SIS composite blends using simple analytical micromechanics models. *Compos. Sci. Technol.*, *142*, 302–310, 2017, https://doi.org/10.1016/j.compscitech.2017.02.021.
33. Vural, K.B., Furtana, Y.B., Yilmaz, M., Yalcin, E., Performance evaluation of bitumen modified with styrene–isoprene-styrene and crumb rubber compound. *Constr. Build. Mater.*, *344*, 128304, 2022, https://doi.org/10.1016/j.conbuildmat.2022.128304.
34. Han, H., Tian, G., Gao, Q., Hu, H., Zhao, J., Li, J., Wall slip of styrene-isoprene-styrene (SIS) triblock copolymer induced by micro elastic phase. *Polymer*, *209*, 122990, 2020, https://doi.org/10.1016/j.polymer.2020.122990.
35. Zhang, F., Hu, C., Zhang, Y., The research for SIS compound modified asphalt. *Mater. Chem. Phys.*, *205*, 44–54, 2018, https://doi.org/10.1016/j.matchemphys.2017.10.023.
36. Luo, Y., Wu, Y., Luo, K., Cai, F., Zhai, T., Wu, S., Structures and properties of alkanethiol-modified graphene oxide/solution-polymerized styrene butadiene rubber composites: Click chemistry and molecular dynamics simulation. *Compos. Sci. Technol.*, *161*, 32–38, 2018, https://doi.org/10.1016/j.compscitech.2018.03.036.
37. Zhu, J., Zhao, X., Liu, L., Yang, R., Song, M., Wu, S., Thermodynamic analyses of the hydrogen bond dissociation reaction and their effects on damping and compatibility capacities of polar small molecule/nitrile-butadiene rubber systems: Molecular simulation and experimental study. *Polymer*, *155*, 152–167, 2018, https://doi.org/10.1016/j.polymer.2018.09.040.
38. Zainal, A.A., Zango, Z.U., Akhdar, H., Aldaghri, O., Ibnaouf, K.H., Sulieman, A., Lim, J.W., Ng, H.S., Khoo, K.S., Sustainable grafting of poly(3-hydroxy-

butyrate-co-3-hydroxyhexanoate) and natural rubber into a new thermoplastic elastomer film. *Polymer*, *289*, 126476, 2023, https://doi.org/10.1016/j.polymer.2023.126476.
39. Qiao, B., Zhao, X., Yue, D., Zhang, L., Wu, S., A combined experiment and molecular dynamics simulation study of hydrogen bonds and free volume in nitrile-butadiene rubber/hindered phenol damping mixtures. *J. Mater. Chem.*, *22*, 24, 12339, 2012, https://doi.org/10.1039/c2jm31716h.
40. Suzuki, T., Osaka, N., Endo, H., Shibayama, M., Ikeda, Y., Asai, H., Higashitani, N., Kokubo, Y., Kohjiya, S., Nonuniformity in Cross-Linked Natural Rubber as Revealed by Contrast-Variation Small-Angle Neutron Scattering. *Macromolecules*, *43*, 3, 1556–1563, 2010, https://doi.org/10.1021/ma9019416.
41. Ooi, Z.X., Ismail, H., Abu Bakar, A., Study on the ageing characteristics of oil palm ash reinforced natural rubber composites by introducing a liquid epoxidized natural rubber coating technique. *Polym. Test.*, *37*, 156–162, 2014, https://doi.org/10.1016/j.polymertesting.2014.05.003.
42. Yin, C., Zhao, X., Zhu, J., Hu, H., Song, M., Wu, S., Experimental and molecular dynamics simulation study on the damping mechanism of C5 petroleum resin/chlorinated butyl rubber composites. *J. Mater. Sci.*, *54*, 5, 3960–3974, 2019, https://doi.org/10.1007/s10853-018-3134-2.
43. Li, J., Han, X., Huyan, C., Liu, D., Li, X., Li, H., Chen, F., Hindered phenolic antioxidant modified waterborne polyurethane films with excellent thermal stability and anti-aging performance at high temperature. *Prog. Org. Coat.*, *176*, 107398, 2023, https://doi.org/10.1016/j.porgcoat.2022.107398.
44. Zeng, X., Zhou, J., Peng, J., Zhang, C., Wang, D., Song, Y., Zheng, Q., Influence of polyethylene glycol-based deep eutectic solvent on vulcanization and mechanical properties of natural rubber gum vulcanizates. *Polymer*, *290*, 126521, 2024, https://doi.org/10.1016/j.polymer.2023.126521.
45. Sarkar, P. and Bhowmick, A.K., Sustainable rubbers and rubber additives. *J. Appl. Polym. Sci.*, *135*, 24, 45701, 2018, https://doi.org/10.1002/app.45701.
46. Zhang, H., Zhao, D., Yin, G., Ni, C., Liu, Y., Shi, T., Niu, X., Zong, Q., Ren, S., Wang, H., Du, Y., Zhang, K., Xiao, F., Effect of SIS block copolymers on damping properties of natural rubber/AO-80 and the performance enhancement mechanism: experimental study and molecular dynamics simulation. *Colloids Surf., A*, *672*, 131705, 2023, https://doi.org/10.1016/j.colsurfa.2023.131705.
47. Miyajima, A., Sakemi-Hoshikawa, K., Usami, M., Mitsunaga, K., Irie, T., Ohno, Y., Sunouchi, M., Thyrotoxic rubber antioxidants, 2-mercaptobenzimidazole and its methyl derivatives, cause both inhibition and induction of drug-metabolizing activity in rat liver microsomes after repeated oral administration. *Biochem. Biophys. Res. Commun.*, *492*, 1, 116–120, 2017, https://doi.org/10.1016/j.bbrc.2017.08.024.
48. Fang, M., Wang, B., Qu, X., Li, S., Huang, J., Li, J., Lu, S., Zhou, N., State-of-the-art of biomass-derived carbon dots: Preparation, properties, and

applications. *Chin. Chem. Lett.*, *35*, 1, 108423, 2024, https://doi.org/10.1016/j.cclet.2023.108423.

49. Wang, B., Waterhouse, G.I.N., Lu, S., Carbon dots: mysterious past, vibrant present, and expansive future. *Trends Chem.*, *5*, 1, 76–87, 2023, https://doi.org/10.1016/j.trechm.2022.10.005.
50. Wei, X., Li, L., Liu, J., Yu, L., Li, H., Cheng, F., Yi, X., He, J., Li, B., Green Synthesis of Fluorescent Carbon Dots from Gynostemma for Bioimaging and Antioxidant in Zebrafish. *ACS Appl. Mater. Interfaces*, *11*, 10, 9832–9840, 2019, https://doi.org/10.1021/acsami.9b00074.
51. Wu, S., Qiu, M., Guo, B., Zhang, L., Lvov, Y., Nanodot-Loaded Clay Nanotubes as Green and Sustained Radical Scavengers for Elastomer. *ACS Sustain. Chem. Eng.*, *5*, 2, 1775–1783, 2017, https://doi.org/10.1021/acssuschemeng.6b02523.
52. Balan, A.V. and Shivasankaran, N., Noise Control using Waste Materials Reinforced Composites. *London J. Res. Sci. Nat. Formal*, *19*, 2, 47–53, 2019.
53. Arroyo, M., López-Manchado, M.A., Herrero, B., Organo-montmorillonite as substitute of carbon black in natural rubber compounds. *Polymer*, *44*, 8, 2447–2453, 2003, https://doi.org/10.1016/S0032-3861(03)00090-9.
54. Mkhari, O., Ntuli, T.D., Coville, N.J., Nxumalo, E.N., Maubane-Nkadimeng, M.S., Supported carbon-dots: A review. *J. Lumin.*, *255*, 119552, 2023, https://doi.org/10.1016/j.jlumin.2022.119552.
55. Hu, Z., Chen, Z., Meng, F., Zhang, Y., Jia, Y., Fei, H., Li, S., Yuan, X., Novel hydrophobic butyl rubber damping composites modified with bio-based PF/DBA *via* the construction of a three-dimensional network. *Sci. Rep.*, *14*, 1, 5007, 2024, https://doi.org/10.1038/s41598-024-55823-x.
56. Fan, Y., Fowler, G.D., Zhao, M., The past, present and future of carbon black as a rubber reinforcing filler – A review. *J. Clean. Prod.*, *247*, 119115, 2020, https://doi.org/10.1016/j.jclepro.2019.119115.
57. Dwivedi, C., Manjare, S., Rajan, S.K., Recycling of waste tire by pyrolysis to recover carbon black: Alternative & environment-friendly reinforcing filler for natural rubber compounds. *Composites, Part B*, *200*, 108346, 2020, https://doi.org/10.1016/j.compositesb.2020.108346.
58. Jiang, C., Bo, J., Xiao, X., Zhang, S., Wang, Z., Yan, G., Wu, Y., Wong, C., He, H., Converting waste lignin into nano-biochar as a renewable substitute of carbon black for reinforcing styrene-butadiene rubber. *Waste Manage. (Oxford)*, *102*, 732–742, 2020, https://doi.org/10.1016/j.wasman.2019.11.019.
59. Rathour, R., Kumar, H., Prasad, K., Anerao, P., Kumar, M., Kapley, A., Pandey, A., Kumar Awasthi, M., Singh, L., Multifunctional applications of bamboo crop beyond environmental management: an Indian prospective. *Bioengineered*, *13*, 4, 8893–8914, 2022, https://doi.org/10.1080/21655979.2022.2056689.
60. Hien, T.T.T., Tsubota, T., Taniguchi, T., Shinogi, Y., Enhancing soil water holding capacity and provision of a potassium source *via* optimization of

the pyrolysis of bamboo biochar. *Biochar*, *3*, 1, 51–61, 2021, https://doi.org/10.1007/s42773-020-00071-1.
61. Dong, Q., Li, H., Niu, M., Luo, C., Zhang, J., Qi, B., Li, X., Zhong, W., Microwave pyrolysis of moso bamboo for syngas production and bio-oil upgrading over bamboo-based biochar catalyst. *Bioresour. Technol.*, *266*, 284–290, 2018, https://doi.org/10.1016/j.biortech.2018.06.104.
62. Gu, X., Wang, Y., Lai, C., Qiu, J., Li, S., Hou, Y., Martens, W., Mahmood, N., Zhang, S., Microporous bamboo biochar for lithium-sulfur batteries. *Nano Res.*, *8*, 1, 129–139, 2015, https://doi.org/10.1007/s12274-014-0601-1.
63. Tang, D., Zhang, X., Hu, S., Liu, X., Ren, X., Hu, J., Feng, Y., The reuse of red brick powder as a filler in styrene-butadiene rubber. *J. Clean. Prod.*, *261*, 120966, 2020, https://doi.org/10.1016/j.jclepro.2020.120966.
64. Song, M., Yue, X., Wang, X., Huang, M., Ma, M., Pan, W., Qin, Q., Improved high-temperature damping performance of nitrile-butadiene rubber/phenolic resin composites by introducing different hindered amine molecules. *E-Polymers*, *20*, 1, 482–490, 2020, https://doi.org/10.1515/epoly-2020-0054.
65. Xia, L., Li, C., Zhang, X., Wang, J., Wu, H., Guo, S., Effect of chain length of polyisobutylene oligomers on the molecular motion modes of butyl rubber: Damping property. *Polymer*, *141*, 70–78, 2018, https://doi.org/10.1016/j.polymer.2018.03.009.
66. Jong, L., Peterson, S.C., Jackson, M.A., Utilization of Porous Carbons Derived from Coconut Shell and Wood in Natural Rubber. *J. Polym. Environ.*, *22*, 3, 289–297, 2014, https://doi.org/10.1007/s10924-013-0637-4.
67. Masek, A., Cichosz, S., Piotrowska, M., Biocomposites of Epoxidized Natural Rubber/Poly(lactic acid) Modified with Natural Fillers (Part I). *Int. J. Mol. Sci.*, *22*, 6, 3150, 2021, https://doi.org/10.3390/ijms22063150.
68. Zhang, K., Zheng, X., Wang, Q., Xu, M., Aramid nanofiber as a polyfunctional crosslinker for epoxidized natural rubber with robust and recyclable properties. *Polymer*, 305, 127195, 2024. https://doi.org/10.1016/j.polymer.2024.127195.
69. Noguchi, T., Bamba, Y., Odaka, I., Iwamoto, R., Endo, M., Isogai, A., Natural rubber composites with high strength, modulus, water-resistance, and thermal stability, prepared with cellulose nanofibrils and sodium methacrylate, *Compos. Part A: Appl. Sci. Manuf.*, 173, 107665, 2023, https://doi.org/10.1016/j.compositesa.2023.107665.
70. Li, Q., Meng, H., Song, Y., Zheng, Q., Performance enhancement of silica filled natural rubber nanocomposites using organic deep eutectic solvent, *Compos. Sci. Technol.*, 256, 110744, 2024, https://doi.org/10.1016/j.compscitech.2024.110744.
71. Ye, Z., Zhao, D., Liu, F., Luo, J., Liu, X., Zhao, W., Flexible and highly conductive Ti3C2Tx/natural rubber composites with interconnected networks for high-performance electromagnetic interference shielding, *Compos. Part A: Appl. Sci. Manuf.*, 180, 108067, 2024, https://doi.org/10.1016/j.compositesa.2024.108067.

72. Duan, X., Tao, R., Chen, Y., Zhang, Z., Zhao, G., Liu, Y., Cheng, S., Improved mechanical, thermal conductivity and low heat build-up properties of natural rubber composites with nano-sulfur modified graphene oxide/silicon carbide, *Ceram. Int.*, 48, 15, 22053–22063, 2022, https://doi.org/10.1016/j.ceramint.2022.04.196.
73. Ismail, H., Edyham, M.R., Wirjosentono, B., Bamboo fibre filled natural rubber composites: the effects of filler loading and bonding agent, *Polym. Test.*, 21, 2, 139–144, 2002, https://doi.org/10.1016/S0142-9418(01)00060-5.
74. Onyeagoro, G.N., Cure Characteristics and Physico-Mechanical Properties of Carbonized Bamboo Fibre Filled Natural Rubber Vulcanizates, *Int. J. Modern Eng. Res. (IJMER)*, 2, 6, 4683–4690, 2012.

3

Natural Rubber-Based Bio-Composites and Bio-Nanocomposites

Jabulani I. Mnyango[1]*, Mandla V. Khumalo[2] and Shanganyane P. Hlangothi[1]

[1]Centre for Rubber Science and Technology, Department of Chemistry, Nelson Mandela University, Port Elizabeth, South Africa
[2]Department of Chemistry, Nelson Mandela University, Port Elizabeth, South Africa

Abstract

An emphasis on eco-friendly, easily accessible, and renewable materials has grown because of heightened concerns about the environment, global warming, and the economy, which have been affected using non-renewable resources. While natural fibers are derived from residues left after essential oils have been extracted or food has been consumed, natural rubber (NR) comes from a natural source. Thus, NR and natural fibers are essential in developing eco-friendly, inexpensive, and renewable materials, such as natural fiber composites and NR-based nanocomposites. The existence of these NR-based bio-composites and bio-nanocomposites results from substantial research in the composites sector. Natural fibers in these composites serve the same purpose as synthetic reinforcing materials, i.e., they reinforce the matrix (NR), improving its curing, physical, mechanical, and thermal properties. These properties, along with non-toxicity and affordability, drive the widespread use of NR/natural fiber bio-composites and bio-nanocomposites in various industries. This chapter provides an overview of these composites, with special attention to their fabrication, characteristics, and current and potential applications. Notably, different natural fibers have a similar influence on NR's properties. The automotive industry has been a frequent user of NR/natural fiber bio-composites.

***Keywords*:** Natural rubber, natural fiber, bio-composites, bio-nanocomposites, rubber properties

**Corresponding author*: s216281547@mandela.ac.za.

Visakh. P.M. and Shiv Kumari Panda (eds.) Natural Rubber-Based Composites and Nanocomposites, (63–116) © 2025 Scrivener Publishing LLC

3.1 Introduction

The increasing environmental awareness around the world has encouraged humans to act on the myriad threats that are disruptive to the conservation of the natural environment. Natural resource depletion as well as water and air pollution, where the latter is commonly known to promote global warming, are examples of these threats [1]. As a result, there has been a gradual transition toward other forms of development and behavior. This is through several regulations aimed at strengthening sustainability, which draws individuals' attention to the importance of committing to decisions that benefit both present and future generations with respect to economic, environmental, and social resources [1, 2]. Sustainable materials, alternatively known as "green" or eco-friendly materials, are currently receiving a lot of attention and are characterized as inexpensive renewable resource materials that are locally produced without causing any kind of harm to living beings [2]. Composites, thermoplastics, thermosets, and elastomers are the most used materials, which justifies the recent interest in their sustainability [1, 3]. Of these materials, composites are relatively newer, and their development has grown significantly over time. By definition, a composite is a multiphase heterogeneous product consisting of at least two materials [1, 3]. Although many composites show great potential as sustainable materials, e.g., bio-composites and bio-nanocomposites, some are not regarded as sustainable if one or more of their parent materials are regarded as eco-unfriendly or if their production process poses a serious harm to the environment and human health [2, 4]. Examples include composites that contain synthetic fillers like carbon black (CB), silica, and carbon fiber, and/or synthetic polymers like polyethylenes, polybutadienes, polypropylenes, polystyrenes, polyesters, butyls, nitriles, nylons, epoxies, vinyls, and those that are acrylated, chlorinated, fluorinated, and sulfonated [5]. Particulate matter, organics, sulfur compounds, trace elements, and various gases are a few of the harmful pollutants linked to the manufacturing of synthetic fillers [5, 6]. Synthetic elastomers, on the other hand, arise from non-renewable sources (i.e., petroleum product), which means that they are non-biodegradable and lead to landfill accumulation [6]. Additionally, petroleum drilling can substantially weaken the quality of air, water, and land *via* leakage from working wells and processing plants [6, 7]. Another downside of the synthetic fillers and elastomers is their high cost [8]. As a result, traditional composites are starting to be replaced with types that in fact contribute (substantially) to sustainability and are potentially cost-effective. These include elastomer-based bio-composites and bio-nanocomposites, which are types of composites with natural (i.e., bio-based) fiber and matrix as their primary components [1, 3–6]. The size of the dispersed natural fiber particle phase is the fundamental distinction between these composites, in addition to the fact that different natural fibers typically have different geometries and varied chemical components that are crucial for their use in such composites [9, 10]. Natural fibers (primary and secondary types) include those derived from plants (vegetables), which contain the important chemical components (polysaccharides) such as cellulose, hemicellulose, lignin, and chitin; those derived from animals, which contain a variety of proteins; and those derived from minerals, which contain a variety of inorganic natural sources. Inexpensive nature, various types of functional groups that are easy to modify, low density, and greater length-to-diameter ratio are also crucial characteristics of natural fibers, and it is these characteristics that are primarily responsible for their superior mechanical properties [1, 2, 10, 11]. In contrast to bio-composites, which comprise natural fiber particles that are not on the nanoscale, the bio-nanocomposites consist of particles with one, two, or three dimensions of less than 100 nm [12, 13]. Furthermore, composites are generally divided into different classes, with *class 1* comprising composites reinforced with a specific filler and *class 2* encompassing structural composites [13, 14]. Because natural fibers are reinforcing fillers, bio-composites and bio-nanocomposites are examples of *class 1* [2, 15]. *Class 2* examples include materials composed of both composite and homogeneous components (i.e., composite sandwich panel).

Bio-composites and bio-nanocomposites can also be prepared using natural rubber (NR), which is primarily harvested as latex from rubber trees (*Hevea brasiliensis*) and serves as their base polymer. There are two types of natural rubbers: *cis-1,4-polyisoprene* and *trans-1,4-polyisoprene*, with the former serving as a typical bio-based elastomeric matrix for bio-composites and bio-nanocomposites [4]. Various approaches can be used to create NR-based bio-composites and bio-nanocomposites, but caution is typically required because natural fibers and NR have different characteristics from the common fillers and matrices used in traditional composites. Most non-polar matrices are generally incompatible with natural fibers due to their polarity and hydrophilic properties [1, 2]. Therefore, since NR molecules of NR are very hydrophobic and non-polar, the surface of natural fibers used in NR-based bio-composites and bio-nanocomposites is often modified to enhance their adhesion to NR and reduce moisture absorption [1, 16, 17]. This modification also facilitates in-depth exploration of these composites. Additionally, NR could also be modified to enhance compatibility with natural fibers, maximizing the exceptional properties of the resultant NR-based bio-composites and bio-nanocomposites [6, 7]. These composites are currently the focus of research and various industries due to their low density, lightweight, economic viability, availability of renewable natural resources, biodegradability, reusability, and advantageous properties such as high strength (e.g., tensile, impact, and tear), low thermal and electrical conductivity, and high resistance to corrosion, fatigue, wear, and creep [2, 7, 18]. NR-based bio-composites and bio-nanocomposites have also garnered significant attention because, while NR is derived from a natural source, natural fibers are often obtained from residues left after essential oils have been extracted and food has been consumed, depending on the source. Moreover, although these composited are currently in the research and development stage with limited applications, primarily in the automotive sector, they are expected to play a significant role in the future, particularly in building and construction, sports equipment, agriculture, aerospace, marine industries, and other engineering-related sectors [2, 8, 9, 19]. This chapter aims to provide an overview of natural rubber-based bio-composites and bio-nanocomposites, with a particular emphasis on their preparation, properties, and both established and potential applications.

3.1.1 Natural Fibers

Natural fibers belong to the category of fibers that are neither synthetic nor man-made and originate from specific plants, animals, and minerals [10, 20]. These fibers have received significant attention as reinforcing agents or fillers in composites due to the harmful effects of synthetic fibers on the environment and human health, including skin allergies and lung cancer [2, 10–12]. The widespread use of natural fibers in composites is attributed their resource efficiency, affordability, minimal wear on processing machinery, low density, and other beneficial properties [10, 12]. Although asbestos is a type of natural fiber (specifically, a mineral fiber), it is associated with health risks due to its carcinogenic properties [2, 12, 21]. Nevertheless, various plant and animal fibers, along with minerals such as chalk, limestone, talc, silica (silicon dioxide), and montmorillonite, remain widely available (Figure 3.1), and the composites sector continues to extensively benefit from them [2, 10, 22]. In addition to the plant fibers shown in Figure 3.1, other fibers originate from bast (e.g., flax, hemp, kenaf, and isora); fruit (e.g., coconut and assai); seed (e.g., kapok and Brazil nut); leaf (e.g., pineapple, abaca, henequen, curaua, piassava, and fique); and stalk (e.g., barley, reed, wheat straw, sugarcane bagasse, and kernel) [2, 11]. Some plant fibers are typically produced and sold globally, whereas trade regulations

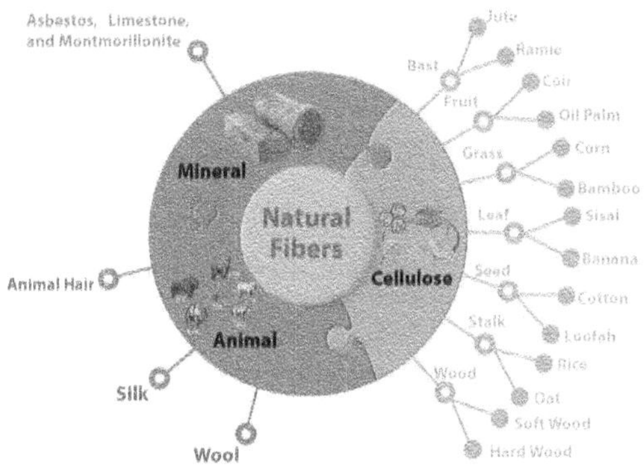

Figure 3.1 Classification of some of the natural fibers based on their sources. Reprinted/adapted with permission from Ref. [2].

for animal fibers are generally stricter, possibly due to the higher likelihood of disease transmission from their sources. Research by Dénes *et al.* [14] supports this finding, demonstrating that examined sheep wool samples contained traces of microbes and nutrients that promote microbial growth, which becomes more pronounced with longer exposure duration.

Animal fibers are often derived from birds, chickens, domestic silk moths (commonly, *Bombyx mori* L., *Antheraea pernyi*, *Antheraea yamamai*, and *Philosamia Cynthia ricini*), alpaca, camel, yak, sheep, horse, Tibetan antelope, Angora rabbit, bison, llama, and various goats, with different types of proteins serving as prime constituents for composite applications [2, 10, 23]. In contrast to the production of avian fibers and silk fibers by birds, chickens, and domestic silk moths, the hair and fur of other mammals serve as sources of cashmere, shearling, persian, shahtoosh, angora, mohair, and bison wool [24, 25]. Among these, cashmere wool is the most popular due to its relatively low density and exceptionally soft, luxurious texture [11, 26]. Eggshells, seashells, and cuttlebone have recently been proposed as potential sustainable sources of animal fibers [13, 15]. Although the chemistry of animal fibers remains uncertain, their proteins—collagen, keratin, and fibroin—are formed through amino acids polymerization *via* peptide bonds into macromolecular chains. Consequently, natural fibers are also referred to as natural protein fibers [26]. Thus, the extension exploration of animal fibers for composite applications may largely depend on the presence of the protein-based constituents [14, 16, 17, 26]. In contrast, polysaccharides, particularly crystalline cellulose, semicrystalline chitin, and amorphous hemicellulose and lignin, are the primary chemical components of plant fibers used in composites [1, 2, 27]. Literature indicates that the existence of cellulose, chitin, hemicellulose, or lignin in plant fibers is the key reason why they are substantially beneficial for composites application [10]. The amounts of these polysaccharides are shown in Table 3.1 for most of the plant fibers. This includes the microfibrillar angle, which

Table 3.1 Essential chemical components of plant fibers for their usage in composites [1, 2].

Fiber	Cellulose (wt. %)	Hemicellulose (wt. %)	Lignin (wt. %)	Microfibrillar angle (°)
Hemp	68–75	1–22.4	3.7–10	2–6.2
Flax	64.1–75	10.4–20.6	2–5	5–10
Jute	61–71	13–20.4	5–13	8
Ramie	61.9–85	3–7.6	0.5–9.1	69–83
Kenaf	31–72	20–33.9	8–21.2	2–6.2
Oil palm	47.9–65	19.1	3.5–4	–
Coir	32–43	0.1–0.3	40–45	30.5
Kapok	35	22	21.5	–
Cotton	82.7–92	–	0	–
Pineapple	70–82	18	5–12.7	–
Sisal	60–78	10–38.2	8–14	10–25
Banana	60–65	10–3	5–10	11–12
Henequen	70–77.6	4–20	8–13.1	–
Abaca	56–63	20–25	7–12.4	20–25
Bagasse	41.1–55.2	16.8–31.8	22.3–25.3	–
Wheat straw	33–45	1–32	12–20	–
Rice straw	36–57	33	8–19	–
Bamboo	22.8–56.7	2.2–43.8	21–31	–
Rice husk	35–45	19–28	20	–
Softwood	30–60	20–30	21–37	–
Hardwood	31–64	25–40	14–34	–

–, Not reported.

is believed to be the primary factor that influences the plant fiber strength characteristics [2, 28]. Moreover, despite lignin being hydrophobic, cellulose and hemicellulose have a high hydrophilicity, making the plant fibers very hydrophilic, high water absorbing and swelling materials [1, 2, 16, 18]. Hence, lignin is often extracted from plant fiber (or its source) and dispersed in the hydrophobic matrix to produce the hydrophobic composites for suitable applications [18]. However, by physically, chemically, or biologically treating the surface (i.e., functionalization or structural modification) of the plant fiber, the hydrophilicity and subsequent swelling of cellulose and hemicellulose can be reduced or avoided [1, 2, 10, 29]. Physical process employs plasma, chemical process uses a variety of chemicals, including alkali, silane, water repellents, peroxides, permanganates, etc., and biological process uses bacteria in anaerobic environments [1, 2, 10, 19]. These processes attach functional groups to the fiber's surface, resulting in the formation of strong covalent bonds with the matrix [1, 2, 10, 19]. Furthermore, surface etching or coarsening resulting from plasma treatment enhances the surface roughness of the fiber, creating a strong interface with the matrix *via* mechanical interlocking. The chemical treatment enhances the fiber/matrix interface by modifying the crystalline structure while reducing or removing weak components like hemicellulose and lignin from the fiber structure, therefore creating the strong chemical bonding between fiber and the matrix [1, 2, 10, 19]. Biological treatment, on the other hand, uses a fermentation process to deposit bacterial cellulose on the fiber surface, therefore improving interfacial adhesion with a matrix significantly [1, 2, 10, 29]. Overall, the presence and functionalization of essential chemical components ultimately determine the desired composite properties and performance [1, 2, 27, 29].

Natural fibers are typically divided into two groups (i.e., primary and secondary groups) from a commercial perspective. Plant fibers in the first group—i.e., jute, hemp, kenaf, and sisal—are primarily cultivated for fiber production, whereas those in the second group—i.e., pineapple, oil palm, and coir—are used as by-products [2, 20, 30]. However, animal fibers differ from plant fibers because they can be sourced from animals raised exclusively for their skin, hair, fur, or feathers (primary category), or from animals slaughtered for food (secondary category) [1, 2, 31, 32]. Additionally, natural fibers are obtained based on the nature of their sources. Shearing and plucking (either manually or using a tumbling machine) are used to separate animal fibers from specific animals [1, 2, 30, 33]. Plant fibers, however, are extracted using methods that depend largely on the benefits and limitations of the fiber manufacturing process, as well as the fiber's intended use [2, 21, 34, 35]. These methods, an essential stage in the creation of plant fiber-based composite materials, include the following [2, 21, 22, 36]:

- The decortication method, often used to separate plant fibers from sources with fibrous stems, such flax, hemp, and jute, involves removing the stem's exterior portion (i.e., bark or skin) to access the interior fibers. This method, which may be mechanical or chemical, produces high-quality fibers with low impurity content.
- The blow molding method, which is commonly used to separate fibers from bamboo, involves heating and softening bamboo fibers using a screw extruder, followed by cooling in a chamber to produce continuous high-quality bamboo fiber samples.
- The enzymatic method, often used to extract high-quality fibers from sisal and kenaf, involves the use of specific enzymes that selectively break down the non-fibrous components of the fiber's source, facilitating the retrieval of fibrous materials.

Methods used to extract natural fibers from their respective sources are equally important to their final properties and the performance of their composite materials [10, 18, 37–40]. Table 3.2 lists the typical natural fibers' mechanical properties, which contribute to their use in composite materials [2, 10, 41, 42]. However, to thoroughly explore these properties and achieve consistency, quality, and reproducibility in composite results, natural fibers must meet a set of standards or performance criteria [2, 10, 23, 43–47]. In addition to the principal chemical composition (i.e., polysaccharides) and the surface treatment of polysaccharides, the general factors considered in the performance criteria for natural fibers, especially in composite materials, are as follows [1, 2, 21, 24, 48]:

- Classification. Natural fibers are classified into three categories based on their source: plant (such as cellulosic, hemicellulosic, lignosic, and lignocellulosic), animal (such as protein-based), and mineral (such as asbestos). Understanding of the fibers' fundamental characteristics and prospective uses is facilitated by their classification.
- Physical properties. These imclude crucial parameters such as density ($g{\cdot}cm^3$), diameter (range: $\mu m - nm$), length (range: $mm - \mu m$), and aspect ratio, all of which are

Table 3.2 Typical mechanical properties of most natural fibers [1, 2, 10].

Fiber	Ultimate tensile strength, MPa	Tensile (Young's) modulus, GPa	Elongation at break, %
Flax	88–1,600	3–80	1.2–10
Hemp	310–900	30–80	1.6–6
Jute	385–850	10–55	1.2–8
Ramie	348–938	3.5–128	1.2–8
Kenaf	216.4–1,191	2.86–60	1.6–6.9
Oil palm	–	1–9	8–25
Coir	106–593	1.3–6	14.2–59.9
Kapok	93.3	–	–
Cotton	45.5–1,000	5.5–12.6	3–10
Pineapple	20–1,672	60–82	1-–14.5
Sisal	80–840	9–38	2–25
Banana	355–980	12–33.8	3–53
Henequen	355–580	10.1–16.3	3–4.7
Abaca	220–980	3–12	3–10
Bagasse	20–40	2–27.1	1.1
Wheat straw	21.2–31.2	4.76–6.58	–
Rice straw	450	26	2.2
Rice husk	19–135	0.3–2.6	–
Bamboo	140–230	11–2	–
Softwood	600–1,000	3.6–40	4.4
Hardwood	–	5.2–37.9	–

–, Not reported.

frequently used to assess the viability of a fiber for a particular application. This is an example of how performance criteria may be established to meet specific requirements. The diameter of fibers can be modified, and the resulting particles can be used to produce both bio-composites and bio-nanocomposites, which are identical materials differing only in the size of the fiber used in their matrix.

- Mechanical properties. These include ultimate tensile strength (MPa), tensile (Young's) modulus (GPa), flexural strength (MPa), elongation at break (%), toughness (MJ/m^3), and impact strength (kJ/m^2). They are key factors in determining the state or quality for fiber reinforcing in composites. Here, setting performance criteria include adjusting the fiber's loading in the matrix to achieve desired mechanical properties.
- Thermal properties. These involve thermal stability and conductivity (W/(m·K)). Generally, the former refers to the material degradation analyzed through Thermal Gravimetric Analysis (TGA) and glass transition temperature (T_g) measured *via* Differential Scanning Calorimetry (DSC). Thermal properties are extensively studied when a fiber-based composite is intended for applications involving elevated temperatures.
- Moisture content and water absorption. This is particularly important for hydrophobic matrices. Fibers are customarily used when they are completely dried. However, performance criteria may set acceptable moisture content levels and absorption limits based on the intended use of the manufactured fiber-based composite material and their inherent capacity to absorb moisture. To ensure consistent performance, dimensional stability, and fiber durability in composites, moisture levels must generally be kept to a minimum.
- Purity. Fibers must typically be as free of contaminants as possible, as their adhesion to the matrix and performance in a composite would be impaired by any non-fibrous material. This remains true even if a fiber's surface has been modified before being used in a composite. Fiber purity is often ensured by performance criteria, which, in relation to the intended application of the composite, include acceptable impurity levels and guidelines for processing and purification.

The development of performance criteria for natural fibers is managed by ASTM (American Society for Testing and Materials) International, ISO (International Organization for Standardization), and numerous other national materials standards bodies [10]. These organizations collaborate or work independently to develop standards that ensure the effectiveness and consistent quality of natural fibers used in various applications [10].

3.1.2 Natural Rubber

Natural rubber (NR) is a high-molecular-weight, non-polar elastomer composed of repeating linear and unsaturated isoprene units (Figure 3.2). Its chemical name is *cis-1,4-polyisoprene*, which is commonly used because the raw source of NR is a milky colloidal suspension called latex (an emulsion of 35 wt.% *cis-1,4-polyisoprene*, 60 wt.% water, and 6 wt.% other non-rubber constituents (Figure 3.3)) [25, 49]. From a commercial and industrial standpoint, *H. brasiliensis*, a tree species indigenous to the Amazon regions of South America, is the primary source of latex [26, 50, 51]. Moreover, *trans-1,4-polyisoprene* occurs naturally found in gutta-percha and balata, two latexes that, like *cis-1,4-polyisoprene*, are obtained from the tropical rubber tree (*Manikara bidentata* species), native to northern South America and the West Indies [4]. Since both *cis-1,4-polyisoprene* and *trans-1,4-polyisoprene* are obtained from renewable sources, NR is considered a bio-based material and an eco-friendly option for developing sustainable composites [1].

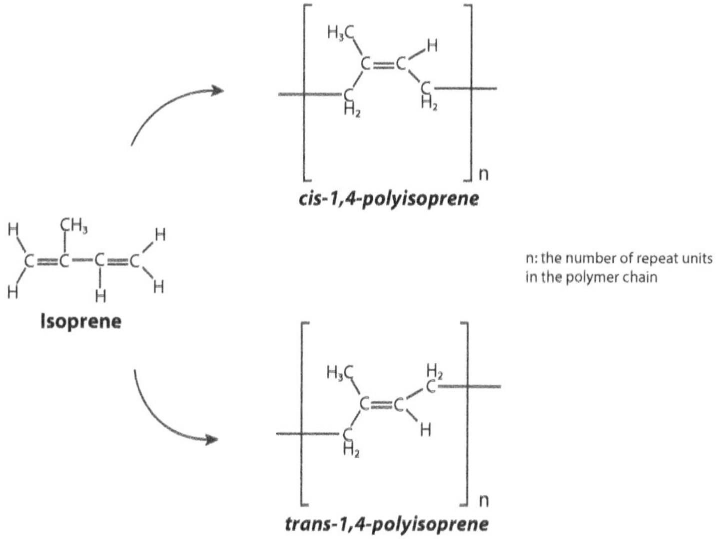

Figure 3.2 Chemical structure and molecular formula of NR: *cis-1,4-polyisoprene* and *trans-1,4-polyisoprene* [4, 31].

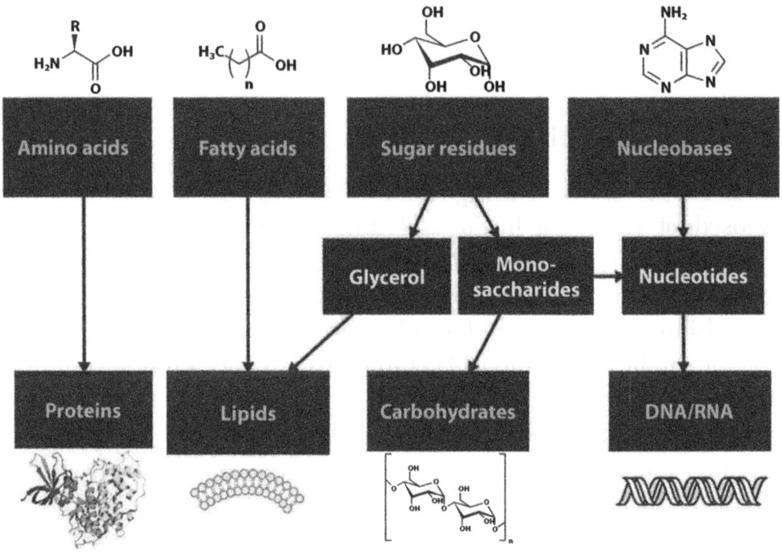

Figure 3.3 Non-rubber constituents (excluding water) present in *cis-1,4-polyisoprene* latex. Reprinted/adapted with permission from Ref. [25].

Generally, the rubber tree's phloem is gently tapped, and the sap (latex) that flows down the side of the tree is collected in a container. A commonly reported for the latex of the *H. brasiliensis* tree, the initial processing step often increases the alkalinity (pH) of the field-collected latex by it with the negatively charged ion or non-ionogenic surfactants (commonly ammonia) [1, 4, 27]. This is typically done to prevent premature coagulation, hinder microbial growth, and preserve or enhance the intrinsic mechanical or colloidal stability of latex [1, 27]. Additionally, treating freshly collected latex with ammonia can denature non-rubber constituents such as proteins and hydrolyze others, like lipids and carbohydrates, into even smaller substances [25, 52]. This is crucial since these constituents have a history of impairing various properties of latex [28, 29, 53]. Similar considerations apply to solid *cis-1,4-polyisoprene*, which is commonly packed and sold in bales of different NR grades, as it is produced by coagulating *H. brasiliensis* latex with acetic or hydrogen carboxylic acid, followed by drying the resulting product [26, 28, 29]. Prior to the production of *cis-1,4-polyisoprene* bales, *H. brasiliensis* latex is concentrated *via* creaming or centrifugation [25, 30, 31, 54]. During this process, a significant amount of non-rubber constituents is removed, while the *cis-1,4-polyisoprene* content is increased to >90 wt.% [25, 30, 31, 55].

The Standard Malaysian Rubber (SMR) system, in use for approximately six decades ago, serves as the basis for various grades of NR (specifically, *cis-1,4-polyisoprene*), which are classified based on processing type and impurity content, including dirt, ash, nitrogen, volatile substances, Wallace rapid plasticity, and plasticity retention index [31, 55]. For example, the order of increasing impurity content in common NR grades is SMR-CV < SMR-L < SMR-5 < SMR-GP ≤ SMR10 < SMR-20 [31]. While the numerals generally indicate impurity content, the letters (i.e., CV, L, and GP) denote controlled viscosity, latex grade, and general purpose, respectively [31]. In contrast to other grades, the SMR-CV and SMR-L are among the highest-quality grades available on the market [4, 25]. Even so, SMR-10 and SMR-20 account for most of the 40% share of the global elastomer market [9]. This is because these grades are widely used in the rubber industry, particularly in the production of vehicles tires [9]. The feasibility of this application, along with the production of various molded and extruded products, depends on NR being compounded with additives and fillers, followed by vulcanization using sulfur (commonly), peroxides, or isocyanates and an accelerator [9, 31–34, 56].

Vulcanized *cis-1,4-polyisoprene* is amorphous when unstretched, but when stress is applied, some of its regions exhibit an oriented crystal structure along the direction of the force (Figure 3.4) [35]. It spontaneously crystallizes when stretched, even at low temperatures, due to its high stereo regularity or tacticity [4, 31, 35, 57]. Unlike *cis-1,4-polyisoprene*, *trans-1,4-polyisoprene* is exceptionally crystalline, which accounts for the tougher, more durable, and leathery properties of gutta-percha and balata [4]. Nonetheless, the maximum crystallinity of *cis-1,4-polyisoprene* is typically approximately 25–30% [31]. Its rapid crystallization during stretching often yields materials with exceptionally high tensile and tear strengths,

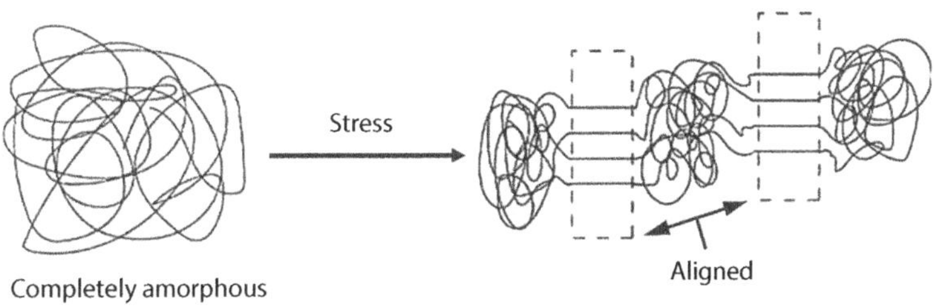

Figure 3.4 Schematic diagram showing *cis-1,4-polyisoprene* under strain-induced crystallization. Reprinted/adapted with permission from Ref. [35].

whether in the form of filled or unfilled vulcanizates [31]. These properties, along with others in Table 3.3, also arise from the fact that *cis-1,4-polyisoprene* molecules are highly intertwined and connected, leading to a strong intermolecular interactions [31]. Table 3.3 also presents the negative aspects of *cis-1,4-polyisoprene*. Due to advantages and disadvantages

Table 3.3 Common property-based advantages and disadvantages of unvulcanized and vulcanized *cis-1,4-polyisoprene* [3, 9, 26, 31, 35].

Advantages	Disadvantages
• *Less potent odor and highly versatile.	• Poor resistance to abrasion.
• *Good behavior and interaction with additives and fillers during compounding, extrusion, calendaring, and molding processes.	• Low resistance to weather-, ozone-, and light-induced environmental deterioration.
• *Highly tacky. Thus, its wide use in industrial applications because it adheres, with ease, to itself as well as other rubbers and materials like brass-plated steel cable.	• Low resistance to oil and fuel.
• Low dissipation factor or hysteresis. Thus, under dynamic stress, there is minimal build-up of heat. This normally maintains the service integrity of new tyres and elevates retreadability.	• Poor resistance to heat. Typical working temperature range: between −55 °C and +82 °C.
• Low rolling resistance and enhances fuel efficiency.	• *Poor resistance to organic liquids.
• Excellent elasticity or stretchability (i.e., high elongation), flexibility, and compressibility.	
• Outstanding resistance to fatigue derived from wear, including chipping, cutting, etc.	
• Excellent resistance to cold.	
• Good electrical insulator.	

*, Unvulcanized.

of *cis-1,4-polyisoprene* based on its physical and mechanical properties, researchers have explored *cis-1,4-polyisoprene*-based composites to expand its applications [4, 31, 35, 58]. Additionally, to enhance the properties of NR and to meet the special needs of NR-based product manufacturers, it can be modified to produce 20–30% oil-extended, epoxidized, or methacrylate-grafted NR [5, 6, 31, 59].

3.2 Natural Rubber-Based Bio-Composites

Rubber composites, whether they are liquid or solid matrix-based, contain a variety of necessary ingredients, such as activators, processing aids, additives, reinforcing agents (i.e., fillers), and curatives [1, 36, 37]. Since each of these ingredients plays an essential role in enhancing the rubber composites' compounding processes, the reinforcing fillers, in particular, further enhance the properties of these composites (Table 3.4) while also reducing the cost of the final product [1]. Natural rubber (NR)-based bio-composites are an important type of rubber composite that has recently attracted considerable interest in both academia and industry, particularly in terms of properties, performance, and applications, and sustainability [1, 38–40, 60]. These composites, in which NR is primarily *cis-1,4-polyisoprene*, are generally referred to as "green" or eco-friendly materials, in which a mechanically and/or chemically transformed natural fiber serves as a reinforcing filler and is dispersed efficiently in a solid or latex NR matrix [1, 2, 61]. Both NR and natural fibers, which are derived from natural systems, contribute to the outstanding properties and high performance of NR-based bio-composites [1, 62, 63]. These renewable natural resources serve as alternatives to their synthetic counterparts due to their advantageous properties, such as low cost, non-toxicity, low density, length, and aspect ratio, and strong compatibility or interfacial adhesion between their phases [1, 64, 65].

Natural fibers utilized in NR-based bio-composites are generally short but do not fall into the nanoscale region [1, 2]. As shown in Figures 3.5(a,b), they are twisted materials that produce continuous threads with sufficient mechanical integrity to withstand stresses exerted during composite preparation or service [2, 9, 10, 18, 66]. NR binds these natural fibers, transferring loads between them while also providing the bio-composites their net shape, high surface quality, and excellent dimensional stability [1, 2, 67]. As a result, NR-based bio-composites possess novel properties superior to those of NR and natural fibers individually [1, 2, 10, 18, 68]. Moreover, since various natural fibers and rubbers have different properties, the

Table 3.4 A variety of common properties of rubber composites essential for industrial applications [1, 64, 65].

Property type	Subtype
Curing	Scorch and optimum cure time
	Minimum and maximum torque
	Torque difference
	Cure rate index
Mechanical	Tensile (i.e., tensile strength, elongation at break, and tensile moduli)
	Abrasion and tear resistance
	Hardness
	Rebound resilience
	Flexural (i.e., bending or modulus of rupture) strength
Dynamic mechanical	Storage and loss modulus
	Loss or damping factor
	Glass transition temperature (T_g)
Electrical	Dielectric strength
Thermal	Heat resistance (i.e., thermal stability) and T_g

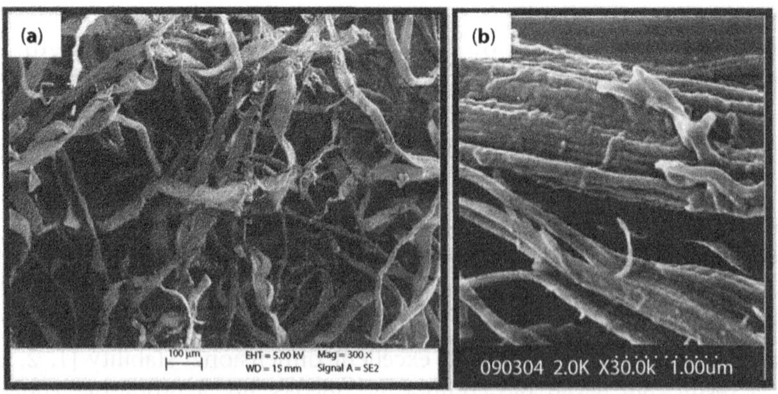

Figure 3.5 SEM micrographs of natural fibers: (a) plant (kenaf bast) fiber and (b) animal (*Antheraea pernyi*) fiber. Reprinted/adapted with permission from Ref. [9].

properties (and performances) of NR-based bio-composites are expected to vary among different NR-based bio-composites. This is because the way the matrix transfers stress to the fiber determines the mechanical properties of NR-based bio-composites [1, 69]. The interfacial adhesion between NR and natural fiber—commonly visualized through scanning electron microscopy (SEM)—also influences the study of the properties and performance of NR-based bio-composites [9, 70–74]. For example, Figures 3.6(a–d), which show micrographs of an NR-based bio-composite composed of unfunctionalized horsetail (*Equisetum arvense*) fibers, demonstrate that the vulcanizates studied had good properties due to strong interfacial adhesion of horsetail fibers to the NR matrix (as described briefly in Table 3.5, particularly under reference [38]). Although such interfacial adhesion was achieved despite the presence of larger size particles, aggregates, and microdefects, the functionalization (i.e., treatment of the surface and subsequent modification) of most natural fibers (especially plant fibers) is typically performed, as it determines the extent to which they adhere to the NR matrix [18, 74]. Several solutions

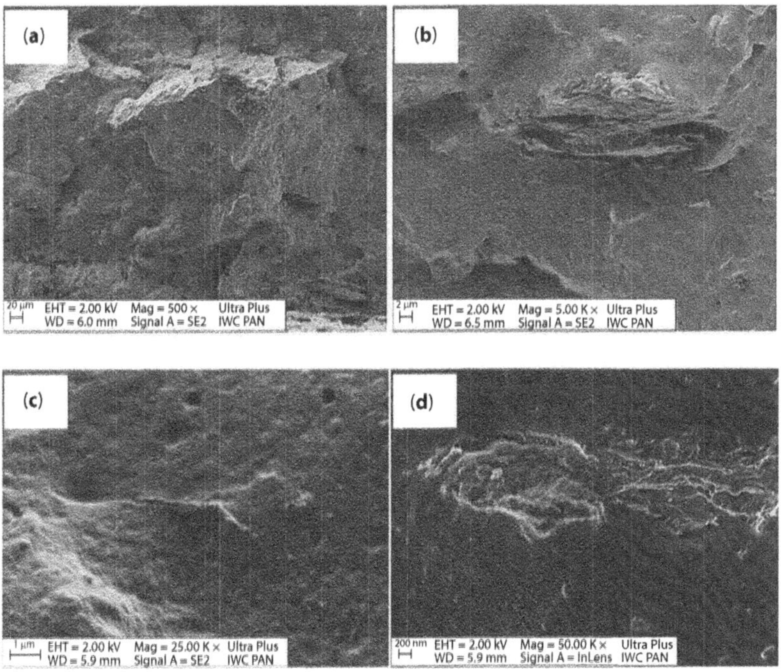

Figure 3.6 SEM micrographs of NR-based bio-composite (with 30 phr of *Equisetum arvense* fiber) under different magnifications: (a) 500 X, (b) 5.00 KX, (c) 25.00 KX, and (d) 50.00 KX. Reprinted/adapted with permission from Ref. [38].

Table 3.5 Selected studies that have investigated the properties of bio-composites that are based on NR (*cis-1,4-polyisoprene*).

Source of filler	Properties of NR-based bio-composites whose filler's source surface was modified using solutions comprised of common inorganic compounds	Study
Pineapple leaf	Loading of 20 wt.% pineapple leaf-sourced filler resulted in the highest ultimate tensile strength (i.e., tensile stress, UTS), hardness, elongation at break (tensile strain, ε_b), and tensile modulus at 100% (M100) and 300% (M300) elongation.	[39]
Pawpaw (*Carica papaya*) seeds	Upon adding 10 phr pawpaw (*Carica papaya*) seeds-sourced filler and increasing it to 50 phr, the scorch time (t_{s5}) and cure time (t_{c90}) of the resultant bio-composites decreased, while the maximum torque (M_H) increased. From 10 to 40 phr filler loading, ε_b decreased, while the UTS and M100 both increased. While the compression set and swelling % decreased with an increase in filler loading, both the abrasion resistance and stiffness (i.e., hardness) of the bio-composites increased.	[40]
Common Nettle (*Urtica dioica L.*)	The t_{c90}, M_H, and crosslink density (v_e) of the bio-composites increased when the loading of 10 phr common nettle-sourced filler and its subsequent increased to 30 phr. The filler resulted in NR-filler and filler–filler interactions, according to the Payne effect, which increased with increasing filler loading. Characterization of the bio-composites with FT-IR led to the identification of flavonols, glycosides, and phenolic acids, which are generally confirmed by their high antioxidant activity, and thus their anti-aging effect on the bio-composites studied. The increase in filler loading increased the resistance to thermal degradation, UTS, ε_b, tensile modulus, and gas (air) barrier properties of the bio-composites.	[42]

(*Continued*)

Table 3.5 Selected studies that have investigated the properties of bio-composites that are based on NR (*cis-1,4-polyisoprene*). (*Continued*)

Source of filler	Properties of NR-based bio-composites whose filler's source surface was modified using solutions comprised of common inorganic compounds	Study
Sisal/Oil palm hybrid	An increased loading of a filler derived from sisal/oil palm hybrid decreased the UTS and tear strength of the resultant bio-composites but increased tensile modulus.	[43]
Rubberwood sawdust	The thermal stability and modulus of rupture (MOR) (i.e., flexural strength) of the NR-based bio-composites comprised of rubberwood sawdust-sourced filler decreased while the UTS, tensile modulus, modulus of elasticity (i.e., elastic modulus), and hardness increased with the addition of 30 wt.% filler and its subsequent increase to 60 wt.%. Even though NR is hydrophobic, it was discovered that the hydrophilic nature of the filler caused the water absorption of bio-composites to increase gradually as the filler loading increased.	[44]
*Sheep wool	The NR/sheep-sourced wool bio-composites had higher thermal resistance (m^2K/W) and higher sound (acoustic) absorption (at higher frequencies) than the resin/sheep-sourced wool composite, but lower thermal conductivity ($W/(m{\cdot}K)$), water absorption, and water vapor diffusion resistance.	[14]
Rice husk	The curing characteristics (i.e., t_{s5}, t_{c90}, minimum torque (M_L), M_H, and torque difference ($M_H - M_L$) and v_e of the NR-based bio composite sample comprising 30 phr rice husk flour (RHF)-sourced filler were the same as those of the reference sample (i.e., NR without RHF filler), but the mechanical properties of the former were different from those of the latter. UTS, ε_b, and tear strength of the bio-composite sample were found to be significantly lower than those of the NR alone, but the M100 and M300 were higher.	[45]

(*Continued*)

Table 3.5 Selected studies that have investigated the properties of bio-composites that are based on NR (*cis-1,4-polyisoprene*). (*Continued*)

Source of filler	Properties of NR-based bio-composites whose filler's source surface was modified using solutions comprised of common inorganic compounds	Study
*Chitin	The increased loading of chitin-sourced filler in the mass of the NR latex foam (NRLF) decreased several air bubble volumes, consequently increasing the spaces for the filler, which increased the density, hardness, compressive strength, and antibacterial activity of the resultant bio-composites, but decreased the UTS, ε_b, and swelling %.	[46]
*Oil palm empty fruit bunch	When the loading of oil palm empty fruit bunch (OPEFB)-sourced filler was increased in the bio-composites, the UTS, Young's modulus, and storage modulus (E') increased, while ε_b, damping factor $(\tan\delta)$, and glass transition temperature (T_g) decreased. Both the filler incorporation and increase had no effect on the thermal degradation of the bio-composites.	[47]
Banana	Increased loading of the filler derived from banana resulted in increased hardness and tensile moduli of the resultant bio-composites but decreased UTS and ε_b.	[48]
Spirulina Platensis	As the loading of the *Spirulina platensis*-sourced filler was increased, the t_{s2}, t_{c90}, M_L, and M_H of the resultant bio-composites decreased while cure rate index (CRI) increased. Increasing the filler loading also increased the hardness, E', and loss modulus (E''), but decreased the UTS, ε_b, $\tan\delta$, and T_g. Filler's incorporation and subsequent increase improved the thermal stability of the bio-composites.	[49]

(*Continued*)

Table 3.5 Selected studies that have investigated the properties of bio-composites that are based on NR (*cis-1,4-polyisoprene*). (*Continued*)

Source of filler	Properties of NR-based bio-composites whose filler's source surface was modified using solutions comprised of common inorganic compounds	Study
Jute	Increasing the loading of jute-sourced filler in the NR/Jute-sourced filler bio-composites decreased the UTS and ε_b but increased the Young's modulus and hardness.	[51]
Chicken feather	Increased loading of a filler sourced from the chicken feathers increased tensile moduli but decreased swelling %, UTS, and ε_b.	[52]
Field horsetail (*Equisetum arvense*)	Reinforcing the NR-based bio-composites by increasing the loading of field horsetail-sourced filler resulted in increased t_{c90}, M_L, M_H, and $M_H - M_L$. However, the thermal stability of the bio-composites decreased as the filler loading increased. The v_e and moduli (M100, M200, and M300) both increased while UTS and ε_b decreased as the filler loading was increased. Moreover, UV, weather, and thermo-oxidative aging processes of bio-composites all led to an additional rise in v_e values, but a decrease in UTS and ε_b values. However, it was only the UV radiation and weather aging parameters that caused significant color changes on the bio-composites.	[38]
Bagasse	When the loading of SB-sourced filler was increased in NR matrix, the t_{s5}, t_{c90}, $M_H - M_L$, and v_e of the resultant bio-composites also increased. The hardness, M100, and compression set increased while UTS and ε_b decreased as the filler loading was increased. But there was no change in lightness (i.e., color and gloss) of the bio-composites when the filler loading was increased.	[53]

(*Continued*)

Table 3.5 Selected studies that have investigated the properties of bio-composites that are based on NR (*cis-1,4-polyisoprene*). (*Continued*)

Source of filler	Properties of NR-based bio-composites whose filler's source surface was modified using solutions comprised of common inorganic compounds	Study
*Cotton	The UTS, ε_b, hardness, and v_e of the NR-based bio-composites increased when the loading of the filler that was sourced from the buffing dust-jeans cotton was increased. Increasing the filler loading also increased the thermal stability and gas (oxygen) barrier properties of the bio-composites.	[16]
Kaolin	Incorporation of the kaolin-sourced filler in the NR matrix contributed to the significant increase in CRI, UTS, ε_b, and moduli (M100 and M200) of the resultant bio-composite.	[54]
Oil palm/Eggshell hybrid	An increase in the loading of a filler sourced from oil palm ash (OPA)/eggshell powder (ESP) hybrid resulted in an increase in UTS, ε_b, and tensile modulus of the resultant bio-composites, but a decrease in hysteresis loss ratio. Stress relaxation response during uploading of the bio-composites at lower and higher strain levels was almost the same as that of the unreinforced NR vulcanizate. However, during unloading, the bio-composites showed a clear decrease in stress as the filler loading was increased. Increasing the filler loading also increased the dynamic stiffness, phase angle, damping coefficient of the bio-composites.	[55]
Kenaf	Increased loading of kenaf-sourced filler in the NR/kenaf-sourced filler bio-composites increased the UTS and elasticity modulus while the ε_b decreased. However, at the highest filler loading, both the UTS and elasticity modulus decreased due to the resultant discontinuity of the matrix.	[56]

(*Continued*)

Table 3.5 Selected studies that have investigated the properties of bio-composites that are based on NR (*cis-1,4-polyisoprene*). (*Continued*)

Source of filler	Properties of NR-based bio-composites whose filler's source surface was modified using solutions comprised of common inorganic compounds	Study
Rice husk and Rice straw	Separately increasing the loadings of RH-sourced and rice straw (RS)-sourced fillers decreased UTS and increased thermal stability, Young's modulus, ε_b, water absorption, and oil absorption of their respective resultant bio-composites.	[57]
*Silk fabric	UTS, modulus of elasticity, thermal stability, and v_e of the NR/Silk fabric-sourced filler bio-composite increased when a filler was added to NR matrix, but the ε_b decreased.	[17]
Field horsetail and Common wheat	Incorporating the field horsetail (HT)-sourced and common wheat (WS)-sourced fillers on the two separate 100 phr NR compounds resulted in increased thermal stability of their respective resultant bio-composites. This also resulted in increased t_{c90}, t_{s5}, and v_e, but decreased M_L, M_H, $M_H - M_L$. UTS and ε_b, and moduli (M5, M50, M100, M200, and M300) of the bio-composites increased when the fillers were added. Both fillers did not have any effect on the $\tan\delta$, and T_g of their respective bio-composites.	[58]

(*Continued*)

Table 3.5 Selected studies that have investigated the properties of bio-composites that are based on NR (*cis-1,4-polyisoprene*). (*Continued*)

Source of filler	Properties of NR-based bio-composites whose filler's source surface was modified using solutions comprised of common inorganic compounds	Study
Cereal straw, Chalk, Talc, and Silicon dioxide (Silica)	Based on oil absorption test, the following order of increase in oil absorption was reported: NR/Talc-sourced filler bio-composite <NR/chalk-sourced filler bio-composite <NR/cereal straw-sourced filler bio-composite <NR/silica-sourced filler bio-composite. Increased loading of the fillers increased the t_{s2}, t_{c90}, M_L, M_H, and $M_H - M_L$ of their respective resultant bio-composites. Following is order of increase in these curing properties, based on filler source used: NR/cereal straw-sourced filler bio-composite <NR/chalk-sourced filler bio-composite <NR/talc-sourced filler bio-composite <NR/silica-sourced filler bio-composite. Increased loading of the fillers also increased the v_e, UTS, moduli (M100 and M200), and hardness of their respective bio-composites but decreased the tear strength, ε_b, and gas (air) barrier properties. Aging the bio-composites resulted in decreased UTS and ε_b values. Following is order of decrease in the UTS and ε_b values after aging: NR/chalk-sourced filler bio-composite >NR/talc-sourced filler bio-composite >NR/cereal straw-sourced filler bio-composite >NR/silica-sourced filler bio-composite.	[59]

(*Continued*)

Table 3.5 Selected studies that have investigated the properties of bio-composites that are based on NR (*cis-1,4-polyisoprene*). (*Continued*)

Source of filler	Properties of NR-based bio-composites whose filler's source surface was modified using solutions comprised of common inorganic compounds	Study
Rye, Oat, and Triticale straws	Based on oil absorption study, the following order of increase in oil absorption based on the filler used was reported: NR/oat-sourced filler bio-composite <NR/triticale-sourced filler bio-composite <NR/rye-sourced filler bio-composite. Increased loading of the fillers increased the v_e, t_{c90}, M_L, and M_H of their respective resultant bio-composites. Increased loading of the fillers also increased the UTS, M100, hardness, $\tan\delta$, E', E'', and flammability of their respective bio-composites but decreased the tear strength, ε_b and gas (air) barrier properties.	[3]
Mengkuang	An increased loading of Mengkuang (MK)-sourced filler in the NR matrix decreased UTS, ε_b, and impact strength of resultant bio-composites but increased tensile modulus.	[61]
*Coffee grounds	Increased loading of the spent coffee ground (SCG)-sourced filler in the NR matrix increased both the hardness and compression set of resultant bio-composites but UTS and ε_b decreased.	[63]
Chitosan	The chitosan-sourced filler did not have any effect on the specific gravity of its bio-composites but decreased the UTS, M300, M400, M500, tear strength, and ε_b as its loading was increased.	[64]

(*Continued*)

Table 3.5 Selected studies that have investigated the properties of bio-composites that are based on NR (*cis-1,4-polyisoprene*). (*Continued*)

Source of filler	Properties of NR-based bio-composites whose filler's source surface was modified using solutions comprised of common inorganic compounds	Study
Coir	Upon adding a coir-sourced filler in a mass of 100 phr NR compound resulted in increased UTS and ε_b of the resultant bio-composite. An increase in filler loading also increased ε_b but decreased UTS of the composites. Cold aging the bio-composites resulted in increased UTS and ε_b of the bio-composites, but the water and heat aging decreased the UTS while increasing the ε_b.	[65]
*Limestone	Incorporating the limestone-sourced filler in a mass of 100 phr NR compound resulted in increased UTS, Young's modulus, and thermal stability of the resultant bio-composite but decreased the ε_b. Increasing filler loading, however, decreased UTS, ε_b, and Young's modulus of the bio-composites.	[36]
Coffee grounds	Increased loading of the spent coffee ground (SCG)-sourced filler in NR matrix contributed to an increase in t_{s2} and t_{c90} of resultant bio-composites. UTS and tear strength of these bio-composites decreased, but both hardness and odor absorption increased while M300 did not change significantly as the filler loading was increased.	[66]
Poly(lactic acid)	Temperature higher than poly(lactic acid) (PLA)'s T_g causes abiotic factors and biodegradation of this material to increase because of its increased flexibility and water absorption, which speed up microbial attachment and hydrolysis. However, loading NR matrix (i.e., amorphous phase) to PLA, and increasing this loading, modifies the chemical structure of the resultant NR/PLA bio-composites, leading to a heterogenous system, and hence increased degree of biotic degradation of PLA.	[67]

(*Continued*)

Table 3.5 Selected studies that have investigated the properties of bio-composites that are based on NR (*cis-1,4-polyisoprene*). (*Continued*)

Source of filler	Properties of NR-based bio-composites whose filler's source surface was modified using solutions comprised of common inorganic compounds	Study
Hemp	Increased loading of hemp-sourced filler in NR matrix increased the t_{s2}, t_{c90}, M_L, M_H, and $M_H - M_L$ of resultant NR/hemp-sourced filler bio-composites. v_e, hardness, and M100 of the bio-composites studied increased with increased filler content, while UTS and ε_b decreased. The increase in filler loading also increased the E' and T_g of the bio-composites but decreased the $\tan\delta$.	[68]
Oil palm	v_e, thermal stability, UTS, and M100 of the NR/OPA-sourced filler bio-composites increased, while ε_b decreased, as the filler loading was increased. Aging the bio-composites did not significantly change their UTS but decreased their ε_b while increasing the M100. E' and T_g of the bio-composites increased but $\tan\delta$ decreased as the filler loading was increased.	[71]
Bioplastic	Increased loadings of poly(butylene adipate-co-terephthalate), polycaprolactone, and poly (butylene succinate) in the NR matrix resulted in increased UTS and tension set of their respective resultant NR-based bio-composites.	[72]

(*Continued*)

Table 3.5 Selected studies that have investigated the properties of bio-composites that are based on NR (*cis-1,4-polyisoprene*). (*Continued*)

Source of filler	Properties of NR-based bio-composites whose filler's source surface was modified using solutions comprised of common inorganic compounds	Study
*Fuit (Mangosteen, Pomelo, and Durian) peels	Separately studied fruit peel powder (FPP)-sourced fillers were the mangosteen peel powder (MPP)-sourced filler, pomelo peel powder (PPP)-sourced filler, and durian peel powder (DPP)-sourced filler. Obtained higher values of aging (thermal or 25 kGy gamma) coefficients for all NR/FPP bio-composites indicated that FPPs were not only bio-fillers but also natural antioxidants. When the FPPs were added to the NR matrix, the colors of resultant bio-composites changed, and the degree of color change increased with higher FPP loadings. The densities of the bio-composites also increased with increased loadings of fillers. Thermal stability of the NR/PPP-sourced filler bio-composites and NR/MPP-sourced filler bio-composites was higher than that of NR/DPP-sourced filler bio-composites and pristine NR. Tensile moduli of the bio-composites increased when the loadings of fillers increased, but UTS and ε_b decreased.	[73]
*Chitosan	The chitosan-sourced filler increased the Young's modulus of the NR/Chitosan-sourced filler bio-composites, while decreasing the swelling %, UTS, and ε_b, as its loading was increased. Due to aging, the swelling % and moduli of the bio-composites increased while those of UTS and ε_b decreased. Increasing filler loading also increased the degree of biodegradation of bio-composites.	[74]
Kraft pulp	Increasing the loading of bleached kraft pulp-sourced filler in NR matrix resulted in increased E' (above T_g), but decreased E'', UTS, and ε_b of the resultant bio-composites.	[75]

(*Continued*)

Table 3.5 Selected studies that have investigated the properties of bio-composites that are based on NR (*cis-1,4-polyisoprene*). (*Continued*)

Source of filler	Properties of NR-based bio-composites whose filler's source surface was modified using solutions comprised of common inorganic compounds	Study
#Rice husk	When the loading of RHA-sourced filler was increased in ENR matrix, the t_{s2} and t_{c90} of the resultant bio-composites decreased but the M_L, M_H, and $M_H - M_L$ increased. Increased filler loading also increased the v_e, UTS and moduli of the bio-composites but decreased the ε_b. E' and E'' of bio-composites increased while the $\tan\delta$ decreased when the filler loading increased.	[5]

*, Its matrix was NR latex in place of solid NR; #, Its matrix was modified (i.e., epoxidized) NR.

and or methods have been developed for functionalization, comprising one or more of the common inorganic compounds—sodium hydroxide (NaOH), potassium hydroxide (KOH), water (H_2O), hydrazine hydrate ($NH_2NH_2 \cdot xH_2O$), calcium nitrate ($Ca(NO_3)_2$), sodium sulfite (Na_2SO_3), calcium carbonate ($CaCO_3$), potassium permanganate ($KMnO_4$), sodium chlorite ($NaClO_2$), ammonium fluoride (NH_4F), sulfuric acid (H_2SO_4) and hydrochloric acid (HCl) (Table 3.5). This is primarily due to their affordability, reliability, and user-friendliness [12, 19, 27, 36, 75]. Furthermore, besides natural fiber functionalization, the adhesion of natural fiber to NR during the creation of NR-based bio-composites could also be enhanced by using oil-extended or methacrylate-grafted NR. This approach is viable since these NR varieties have long been established and explored; however, their composites, particularly biobased types, have yet to be examined. The existence of epoxidized NR (ENR)-based bio-composites further indicates that producing bio-composites based on modified NR is feasible [5, 6, 76].

The current state of research on NR-based bio-composites is summarized in Table 3.5, which primarily addresses the effect of increased natural material-sourced fillers on various properties of NR-based bio-composites. These properties, analyzed using SEM morphology, which displays physical interactions between fiber and the NR matrix, are equally significance to the intended applications of these bio-composites. Moreover, the studies summarized in Table 3.5 revealed that the applications of these composites typically depend on the properties achieved with the filler type. For instance, if a filler enhances a composite material's thermal stability, that composite material would be used in applications requiring high thermal performance. Furthermore, NR-based bio-composites composed of plant fibers appear to be studied more frequently than those containing animal fibers, minerals, or mineral fibers. Also, solid NR has been studied more extensively in NR-based bio-composites, compared to NR latex. As a result, further research on reinforcing NR-based bio-composites using mineral- and animal-fiber-sourced fillers, as well as related studies on NR latex, is necessary.

Nevertheless, it has been demonstrated that after functionalization, the typical procedure for preparing the bio-filler sample is to freeze the corresponding source in liquid nitrogen and then mechanically shear (e.g., *via* planetary ball mill) the resultant sample into smaller particles, which are then run through a mesh sieve with a corresponding micron-sized number [1, 2]. However, there are instances when sample freezing is not required because of the sample's nature, as in the case with samples such as palm kernel shells from palm plants or trees, eggshells from chicken eggs, and clay from minerals. Sufficient drying of the bio-based filler sample is then required to achieve an acceptably low moisture level, as specified by ASTM D2654. Moreover, as would be anticipated, the type of matrix, i.e., whether it is solid or liquid, determines how the NR-based bio-composites are prepared. Both NR latex- and solid NR-based bio-composites typically use conventional vulcanization (CV), efficient (EV), or semi-EV systems; however, the former utilizes a melt compounding method, while the latter uses an internal mixer and/or two-roll mill mixing method [1, 3, 5, 8, 50, 54, 70]. Furthermore, using the ASTM D2084-19a standard method, NR latex- and solid NR-based bio-composites are often sulfur-vulcanized; for the former, an oven or radiation is utilized, and for the latter, a hot vulcanizing press machine is employed after their optimum curing times have been determined using an Oscillating Disc Rheometer.

Several references in Table 3.5 suggest that, to date, NR-based bio-composites comprising rice husk (RH)-sourced fillers are the most widely researched materials. This is because RH is an abundantly available waste product in all rice-producing countries that also have higher agricultural nominal Gross Domestic Product (GDP)—the United States, China, Japan, Brazil, India, Indonesia, Bangladesh, and Malaysia [1, 4, 77]. Currently, the second most researched bio-filler sources for NR-based bio-composites are oil palm and chitosan. Sugarcane bagasse (also known as megass or bagasse) is another source of key fibers that are widely used as bio-fillers around the world, with nearly 23,000 and 48.84 lakh of sugarcane growers in South Africa and India, respectively [2, 30]. The general synopsis of NR-based bio-composites, as shown in Table 3.5, should encourage countries to closely monitor their most important agricultural commodities and investigate their respective waste for possible NR-based composite applications to maintain or improve sustainability.

The major factors contributing to NR-based bio-composites' numerous applications in a variety of sectors are their loosely packed fiber particles in the NR matrix (which makes the end product lightweight), environmentally beneficial, and affordable [1, 76]. The long-established and potential applications of NR-based bio-composites include soles, belts, straps, gloves, hoses, furniture, caskets, sports equipment, brush bristles, architectural elements, and point-of-sale displays, as well as various storage, packaging and gadgets' materials [1, 46, 77]. Nonetheless, research that has been published thus far makes it abundantly evident that the automotive sector is comparatively the one that has made the greatest use of NR-based bio-composites [2, 10, 18, 19, 78]. In addition to the benefits mentioned above for NR-based bio-composites, the extensive use of these composites in this industry is because natural fiber textiles, which offer enhanced insulation, have long been employed to reinforce various polymers used in vehicles [2, 10, 18, 19, 78]. For instance, the utilization of natural fibers has increased by 73% at Mercedes-Benz [1, 2], in addition to approximately 10,000 tons of bio-composites that Bayerische Motoren Werke (BMW) utilized in 2004 [2, 74–79]. The C-, S-, E-, and A-class cars of the 164/251 2004 model from Mercedes-Benz had dashboard support as well as seats and backrest surfaces made of NR/coconut fiber bio-composites [1, 2]. Moreover, Shahril *et al.* [56] suggested using NR/kenaf fiber bio-composites to create rubber mounts for automotive engines, while Patil *et al.* [65] suggested the use of NR/coir bio-composites to manufacture insulation, mattresses, toys, and automobile tires. According to a study by Zhang *et al.* [46], NR latex foam/chitin bio-composites offer strong practical value for future applications and may be broadly utilized in home products, including pillows, beds, and cushions. Thumwong *et al.* [73] reported that fruit peels, namely the powders of durian, pomelo, and mangosteen, can be used as reinforcing bio-fillers and natural antioxidants in the manufacturing of high-quality NR latex gloves. Samsir *et al.* [80] proposed using carrageenan seaweed-sourced filler in the NR-based bio-composites for biodegradable applications, like polybags for pottery and plants, while Toda *et al.* [81] suggested the possibility of using NR/bamboo fiber bio-composites for icy or frozen roads. Tang *et al.* [82], in contrast, suggested the utilization of NR/kapok fibers bio-composites for pressure sensor applications. The additional uses of various natural fiber composites, particularly in the automotive sector, were discussed in detail by Elfaleh *et al.* [2], Wazeer *et al.* [78], and Zwawi *et al.* [83].

3.3 Natural Rubber-Based Bio-Nanocomposites

3.3.1 Nanocomposites Definition

"Nano" refers to items at the nanoscale (10^{-9} m), which corresponds to one billionth of a meter. The nanoscale range encompasses particle sizes larger than several atoms, but smaller than the wavelength range of visible radiation [84]. Growing attention to working with materials at the nanometric scale has emerged since Richard Feynman introduced the concept of nanotechnology in 1959. This has contributed to the expansion of innovative materials known as *nanocomposites*, with NR-based nanocomposites now referred to as those comprising fillers with at least one dimension of <100 nm. Compared with common composites, NR-based nanocomposites typically contains a small amount of well-dispersed nanofillers [85]. Nanoparticles offer opportunities to develop high-performance multifunctional composites without the need for high filler quantities to achieve superior mechanical properties [86].

The renewed interest in nanocomposites is driven by their significant industrial impact, has enabled the design of new materials with unique physical properties. Owing to their near-perfect quality, nanofillers can overcome a way around the drawbacks of conventional micrometer-scale materials. The large specific surface area of nanocomposites results in a substantial matrix-filler interfacial area, altering the molecular mobility, relaxation behavior, and subsequent thermal and mechanical properties [87, 88]. Recently, attention has shifted towards understanding and exploiting the unique physical properties of NR nanocomposites, moving beyond simple formulations to engineered and functional nanocomposites. Additionally, there is growing interest in natural fiber composites or bio-composites, driven by factors such as increased fuel efficiency, cost-effective building materials, and environmental preservation. The emerging category of bio-based nanocomposites, known as bio-nanocomposites, involves renewable nanomaterials, including cellulose, hemicellulose, and lignin, combined with petroleum-derived or biopolymers (e.g., PLA), polyhydroxyalkanoate (PHA)), as well as synthetic and/or inorganic nanometer fillers (i.e., carbon nanotubes and nanoclay). The bio-composite and bio-nanocomposite markets are rapidly growing owing to increased demand and eco-friendly properties. Industries are now competing to design superior strength bio-composites and bio-nanocomposites with commercial importance, and research on biopolymers derived from renewable resources is ongoing [89].

Rubber-based bio-nanocomposites, in which rubber serves as the matrix, are crucial given rubber's versatility and application levels, driven by its superior strength, elasticity, flexibility, resilience, and abrasion resistance. The development of natural-based bio-nanocomposites, incorporating nanometer-scale reinforcing bio-fillers such as nanocelluloses, shows their potential to impart improved properties at lower volume fractions compared to macro- or micro-sized fillers [90, 91]. It provides an excellent matrix for studying filler reinforcement effects, allowing the tailoring of properties through the addition of reinforcing fillers with diverse surface chemistries and aggregate size/aspect ratios [92, 93].

Silica and carbon black (CB) are common reinforcing agents in the rubber industry; however, their processing contributes to high energy consumption and ecological pollution. Cellulose nanocrystals (CNCs) offer a potential alternative as aqueous suspensions, allowing their easy incorporation into NR processing routes. The use of CNCs and other bio-nanoparticles in NR facilitates the development of bio-based green nanocomposites [94]. CNCs from various sources, including capim dourado [95], rachis of palm date trees [96], bagasse [97], sisal [98], and bamboo [99], have been studied as nano-reinforcements in NR-based bio-nanocomposites.

3.3.2 NR–Biopolymer-Based Nanocomposites

Bio-based polymer nanocomposites are a novel class of sustainable materials made from bio-renewable resources. In the field of research, there is an increasing need for reinforcing fillers made from renewable resources because of their low density, biodegradability, superior mechanical properties, affordability, environmental friendliness, biocompatibility, and ease of processing [1]. Scientists have been paying close attention to bio-fillers such as chitin, cellulose, and lignin, as they work to create sustainable polymer nanocomposites. As previously mentioned, NR is a biopolymer, and its combination with bio-derived nanofillers has the potential to create eco-friendly polymer nanocomposites with remarkable qualities.

3.3.3 NR–Cellulose Nanocomposites

Bio-based nanocomposites have been created through a multitude of research endeavors aimed at utilizing biomaterials as reinforcing agents in polymeric matrices. The eco-friendliness, biocompatibility, and biodegradability of these materials are generating a lot of significant interest. Among the renewable biomaterials, cellulose, which is abundant and renewable, has emerged as a key player in this field [100, 101]. The reinforcing capabilities of cellulose nanoparticles in nanocomposites were explored approximately one year ago [102]. Owing to its excellent mechanical properties, biodegradability, low density, and accessibility, nanocellulose, which comes from a variety of renewable sources, is a perfect green filler for polymers [103].

Because NR is a biopolymer, the use of bio-nanofillers greatly enhances its potential to create bio-based and environmentally friendly nanocomposites. Researchers have isolated cellulose whiskers (cellulose nanocrystals) with a length of 84–102 nm from sugar cane bagasse and demonstrated their reinforcing ability in NR matrices. The addition of these whiskers significantly improved the Young's modulus and ultimate tensile strength (UTS), which is attributed to the development of a continuous network through hydrogen bonding in cellulosic nanoparticles [104]. Similar positive effects on the thermal stability, UTS, E', and tan delta peak were observed when cellulose nanowhiskers (CNWs) from bamboo pulp were introduced into NR as shown in Figure 3.7 [105].

The transport properties of polymeric samples are crucial for understanding their morphologies and internal structures. Cellulose nanofibers

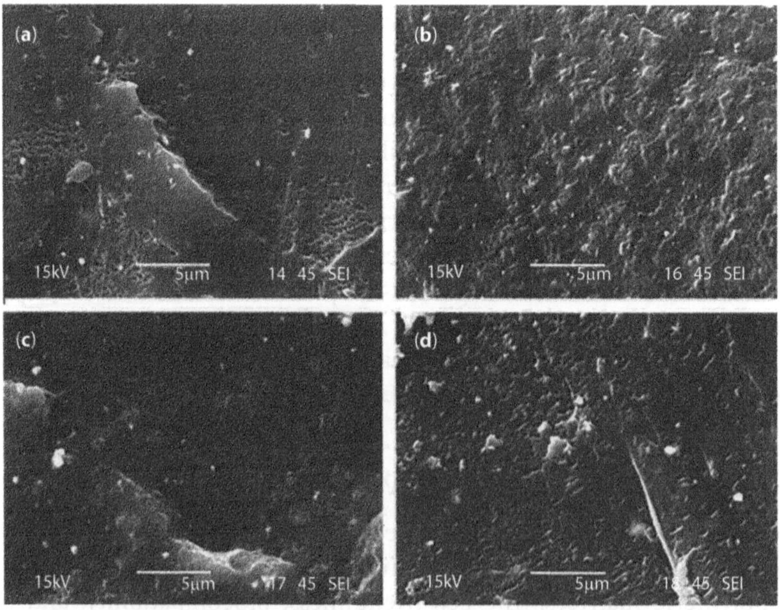

Figure 3.7 SEM micrographs of NR/cellulose nanowhiskers nanocomposites: (a) untreated NR, (b) NR treated with CNW2.5, (c) NR treated with CNW5, and (d) NR treated with CNW10. Reprinted/adapted with permission from Ref. [105].

(CNFs) and CNWs found in NR-based bio-nanocomposites made from bamboo pulp demonstrated their ability to stop molecular motion and polymer chain relaxation by reducing solvent uptake. CNFs, due to their tangling effect, showed a more significant decrease in solvent uptake than CNWs. The order of solvent reduction was benzene > toluene > p-xylene, which was attributed to differences in bulkiness and molecular weight. Banana fiber-derived CNFs, when dispersed in NR matrices, demonstrated improved stress levels in nanocomposites owing to the reaction between zinc cellulose and CNF [106]. CNFs were extracted from banana fiber through steam explosion and dispersed in the NR matrix, as confirmed by atomic force microscopy (AFM) and SEM, as shown in Figures 3.8 and 3.9, respectively. According to the X-ray diffraction (XRD) results, the crystalline peaks of CNFs were absent from all the NR-based bio-nanocomposites. This was because NR and the crosslinking agents formed additional chemical interactions with the CNF. An important factor in the improved stress level of the synthesized nanocomposites id the reaction between zinc cellulose and CNF. A reaction mechanism is proposed, as shown in Figure 3.10.

NR applies force to the CNFs when a load is applied, and CNF are an extremely strong material. The improved mechanical properties were a result of the synergistic interactions between rubber and cellulose, rubber and rubber, cellulose and cellulose, and Zn and cellulose [107]. In a different study, the authors [107] investigated the transport properties of NR/nanocellulose nanocomposites using p-xylene and toluene. At low filler loadings, the nanofillers were uniformly dispersed, as shown in the SEM and TEM micrographs. The development of a double network between

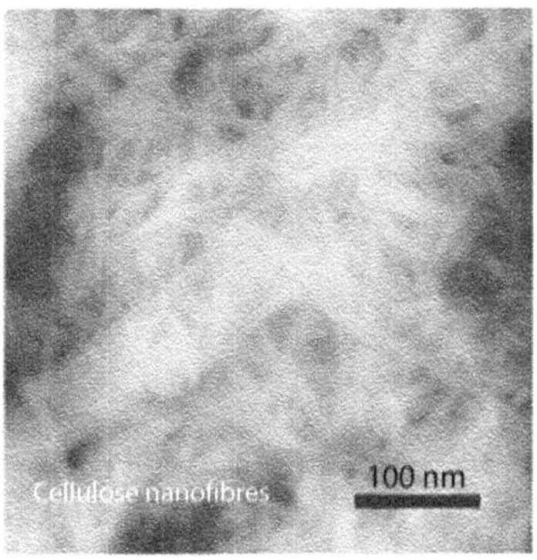

Figure 3.8 AFM micrograph of the cellulose nanofiber. Reprinted/adapted with permission from Ref. [106].

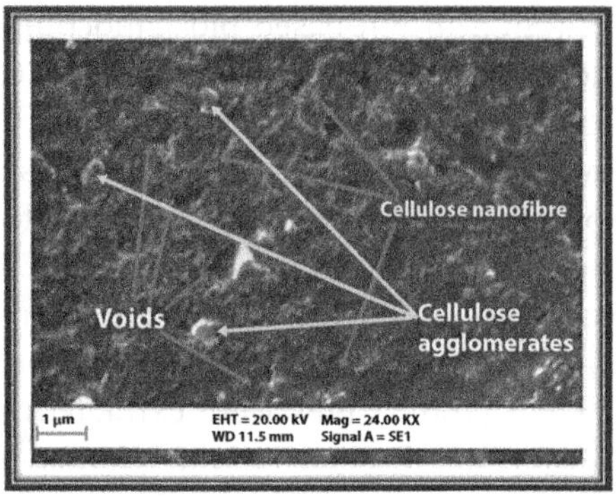

Figure 3.9 SEM micrograph of the tensile fracture surface of nanocomposite reinforced with 2.5% CNF. Reprinted/adapted with permission from Ref. [106].

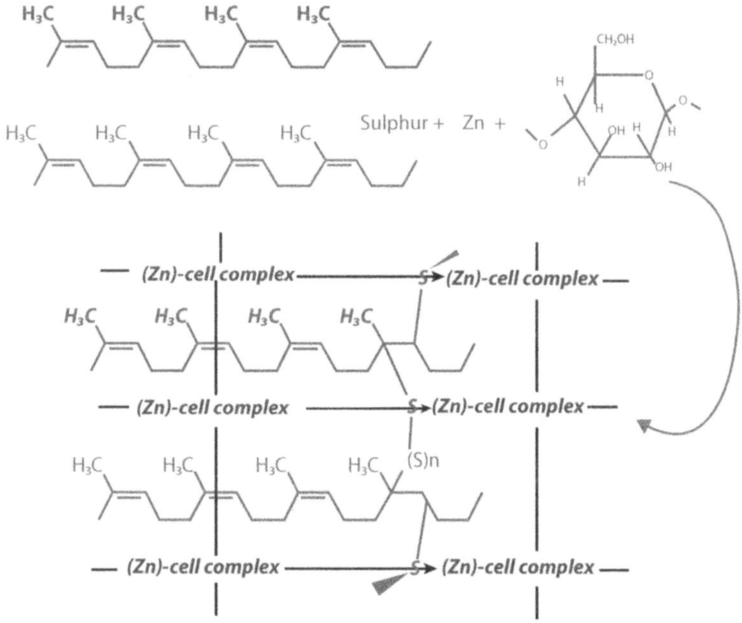

Figure 3.10 A suggested process by which the cross-linked NR/CNF nanocomposite interacts. Reprinted/adapted with permission from Ref. [106].

rubber–rubber and cellulose–cellulose helped lower the rate of solvent absorption. When the nanofiller was inserted, an increase in solvent resistance was observed because the hydrophilic OH groups prevented the organic solvents from becoming hydrophobic [108]. The UTS and Young's modulus increased when 50 nm-diameter nanocellulose fibers were separated from the jute fibers using the steam explosion method and then inserted into the NR matrix. This suggests that nanocellulose in NR has reinforcing properties. Zn/cellulose and cellulose/cellulose networks, among other three-dimensional networks of nanocellulose fibers in the rubber matrix, lead to remarkable enhancements in the properties of nanocomposites [109–111]. Similarly, nanocellulose fibers from jute fibers, corncobs, pineapple leaf fibers, and rice husks have been explored for their reinforcing abilities in NR matrices. These studies revealed enhancements in the UTS, Young's modulus, tear strength, and other mechanical

Table 3.6 Multiple approaches for producing nanocellulose.

No	Source of filler	Method	Filler	Size, nm	Study
1	Bagasse	Acidic hydrolysis	Cellulose nanocrystals	84–102	[103]
2	Bamboo pulp	Acidic hydrolysis	Cellulose nanocrystals	14	[105]
3	Bamboo pulp	Mechanical fibrillation for CNFs, and acidic hydrolysis for CNWs	Cellulose nanometer fibers and cellulose nanocrystals	23–42 (CNFs) and 5–14 (CNWs)	[106]
4	Banana fiber	Thermomechanical treatment	Cellulose nanometer fibres	10–60	[107]
5	Jute fiber	Thermomechanical treatment	Nanocellulose fibres	50	[108]
6	Corn cob	Acidic hydrolysis	Cellulose nanocrystals	2–5	[109]
7	Pineapple leaf fiber	Acidic hydrolysis	Nano-crystalline cellulose	30–50	[110]
8	Rice husk	Alkali treatment, bleaching, thermomechanical treatment, and acidic hydrolysis	Cellulose nanofibrils	45	[60]

properties, showcasing the potential of nanocellulose as a versatile and effective reinforcing agent in the development of bio-based nanocomposites. The different methods for producing nanocellulose are listed in Table 3.6.

3.3.4 NR–Lignin Nanocomposites

Of all the biopolymers found in nature, lignin is the most abundant because of its highly branched and renewable nature. Lignin is defined by functional groups that are carboxyl, methoxyl, hydroxyl, and carbonyl. It has several useful qualities, such as UV absorption, biodegradability, stabilizing power [112, 113], reinforcing, and anti-fungal capabilities [114, 115] because of its diverse functional groups and network structure. When incorporated into rubber, lignin imparts nonconductivity, lower density, and lighter color than CB. However, its reinforcement in nonpolar rubbers is comparatively low [116, 117]. Despite the abundance of lignin, reports on nanolignin reinforcement in NR are limited.

In one study, colloidal lignin-poly (diallyldimethylammonium chloride) (PolyDADMAC) complexes (LPCs) obtained by self-assembly were combined with NR latex to create NR/lignin bio-nanocomposites using the co-precipitation method. The reaction between PolyDADMAC and lignin, facilitated by cation $-\pi$ and $\pi-\pi$ interactions, enhances the compatibility of LPC with NR [118]. As demonstrated in Figure 3.11, this led to the uniform dispersion of LPC in NR, enhancing the mechanical, thermo-oxidative, and thermal stability of the NR/LPC bio-nanocomposites. These LPCs also hastened vulcanization of the NR/LPC bio-nanocomposite. Another strategy was to use a soft processing method to add different concentrations of sodium lignosulfonate to NR. As the amount of lignin loading increased, the resulting bio-composite's tensile stress increased. Additionally, the composite exhibited a shorter dissipative loss with a lower glass transition temperature (T_g) and higher storage modulus as the lignin concentration increased. Laser scanning confocal microscopy (LSCM) revealed that the size of lignin in the NR/lignin bio-nanocomposite was less than 100 nm [119].

Nanoscale organosolv lignin (OSL) has been introduced into NR latex to enhance its mechanical properties. The addition of 5 wt.% of nanoscale OSL improved toughness, Shore A hardness, and UTS while retaining high elongation at break (ε_b). In addition, the effect of the leaching process on the samples' toughness and thermo-oxidative stability was investigated. After leaching, the toughness, ε_b, and UTS increased by 32, 9, and 27%, respectively. In the NR/OSL bio-nanocomposite, superior interparticle

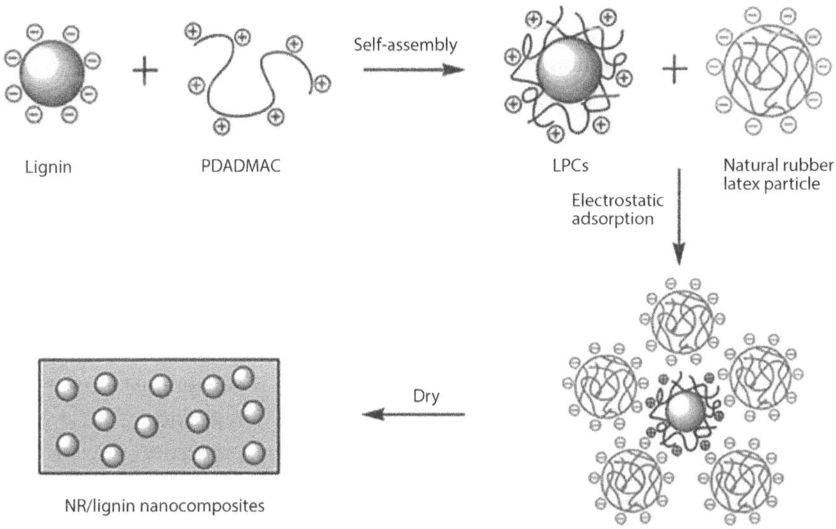

Figure 3.11 Process flow diagram for NR/Lignin nanocomposites. Reprinted/adapted with permission from Ref. [118].

integration resulted from the leaching process's removal of proteins and surplus chemicals, which also enhanced interparticle attraction and cross-linking [120].

3.3.5 NR–Chitin Nanocomposites

Chitin, also referred to as poly-[1→4]-β-N-acetyl-D-glucosamine, is a naturally occurring amino polysaccharide derived from marine biomass that is widely distributed in the Earth's crust. It is hydrophilic and biocompatible and has good biodegradability and desirable qualities [121, 122]. One-dimensional (1D) nanomaterials, such as nanofibers and nanocrystals, have a larger specific surface area, crystallinity, and modulus in chitin [123, 124]. These characteristics make chitin an ideal candidate for applications in optoelectronics and biomedicine, and as a green nanofiller for reinforcing polymer matrices.

In one study, chitin whiskers (with an average length of ~30 nm) were extracted from crab shells and added to colloidal solutions. These whiskers were then mixed with vulcanized and unvulcanized NR latex. Solid composite films were obtained by casting, evaporation, freeze-drying, and hot pressing. This investigation evaluated how the processing methods affected the final nanocomposites' characteristics. It was found that whiskers only

Table 3.7 An overview of different NR nanocomposites.

Nanocomposite	Method of fabrication	Loading, wt. %	Enhanced properties	Study
Cellulose NR/cellulose whiskers	Melt blending	0–10	Biodegradability, thermal resistance, mechanical, and barrier properties	[102]
NR/cellulose nanowhiskers	Melt co-coagulation and mechanical mixing	0–10	Mechanical and thermal resistance properties	[105]
NR/CNWs and NR/CNFs	Melt co-coagulation and mechanical mixing	0–10	Solvent absorption inhibition	[106]
NR/CNF	Melt blending	0–10	Mechanical properties	[107]
NR/ nanocellulose	Melt blending	0–10	Solvent absorption inhibition and thermal resistance properties	[60]
NR/ nanocellulose	Melt blending	0–3	Mechanical and thermal resistance properties	[108]
NR/CNC	Melt blending and γ-ray irradiation	1–5	Mechanical properties	[94]
NR/NCC	Melt blending		Mechanical properties	[110]

(*Continued*)

Table 3.7 An overview of different NR nanocomposites. (*Continued*)

Nanocomposite	Method of fabrication	Loading, wt. %	Enhanced properties	Study
Lignin NR/LPC	Co-precipitation	7*	Thermal stability, thermo-oxidative stability and mechanical properties	[118]
NR/lignin (sodium lignosulfonate)	Melt blending	10–40*	Mechanical properties	[119]
NR/OSL (organosolv lignin)	Melt blending	5	Mechanical and thermal resistance properties	[120]
Chitin NR/chitin	Melt blending	5–20	Solvent absorption inhibition	[121]
NR/chitin	Melt blending	5–20	Mechanical and thermal resistance properties	[122]
NR/ChNFs	Melt blending	0.3–1.0	Mechanical and biocompatibility properties	[124]

*, phr.

formed a rigid network in the evaporated samples, which enhanced the samples' exceptional swelling resistance to organic solvents [125]. The diffusion coefficient values, bound rubber content, and relative weight loss all support the existence of this network.

Another study explored the use of rigid chitin nanofibers (ChNFs) as reinforcing fillers for NR. An α-chitin/KOH/urea aqueous solution was used to self-assemble ChNFs, which demonstrated improved dispersion in NRL at the nanoscale and better compatibility with NRL. With modest loading (0.3–2 wt.%), the NR/ChNFs bio-nanocomposites showed a notable improvement in mechanical properties. In addition, the NR nanocomposite's improved biocompatibility was demonstrated by the promotion of mouse bone mesenchymal stem cells (mBMSCs) through the insertion of ChNFs. This finding suggests potential applications for biomedical materials such as human diaphragms, cosmetic prostheses, and blood vessels [126, 127]. The outcomes of cell growth on the NR surface and its chitin nanofiber nanocomposite at different loadings are displayed, and Table 3.7 lists the reinforcements of the different NR-based bio-nanocomposites.

3.4 Conclusion and Future Perspectives

The heightened awareness of ecological concerns in contemporary times has led to revived interest in materials derived from natural, compostable, and renewable resources. Issues surrounding material disposal and environmental safety are gaining prominence, leading to the exploration of alternatives, such as natural fibers and completely or partially biodegradable polymers. Against this backdrop, the development of bio-composites and bio-nanocomposites based on natural rubber (NR) has emerged as a focal point, garnering attention from both the industrial and academic sectors. This trend reflects a broader effort to create sustainable and environment-friendly materials.

The absolute potential of NR-based bio-composites and bio-nanocomposites is frequently pursued through continuous research and development endeavors aimed at exploring their advanced versatility. These composites offer notable advantages, primarily owing to the availability of natural bio-fillers, including bio-nanofillers, and their cost-effectiveness compared to synthetic nanofillers. Beyond economic considerations, these materials present technical benefits, such as low density, high toughness, acceptable specific strength properties, ease of separation, enhanced energy recovery, carbon dioxide reduction, and biodegradability. These factors collectively contribute to driving the growth of markets centered around rubber-based bio-composites.

Natural fibers, such as jute, rice husk, flax, oil palm, hemp, chitosan, sisal, bagasse, abaca, and coir, play a pivotal role in the paradigm shift towards NR-based bio-composites and bio-nanocomposites by offering the potential to replace synthetic fibers in numerous applications. Moreover, guayule NR, which has a structural backbone composed of 99.9 wt.% *cis-1,4-polyisoprene* units and shares similar molecular weight and physico-mechanical properties with *Hevea* NR, has been used to manufacture NR-based bio-composites using eggshell-sourced fillers. Similarly, it is possible to employ *trans-1,4-polyisoprene* in elastomer-based bio-composites and bio-nanocomposites, and future studies are expected to address this topic. This is because, like those based on unmodified and epoxidized *cis-1,4-polyisoprene* NR, *trans-1,4-polyisoprene*-based bio-composites would have great potential to build a more ecologically sensitive and sustainable future. Furthermore, other bio-filler sources, such as palm kernel shells, peanut shells, ramie fibers, and henequen fibers, should be studied for NR-based bio-composites and bio-nanocomposites. Rubber-based bio-composites and bio-nanocomposites, where synthetic rubber-based variants are partially biodegradable, are poised to play crucial roles in various industries and meet the evolving requirements of our day-to-day lives. This shift towards sustainable and environmentally conscious materials represents a significant stride towards a more responsible and eco-friendly approach to manufacturing and consumption.

Novel and inventive applications of NR (solid and latex)-based bio-composites and bio-nanocomposites in a range of industries, such as biomedicine, building, agriculture, marine, aerospace, and automotive industries, are expected to continue to emerge as scientists and engineers learn more about natural fibers and elastomers, as well as diverse processing methods and properties of such composites. This calls for further research in accomplishing the already achieved and additional curing, physical, mechanical, electrical, and thermal properties of NR-based bio-composites and bio-nanocomposites at lower bio-filler loadings, similar to synthetic fibers (e.g., carbon nanotubes). Shear strength, wear strength, dielectric strength, creep resistance, rebound resilience, durability, and wet traction are additional properties that require further investigation for NR-based bio-composites and bio-nanocomposites.

Acknowledgments

The authors are grateful for funding support provided by the National Research Foundation (NRF).

References

1. Lozada, E.R., Aguilar, C.M.G., Carvalho, J.A.J., Sánchez, J.C., Torres, G.B., Vegetable cellulose fibers in natural rubber composites. *Polymers*, 15, 294, 1–16, 2023.

2. Elfaleh, I., Abbassi, F., Habibi, M., Ahmad, F., Guedri, M., Nasri, M., Garnier, C., A comprehensive review of natural fibers and their composites: An eco-friendly alternative to conventional materials. *Results Eng.*, 19, 101271, 1–31, 2023.
3. Masłowski, M., Miedzianowska, J., Strąkowska, A., Strzelec, K., Szynkowska, M.I., The use of rye, oat and triticale straw as fillers of natural rubber composites. *Polym. Bull.*, 75, 4607–4626, 2018.
4. Szczepanowska, H.M. and Akhmedov, N.G., Gutta percha, natural rubber and balatachemical characterization of polyisoprenes in the context of cultural heritage. *Heritage Sci.*, 11, 125, 1–18, 2023.
5. Pongdong, W., Nakason, C., Kummerlöwe, C., Vennemann, N., Influence of filler from a renewable resource and silane coupling agent on the properties of epoxidized natural rubber vulcanizates. *J. Chem.*, 796459, 1–15, 2015.
6. Zakaria, Z., Islam, M.S., Hassan, A., Haafiz, M.K.M., Arjmandi, R., Inuwa, I.M., Hasan, M., Mechanical properties and morphological characterization of PLA/Chitosan/Epoxidized natural rubber composites. *Adv. Mater. Sci. Eng.*, 629092, 1–7, 2013.
7. Pothan, L.A., Oommen, Z., Thomas, S., Dynamic mechanical analysis of bananas fiber reinforced polyester composites. *Compos. Sci. Technol.*, 63, 283–293, 2003.
8. Zhang, L., The application of composite fiber materials in sports equipment. *5th International Conference on Education, Management, Information and Medicine*, Atlantis Press, pp. 450–453, 2015.
9. Brüning, K., Natural rubber, in: *Encyclopedia of polymeric nanomaterials*, S. Kobayashi, and K. Müllen, (Eds.), pp. 1377–1382, Springer, Berlin, Heidelberg, 2015.
10. Mohammed, M., Oleiwi, J.K., Mohammed, A.M., Jawad, A.J.M., Osman, A.F., Adam, T., Betar, B.O., Gopinath, S.C.B., Dahham, O.S., Jaafar, M., Comprehensive insights on mechanical attributes of natural-synthetic fibres in polymer composites. *J. Mater. Res. Technol.*, 25, 4960–4988, 2023.
11. Kiron, M.I., Mineral fiber properties, types and uses, 2022, Available at https://textilelearner.net/mineral-fiber-properties-types-and-uses/ [Accessed on: 20/02/203].
12. Jawaid, M.H.P.S. and Khalil, H.A., Cellulosic/synthetic fibre reinforced polymer hybrid composites: a review. *Carbohydr. Polym.*, 86, 1, 1–18, 2011.
13. Karine, C., Jean-Paul, J., Moussa, G., Christophe, B., Laurent, B., Joël, B., Morphology and mechanical behavior of a natural composite: The flax fiber. *16th International Conference on Composite Materials, Materials Science, Engineering*, Japan, pp. 1–8, 2007.
14. Dénes, T.-O., Istoan, R., Tămaş-Gavrea, D.R., Manea, D.L., Hegyi, A., Popa, F., Vasile, O., Analysis of sheep wool-based composites for building insulation. *Polymers*, 14, 2109, 1–25, 2022.

15. Barrera, C.S. and Cornish, K., Processing and mechanical properties of natural rubber/waste-derived nano filler composites compared to macro and micro filler composites. *Ind. Crops Prod.*, 107, 22–231, 2017.
16. Islam, A., Molla, Y., Dey, T.K., Jamal, M., Rathanasamy, R., Uddin, M.E., Latex reinforced waste buffing dustjeans cotton composites and its characterization. *J. Polym. Res.*, 28, 322, 1–8, 2021.
17. Kumar, M.T., Shetty, S.S.G., Kalkornsurapranee, E., Songtipya, L., Nakaramontri, Y., Johns, J., Combination of silk fabric and natural rubber for the development of green composites: Influence of curing on mechanical and thermal properties. *Polym. Polym. Compos.*, 4, 9S, S1204–S1215, 2021.
18. Mohammed, M., Jawad, A.J.M., Mohammed, A.M., Oleiwi, J.K., Adam, T., Osman, A.F., Dahham, O.S., Betar, B.O., Gopinath, S.C.B., Jaafar, M., Challenges and advancement in water absorption of natural fiber-reinforced polymer composites. *Polym. Test.*, 13, 108083, 1–33, 2023.
19. Cruz, J. and Fangueiro, R., Surface modification of natural fibers: a review. *Procedia Eng.*, 15, 285–288, 2016.
20. Rohit, K. and Dixit, S., A review-future aspect of natural fiber reinforced composite. *Polym. Renew. Resour.*, 7, 2, 43–60, 2016.
21. Mohanty, A.K., Misra, M., Drzal, L.T., Selke, S.E., Harte, B.R., Hinrichsen, G., Natural fibers, biopolymers, and biocomposites: An introduction, in: *Natural fibers, biopolymers, and biocomposites*, A.K. Mohanty, M. Misra, L.T. Drzal, (Eds.), pp. 1–35, CRC Press, Boca Raton, 2005.
22. Fu, J., Li, X., Gao, W., Wang, H., Cavaco-Paulo, A., Silva, C., Bio-processing of bamboo fibres for textile applications: a mini review. *Biocatal. Biotransform.*, 30, 1, 141–153, 2012.
23. Ahmad, E. and Luyt, A.S., Effects of organic peroxide and polymer chain structure on morphology and thermal properties of sisal fibre reinforced polyethylene composites. *Composites, Part A*, 43, 4, 703–710, 2012.
24. Das, P.P., Chaudhary, V., Ahmad, F., Manral, A., Gupta, S., Gupta, P., Acoustic performance of natural fiber reinforced polymer composites: influencing factors, future scope, challenges, and applications. *Polym. Compos.*, 43, 3, 1221–37, 2022.
25. Perrella, F.W. and Gaspari, A.A., Natural rubber latex protein reduction with an emphasis on enzyme treatment. *Methods*, 27, 77–86, 2002.
26. Ibrahim, S., Riahi, O., Said, S.M., Sabri, M.F.M., Rozali, S., Biopolymers from crop plants, in: *Reference Module in Materials Science and Materials Engineering*, Elsevier, Amsterdam, Netherlands, 2019, https://doi.org/10.1016/B978-0-12-803581-8.123-5.
27. Payungwong, N., Sakdapipanich, J., Wu, J., Ho, C., The interplay of protein hydrolysis and ammonia in the stability of *Hevea* rubber latex during storage. *Polymers*, 1, 4636, 1–15, 2023.
28. Wang, M., Wang, R., Chen, X., Kong, Y., Huang, Y., Lv, Y., Li, G., Effect of non-rubber components on the crosslinking structure and thermo-oxidative degradation of natural rubber. *Polym. Degrad. Stab.*, 196, 109845, 1–11, 2022.

29. Sarkawi, S.S., Dierkes, W.K., Noordermeer, J.W.M., The influence of non-rubber constituents on performance of silica reinforced natural rubber compounds. *Eur. Polym. J.*, 49, 10, 3199–3209, 2013.
30. Linos, A., Berekaa, M.M., Reichelt, R., Keller, U., Schmitt, J., Flemming, H., Kroppenstedt, R.M., Steinbüchel, A., Biodegradation of *cis*-1,4-polyisoprene rubbers by distinct actinomycetes: microbial strategies and detailed surface analysis. *Appl. Environ. Microbiol.*, 66, 4, 1639–1645, 2000.
31. Hanhi, K., Poikelispää, M., Tirilä, H., Elastomeric materials, Plastics and Elastomer Technology (VERT), pp. 1–84, 2007.
32. Abraham, E., Cherian, B.M., Elbi, P.A., Pothen, L.A., Thomas, S., Recent advances in the recycling of rubber waste, in: *Recent developments in polymer recycling*, A. Fainleib, and O. Grigoryen, (Eds.), pp. 47–100, Transworld Research Network, India, Kerala, 2011.
33. Akiba, M. and Hashim, A.S., Vulcanization and crosslinking in elastomers. *Prog. Polym. Sci.*, 22, 475–521, 1997.
34. Kumar, C.S.S.R. and Nijasure, A.M., Vulcanization of rubber: How to alter molecular structure and influence physical properties. *Resonance*, 2, 55–59, 1997.
35. Madubhashini, S., Properties of natural rubber, Rubber Technology | Polymer Science and Technology, pp. 1–5, 2023.
36. Siregar, A.H., Gea, S., Indriani, N., Making composites from mixing limestone with addition of latex, in: *Proceedings of the 1st International Conference on Chemical Science and Technology Innovation (ICOCSTI)*, pp. 165–173, 2019.
37. Joy, J., Chan, C.H., Maria, H.J., Thomas, S., Natural rubber-based composites and nanocomposites: State of the art, new challenges and opportunities, in: *Natural rubber materials*, S. Thomas, C.H. Chan, L. Pothen, J. Joy, H. Maria, (Eds.), pp. 1–33, RSC publishing, London, 2003.
38. Maslowski, M., Miedzianowska, J., Czylkowska, A., Strzelec, K., Horsetail (*Equisetum Arvense*) as a functional filler for natural rubber biocomposites. *Materials*, 13, 1–20, 2020.
39. Mustafa, M.S., Syed Ismail, S.N., Mohd Shaipul Amini, S., Abdul Wahab, N.M., Majid, N.A., Sarip, M.N., Pineapple leaf fibre filled natural rubber composite: The effect of filler loading, in: *Charting the sustainable future of ASEAN in science and technology*, N. Alias, and R. Yusof, (Eds.), pp. 447–454, Springer, Singapore, 2020.
40. Osabohien, E., Soya, U., Ojeifo, N.O., Characteristics of natural rubber-carbonized pawpaw seed composites. *Int. J. Biol. Chem. Sci.*, 14, 8, 451–464, 2020.
41. Aguele, F.O. and Madufor, C.I., Effects of carbonised coir on physical properties of natural rubber composites. *Am. J. Polym. Sci.*, 2, 3, 28–34, 2012.
42. Maslowski, M., Aleksieiev, A., Miedzianowska, J., Strzelec, K., Common nettle (*Urtica dioica L.*) as an active filler of natural rubber biocomposites. *Materials*, 14, 1616, 1–30, 2021.

43. Jacob, M., Thomas, S., Varughese, K.T., Mechanical properties of sisal/oil palm hybrid fiber reinforced natural rubber composites. *Compos. Sci. Technol.*, 64, 955–965, 2004.
44. Homkhiew, C., Rawangwong, S., Boonchouytan, W., Thongruang, W., Ratanawilai, Composites from thermoplastic natural rubber reinforced rubberwood sawdust: Effects of sawdust size and content on thermal, physical, and mechanical properties. *Int. J. Polym. Sci.*, 1–11, 7179527, 2018.
45. Srisuwan, L., Jarukumjorn, K., Suppakarn, N., Effect of silane treatment methods on physical properties of rice husk flour/natural rubber composites. *Adv. Mater. Sci. Eng.*, 4583974, 1–14, 2018.
46. Zhang, N. and Cao, H., Enhancement of the antibacterial activity of natural rubber latex foam by blending it with chitin. *Materials*, 13, 1039, 1–15, 2020.
47. Fiorote, J.A., Freire, A.P., Rodrigues, D., Martins, M.A., Andeani, L., Valadares, L.S., Preparation of composites from natural rubber and oil palm empty fruit bunch cellulose: Effect of cellulose morphology on properties. *BioResources*, 14, 2, 3168–3181, 2019.
48. Wongpreedee, T., Prapruddivongs, C., Kengkhetkit, N., Effect of short banana fiber on hardness and tensile properties of natural rubber composites. *Eng. Mater.*, 904, 232–236, 2021.
49. Głowińska, E., Datta, J., Parcheta, P., Kaźmierczak, N., Novel approaches of using of *Spirulina Platensis* in natural rubber based composites. *J. Renew. Mater.*, 6, 7, 1–8, 2018.
50. Torres, G.B., Dognani, G., Agostini, D.L., Santos, R.J., Cabrera, F.C., Aguilar, C.M.G., Paiva, F.F.G., Teixeira, S.R., Job, A.E., Potential ecofriendly application of sugarcane bagasse ash in the rubber industry. *Waste Biomass Valorization*, 12, 4599–4613, 2021.
51. Correia, C.A. and Valera, T.S., Cellulose nanocrystals and jute fiber-reinforced natural rubber composites: Cure characteristics and mechanical properties. *Mater. Res.*, 22, 1, 1–9, 2019.
52. Divia, P.S., Sujith, A., Rajesh, C., Cure characteristics and mechanical properties of biocomposites of natural rubber reinforced with chicken feather fibre: Effect of fibre loading, alkali treatment, bonding and vulcanizing systems. *Mater. Today Commun.*, 20, 100555, 2019.
53. Huabcharoen, P., Wimolmala, E., Markpin, T., Sombatsompop, N., Purification and characterization of silica from sugarcane bagasse ash as a reinforcing filler in natural rubber composites. *BioResources*, 12, 1, 1228–1245, 2017.
54. Sukumar, R. and Menon, A., Organomodified kaolin as a reinforcing filler for natural rubber. *J. Appl. Polym. Sci.*, 107, 3476–3483, 2008.
55. Kamaruddin, S., Chai, A.B., Lim, C.L., Ho, J., Characterisation of static and dynamic responses of natural rubber bio-composites. *J. Phys. Conf. Ser.*, 908, 012028, 1–13, 2017.

56. Shahril, D.A., Azammi, A.M.N., Zulfadli, S.M., Reinforcement of kenaf fiber in natural rubber composite for automotive engine rubber mounting. *Int. J. Appl. Eng. Res.*, 12, 24, 14490–14494, 2017.
57. Kengkhetkit, N. and Wongpreedee., T., Physical and mechanical properties of natural rubber composites filled with rice husk and rice straw. *SNRU J. Sci. Technol.*, 13, 3, 86–95, 2021.
58. Maslowski, M., Miedzianowska, J., Delekta, M., Czylkowska, A., Strzelec, K., Natural rubber biocomposites filled with phyto-ashes rich in biogenic silica obtained from wheat straw and field horsetail. *Polymers*, 13, 127, 1–24, 2021.
59. Masłowski, M., Miedzianowska, J., Strzelec, K., Natural rubber composites filled with crop residues as an alternative to vulcanizates with common fillers. *Polymers*, 11, 972, 2–18, 2019.
60. Da Costa, H.M., Visconte, L.L.Y., Nunes, R.C.R., Furtado, C.R.G., Mechanical and dynamic mechanical properties of rice husk ash–filled natural rubber compounds. *J. Appl. Polym. Sci.*, 83, 2331–2346, 2002.
61. Mohamed, W.Z.W., Baharum, A., Ahmad, I., Abdullah, I., Zakaria, N.E., Mengkuang fiber reinforced thermoplastic natural rubber composites: Influence of rubber content on mechanical properties and morphology. *Malays. J. Anal. Sci.*, 22, 5, 906–913, 2018.
62. Miedzianowska, J., Masłowski, M., Rybínski, P., Strzelec, K., Modified nanoclays/straw fillers as functional additives of natural rubber biocomposites. *Polymers*, 13, 799, 1–30, 2021.
63. Mitsaichon, V., Ruaesrijan, K., Phinyocheep, P., Amornsakchai, T., Junkong, P., Thanawan, S., Preparation of spent coffee grounds-rubber composites using natural rubber latex as binder. *J. Met. Mater. Miner.*, 32, 4, 161–166, 2022.
64. Mon, M.M., Cho, C., Nu, T., Study on mechanical properties of natural rubber latex chitosan composites. *J. Myanmar Acad. Arts Sci.*, 2, 1B, 1–13, 2019.
65. Patil, Y. and Sharma, S., Effect of coir fiber reinforcement on mechanical properties of vulcanized natural rubber composites. *Sci. Eng. Compos. Mater.*, 25, 3, 517–528, 2018.
66. Somdee, P., Shettar, M., Prasoetsopha, N., Detsunhnoen, S., Matnok, S., Ansari, M.A., Enhancing eco-friendly natural rubber composites through the incorporation of spent coffee grounds, Research Square, pp. 1–19, 2023.
67. Tertyshnaya, Y.V., Podzorova, M.V., Khramkova, A.V., Ovchinnikov, V.A., Krivandin, A.V., Structural rearrangements of polylactide/natural rubber composites during hydro- and biotic degradation. *Polymers*, 1, 1930, 1–15, 2023.
68. Moonart, U. and Utara, S., Effect of surface treatments and filler loading on the properties of hemp fiber/natural rubber composites. *Cellulose*, 26, 7271–7295, 2019.

69. Santos, R.J., Agostini, D.L.S., Cabrera, F.C., Reis, E.A.P., Ruiz, M.R., Budemberg, E.R., Teixeira, S.R., Job, A.E., Sugarcane bagasse ash: New filler to natural rubber composite. *Polímeros*, 24, 6, 646–653, 2014.
70. Surya, I. and Ginting, M., The cure and tensile properties of montmorillonite-natural rubber composites in the incorporation of alkanolamide. *J. Phys. Conf. Ser.*, 1373, 1–7, 2019.
71. Ooi, Z.Z., Ismail, H., Bakar, A.A., The effect of hydrochloric acid treatment on properties of oil palm ash-filled natural rubber composites. *BioResources*, 8, 4, 5133–5144, 2013.
72. Thongpin, C., Buaksuntear, K., Tanprasert, T., Charoenta, J., Biobased thermoplastic vulcanizates from natural rubber (bioplastic/NR). *IOP Conf. Ser.: Mater. Sci. Eng.*, 773, 1–6, 2020.
73. Thumwong, A., Darachai, J., Thamrongsiripak, N., Tokonami, S., Ishikawa, T., Saenboonruang, K., Fruit peel powder as natural antioxidant and reinforcing bio-filler in natural rubber latex gloves: Cases of Mangosteen, Pomelo and Durian. *Antioxidants*, 12, 1119, 1–21, 2023.
74. Thumwong, A., Poltabtim, W., Kerdsang, P., Saenboonruang, K., Roles of chitosan as bio-fillers in radiation-vulcanized natural rubber latex and hybrid radiation and peroxide-vulcanized natural rubber latex: Physical/mechanical properties under thermal aging and biodegradability. *Polymers*, 13, 3940, 1–18, 2021.
75. Yano, S., Stenberg, B., Flink, P., Mechanical properties of natural rubber composites reinforced with cellulose fibers. *J. Jpn. Soc. Rheol.*, 20, 132–140, 1992.
76. Chandrasekar, M., Ishak, M.R., Sapuan, S.M., Leman, Z., Jawaid, M., A review on the characterisation of natural fibres and their composites after alkali treatment and water absorption. *Plast. Rubber Compos.*, 46, 119–136, 2017.
77. Zhou, Y., Fan, M., Chen, L., Zhuang, J., Lignocellulosic fibre mediated rubber composites: An overview. *Composites, Part B*, 76, 180–191, 2015.
78. Wazeer, A., Das, A., Abeykoon, C., Sinha, A., Karmakar, A., Composites for electric vehicles and automotive sector: A review. *Green Energy Intell. Transp.*, 2, 100043, 1–23, 2023.
79. Mohammed, L., Ansari, M.N.N., Pua, G., Jawaid, M., Islam, M.S., A Review on natural fiber reinforced polymer composite and its applications. *Int. J. Polym. Sci.*, 2015, 1–15, 2015.
80. Samsir, N., Rashid, A.A., Azahari, B., The influence of surface structure of low industrial grade seaweed and semi-refined carrageenan on mechanical and physical properties of natural rubber latex composites. *J. Vinyl Addit. Technol.*, 25, 3, 278–286, 2019.
81. Toda, T., Okubo, K., Fujii, T., Hurutachi, H., Yamanaka, Y., Yamamura, H., Development of rubber shoe sole containing bamboo fibers for frozen roads. *Proceedings of the 16th International Conference on Composite Materials*, Kyoto, Japan, pp. 8–13, 2007.

82. Tang, B., Chen, X., He, Y., Zhou, J., Zhao, H., Chen, W., Wang, J., Wang, X., Fabrication of kapok fibers and natural rubber composites for pressure sensor applications. *Cellulose*, 28, 2287–2301, 2021.
83. Zwawi, M.A., Review on natural fiber bio-composites, surface modifications and applications. *Molecules*, 26, 404, 1–28, 2021.
84. Kamel, S., Nanotechnology and its applications in lignocellulosic composites, a mini review. *eXPRESS Polym. Lett.*, 1, 9, 546–75, 2007.
85. Peppas, N.A., Hilt, J.Z., Khademhosseini, A., Langer, R., Hydrogels in biology and medicine: from molecular principles to bionanotechnology. *Adv. Mater.*, 18, 11, 1345–60, 2006.
86. Esteves, A.C., Barros-Timmons, A., Trindade, T., Polymer based nanocomposites: Synthetic strategies for hybrid materials. *Quím. Nova*, 27, 798–806, 2004.
87. De Azeredo, H.M., Nanocomposites for food packaging applications. *Food Res. Int.*, 42, 9, 1240–53, 2009.
88. Schadler, L.S., Brinson, L.C., Sawyer, W.G., Polymer nanocomposites: A small part of the story. *J. Miner. Met. Mater. Soc.*, 59, 53–60, 2007.
89. Siqueira, G., Bras, J., Dufresne, A., Cellulosic bionanocomposites: a review of preparation, properties and applications. *Polymers*, 2, 4, 728–65, 2010.
90. Khalil, H.A., Davoudpour, Y., Islam, M.N., Mustapha, A., Sudesh, K., Dungani, R., Jawaid, M., Production and modification of nanofibrillated cellulose using various mechanical processes: a review. *Carbohydr. Polym.*, 99, 649–65, 2014.
91. Qing, Y., Sabo, R., Zhu, J.Y., Agarwal, U., Cai, Z., Wu, Y., A comparative study of cellulose nanofibrils disintegrated *via* multiple processing approaches. *Carbohydr. Polym.*, 97, 1, 226–34, 2013.
92. Mariano, M., El Kissi, N., Dufresne, A., Cellulose nanocrystal reinforced oxidized natural rubber nanocomposites. *Carbohydr. Polym.*, 137, 174–83, 2016.
93. Rippel, M.M. and Galembeck, F., Nanostructures and adhesion in natural rubber: new era for a classic. *J. Braz. Chem. Soc.*, 20, 1024–30, 2009.
94. Neto, W.P., Mariano, M., da Silva, I.S., Silvério, H.A., Putaux, J.L., Otaguro, H., Pasquini, D., Dufresne, A., Mechanical properties of natural rubber nanocomposites reinforced with high aspect ratio cellulose nanocrystals isolated from soy hulls. *Carbohydr. Polym.*, 153, 143–52, 2016.
95. Siqueira, G., Abdillahi, H., Bras, J., Dufresne, A., High reinforcing capability cellulose nanocrystals extracted from Syngonanthus nitens (Capim Dourado). *Cellulose*, 17, 289–98, 2010.
96. Bendahou, A., Habibi, Y., Kaddami, H., Dufresne, A., Physico-chemical characterization of palm from phoenix dactylifera–L, preparation of cellulose whiskers and natural rubber–based nanocomposites. *J. Biobased Mater. Bioenergy*, 3, 1, 81–90, 2009.
97. Bras, J., Hassan, M.L., Bruzesse, C., Hassan, E.A., El-Wakil, N.A., Dufresne, A., Mechanical, barrier, and biodegradability properties of bagasse cellulose

whiskers reinforced natural rubber nanocomposites. *Ind. Crops Prod.*, 32, 3, 627–33, 2010.

98. Siqueira, G., Tapin-Lingua, S., Bras, J., da Silva Perez, D., Dufresne, A., Mechanical properties of natural rubber nanocomposites reinforced with cellulosic nanoparticles obtained from combined mechanical shearing, and enzymatic and acid hydrolysis of sisal fibers. *Cellulose*, 18, 57–65, 2011.
99. Visakh, P.M., Thomas, S., Oksman, K., Mathew, A.P., Effect of cellulose nanofibers isolated from bamboo pulp residue on vulcanized natural rubber. *Bioresources*, 7, 2, 2156–68, 2012.
100. Oksman, K., Mathew, A.P., Bondeson, D., Kvien, I., Manufacturing process of cellulose whiskers/polylactic acid nanocomposites. *Compos. Sci. Technol.*, 66, 15, 2776–84, 2006.
101. Grunert, M. and Winter, W.T., Nanocomposites of cellulose acetate butyrate reinforced with cellulose nanocrystals. *J. Polym. Environ.*, 10, 27–30, 2002.
102. Favier, V., Chanzy, H., Cavaillé, J., Polymer nanocomposites reinforced by cellulose whiskers. *Macromolecules*, 28, 18, 6365–7, 1995.
103. Abraham, E., Deepa, B., Pothan, L.A., Jacob, M., Thomas, S., Cvelbar, U., Anandjiwala, R., Extraction of nanocellulose fibrils from lignocellulosic fibres: A novel approach. *Carbohydr. Polym.*, 86, 4, 1468–75, 2011.
104. Visakh, P.M., Thomas, S., Oksman, K., Mathew, A.P., Crosslinked natural rubber nanocomposites reinforced with cellulose whiskers isolated from bamboo waste: Processing and mechanical/thermal properties. *Composites, Part A*, 43, 4, 735–41, 2012.
105. Visakh, P.M., Thomas, S., Oksman, K., Mathew, A.P., Cellulose nanofibres and cellulose nanowhiskers based natural rubber composites: diffusion, sorption, and permeation of aromatic organic solvents. *J. Appl. Polym. Sci.*, 124, 2, 1614–23, 2012.
106. Abraham, E., Deepa, B., Pothan, L.A., John, M., Narine, S.S., Thomas, S., Anandjiwala, R., Physicomechanical properties of nanocomposites based on cellulose nanofibre and natural rubber latex. *Cellulose*, 20, 417–27, 2013.
107. Abraham, E., Thomas, M.S., John, C., Pothen, L.A., Shoseyov, O., Thomas, S., Green nanocomposites of natural rubber/nanocellulose: membrane transport, rheological and thermal degradation characterisations. *Ind. Crops Prod.*, 51, 415–24, 2013.
108. Thomas, M.G., Abraham, E., Jyotishkumar, P., Maria, H.J., Pothen, L.A., Thomas, S., Nanocelluloses from jute fibers and their nanocomposites with natural rubber: Preparation and characterization. *Int. J. Biol. Macromol.*, 81, 768–77, 2015.
109. Tappanawatch, W., Prapainainar, P., Sae-Oui, P., Loykulnant, S., Dittanet, P., Effect of gamma radiation on properties of cellulose nanocrystal/natural rubber nanocomposites. *Key Eng. Mater.*, 772, 13–7, 2018.
110. Gao, T.M., Huang, M.F., Xie, R.H., Chen, H.L., Preparation and Characterization of Nanocrystalline Cellulose/Natural Rubber (NCC/Nr) Composites. *Adv. Mater. Res.*, 712, 111–4, 2013.

111. Gosselink, R.J., Abächerli, A., Semke, H., Malherbe, R., Käuper, P., Nadif, A., Van Dam, J.E., Analytical protocols for characterisation of sulphur-free lignin. *Ind. Crops Prod.*, 19, 3, 271–81, 2004.
112. Parvathy, G., Sethulekshmi, A.S., Jayan, J.S., Raman, A., Saritha, A., Lignin based nano-composites: Synthesis and applications. *Process Saf. Environ. Prot.*, 145, 395–410, 2021.
113. De Paoli, M.A. and Furlan, L.T., Sugar cane bagasse-lignin as photo-stabilizer for butadiene rubber. *Polym. Degrad. Stab.*, 11, 4, 327–37, 1985.
114. Buono, P., Duval, A., Verge, P., Averous, L., Habibi, Y., New insights on the chemical modification of lignin: Acetylation versus silylation. *ACS Sustain. Chem. Eng.*, 4, 10, 5212–22, 2016.
115. Kosikova, B. and Gregorova, A., Sulfur-free lignin as reinforcing component of styrene–butadiene rubber. *J. Appl. Polym. Sci.*, 97, 3, 924–9, 2005.
116. Botros, S.H., Eid, M.A., Nageeb, Z.A., Thermal stability and dielectric relaxation of natural rubber/soda lignin and natural rubber/thiolignin composites. *J. Appl. Polym. Sci.*, 99, 5, 2504–11, 2006.
117. Jiang, C., He, H., Jiang, H., Ma, L., Jia, D.M., Nano-lignin filled natural rubber composites: Preparation and characterization. *eXPRESS Polym. Lett.*, 7, 5, 480–493, 2013.
118. Ikeda, Y., Phakkeeree, T., Junkong, P., Yokohama, H., Phinyocheep, P., Kitano, R., Kato, A., Reinforcing biofiller "Lignin" for high performance green natural rubber nanocomposites. *RSC Adv.*, 7, 9, 5222–31, 2017.
119. Hosseinmardi, A., Amiralian, N., Hayati, A.N., Martin, D.J., Annamalai, P.K., Toughening of natural rubber nanocomposites by the incorporation of nanoscale lignin combined with an industrially relevant leaching process. *Ind. Crops Prod.*, 19, 113063, 2021.
120. Yan, N. and Chen, X., Sustainability: Don't waste seafood waste. *Nature*, 524, 7564, 155–7, 2015.
121. You, J., Li, M., Ding, B., Wu, X., Li, C., Crab chitin-based 2D soft nanomaterials for fully biobased electric devices. *Adv. Mater.*, 29, 19, 1606895, 2017.
122. Thakore, S.I., Chitin based rubber nanocomposites, in: *Rubber Based Bionanocomposites: Preparation*, pp. 35–50, 2017.
123. Zeng, J.B., He, Y.S., Li, S.L., Wang, Y.Z., Chitin whiskers: An overview. *Biomacromolecules*, 13, 1, 1, 2012.
124. Nair, K.G. and Dufresne, A., Crab shell chitin whisker reinforced natural rubber nanocomposites. 1. Processing and swelling behavior. *Biomacromolecules*, 4, 3, 657–65, 2003.
125. Nair, K.G. and Dufresne, A., Crab shell chitin whisker reinforced natural rubber nanocomposites. 2. Mechanical behavior. *Biomacromolecules*, 4, 3, 666–74, 2003.

126. Sethulekshmi, A.S., Saritha, A., Joseph, K., A comprehensive review on the recent advancements in natural rubber nanocomposites. *Int. J. Biol. Macromol.*, 194, 819–42, 2022.
127. Ding, B., Huang, S., Shen, K., Hou, J., Gao, H., Duan, Y., Zhang, J., Natural rubber bio-nanocomposites reinforced with self-assembled chitin nanofibers from aqueous KOH/urea solution. *Carbohydr. Polym.*, 225, 11230, 2019.

4

Mechanical/Dynamical Mechanical, Thermal and Electrical Properties of Natural Rubber-Carbon Nanotube Nanocomposites and their Applications

Saheli Roy[1]* and Suneel Kumar Srivastava[2]†

[1]*Department of Chemistry, ICFAI University, Tripura, India*
[2]*Department of Chemistry, Indian Institute of Technology, Kharagpur, India*

Abstract

Natural rubber (NR) is a widely recognized sustainable biopolymer with multifaceted applications owing to its excellent elasticity, flexibility, biodegradability, etc. Despite this, its applications are limited owing to its intrinsic imperfect aging, inferior oil resistance, poor electrical conductivity, low hardness, durability, and tear resistance, which remain major drawbacks of NR, and it is anticipated that improvements in these properties could enable the application of NR from a broader perspective. In this regard, carbon black (CB) has been widely used as a filler in NR matrices, and the emergence of nanomaterial-filled NR composites appear to have superior physical properties. In this regard, carbon nanotubes (CNTs) as sole fillers or with other additive fillers have gained significant attention as conductive fillers for NR. In addition to improvements in the mechanical properties, these NR/CNT nanocomposites exhibited considerably improved electrical properties and thermal stability compared to NR. This chapter focuses on the different techniques utilized in the fabrication of NR/CNT nanocomposites, dispersing CNT in NR, and comprehensively examining the impact of CNT on the properties (mechanical, dynamic mechanical, and electrical) as the thermal stability of NR. In addition, the effect of CNT-based hybrid fillers in reinforcing NR has been reviewed, followed by applications of NR/CNT nanocomposites in recent fields.

**Corresponding author*: saheli.s.roy@gmail.com
†*Corresponding author*: suneelchemkgp@gmail.com

Visakh. P.M. and Shiv Kumari Panda (eds.) Natural Rubber-Based Composites and Nanocomposites, (117–152) © 2025 Scrivener Publishing LLC

Keywords: Natural rubber, carbon nanotubes, nanocomposites, mechanical properties, thermal properties, electrical properties

4.1 Introduction

The latex of the rubber plant (*Hevea brasiliensis*) is comprised of natural rubber (NR) and some non-rubbery impurities. NR was extracted from rubber latex using the acid coagulation method. Subsequently, it is washed to remove impurities, followed by further processing. NR, the biopolymer, contains isoprene units connected in a cis-1,4 arrangement, and almost 93–95% of the isoprene produced is utilized for manufacturing cis-1,4-polyisoprene (Figure 4.1) with a molecular weight in the range of 100,000–1,000,00000) [1]. NR has received significant attention from the scientific and industrial communities because of its superb elasticity, flexibility, exceptional tear and overall fatigue resistance, minimal heat accumulation, high resilience, strength retention at elevated temperatures, and superior mechanical and dynamic characteristics [2]. NR has been consistently utilized in several applications, such as tyres (used in vehicles), sealants (in industries), other consumer products (rubber boots, sports equipment), hygiene, and medical fields [3, 4]. However, its low hardness, durability, tear resistance, imperfect aging, poor weathering, inferior oil resistance, and poor electrical conductivity limit the fulfilment of the demands for broader scale applications. Their high susceptibility to degradation in the presence of light, ozone, radiation, humidity, and heat is another factor [2–5]. Therefore, NR is subjected to different modifications to form epoxidized natural rubber (ENR) and maleated natural rubber (MNR) [6]. NR in the vulcanization process in the presence of sulfur, an accelerator, and other compounding ingredients and exhibits excellent mechanical properties owing to the strain-induced crystallization phenomenon [7]. Alternatively, high-performance NR nanocomposites have been developed from this point of view by incorporating a variety of micro- and nanofillers comprising clay [8, 9], metal oxides (ZnO, TiO_2, SiO_2, Fe_2O_3, Al_2O_3, $Mn_{1-x}Zn_xFe_2O_4$, $BaTiO_3$, $PbTiO_3$, $BaFe_{12}O_{19}$, and $Ni_{1-x}Zn_xFe_2O_4$ [2, 5, 8, 9], and carbonaceous materials (CB, graphene, and CNTs) [10–12].

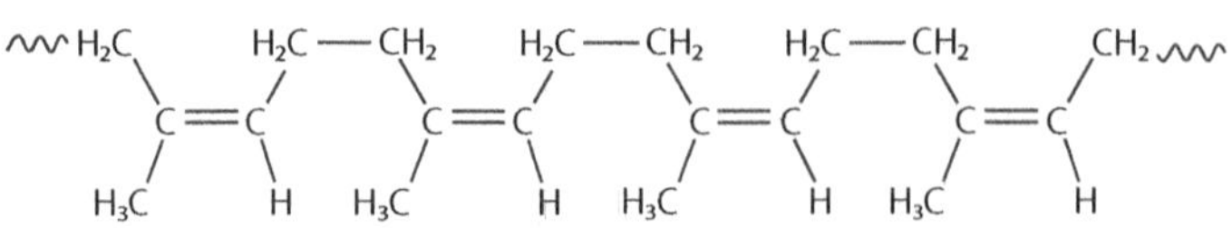

Figure 4.1 Structure of natural rubber (NR).

Recently, compared to different ranges of nanofillers, CNTs have emerged as the most promising nanofillers for fabricating natural rubber/polymer nanocomposites [13–19]. The exceptionally high tensile strength compared to other carbon-based fillers makes them appropriate for the development of high-performance composite materials for their applications as efficient vibration and shock absorbers [20], biomedical applications [21, 22], tires for cars, buses, and even airplanes [23]; automotive parts, such as bearings, seals, o-rings, oil filter gaskets, and manifold gasket brake pads, airbags, drive couplings, springs and window seals, toys, etc. [24].

Hence, this chapter concentrates on the different processing techniques used in the fabrication and dispersion of CNT-filled NR and NR blends. The different properties of NR and its blends are also discussed. Finally, the chapter ends with the application of NR/CNT nanocomposites in multifaceted fields.

4.2 Processing Techniques of NR/CNT Nanocomposites

Uniform dispersion of nanofillers remains one of the most important issues in the preparation of high-performance polymer nanocomposites. To achieve this, it is necessary to prevent agglomeration of the nanofiller in the polymer matrix and improve the matrix–filler interaction [25, 26]. These include solution mixing [27, 28] melt compounding, (dry mixing) [29–31], and latex casting [32–36] (Figure 4.2) have attracted considerable attention as the most popular processing techniques for the preparation of NR nanocomposites. This melt-mixing process is considered to be one of the most widely used techniques for rubber compounding. However, solution-mixing techniques have been found to be more effective than

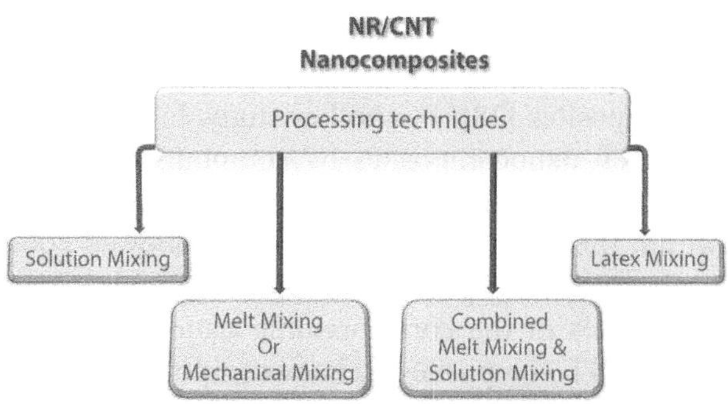

Figure 4.2 Different processing techniques for NR/CNT nanocomposites.

melt-mixing [37, 38]. In addition, a preparative method comprising both solutions and melt mixing was used to achieve excellent CNTs dispersion, as evidenced by its enhanced properties [39, 40].

4.3 Carbon Nanotubes (CNTs), their Functionalization, and Dispersion

The unique features of CNT (MWCNT and SWCNT) arise from the existence of sp^2-hybridized C atoms throughout their molecular framework [41, 42]. Each C atom is linked to three other C atoms through strong covalent bonding, leaving an extra electron on each C atom of the CNT. Consequently, the free delocalized electrons located on the CNT surfaces are attributed to the electrical and thermal conductivities of the CNTs. In addition to the strong covalent bonding present in CNTs, the C–C bond angles can be reversibly mechanically distorted. Furthermore, CNTs are difficult to disperse uniformly in a polymeric medium owing to the strong π–π stacking interaction and their intricate entanglement [43]. Therefore, it is difficult to apply traditional polymer processing to produce high-performance CNT–polymer nanocomposites. The investigations revealed a remarkable enhancement in the properties of NR upon filling it with a low amount of uniformly dispersed CNTs. This was guided by the processing history and morphology of the nanocomposites. To facilitate this, CNTs are subjected to covalent and noncovalent functionalizations [44, 45]. In covalent functionalization, the desired groups are attached to the side walls of the CNT. However, this functionalization method introduces defects in CNTs [46–48]. In contrast, noncovalent modification relies on van der Waals forces and thus does not damage the properties of CNTs. In this regard, the dispersal ability of CNT in the fabrication of NR, NR latex, and ENR nanocomposites is described below.

Guo *et al.* [28] achieved enhanced dispersion of CNT following the ultrasonic grinding and prepared its waterjet-produced rubber powder (WPRP/NR) composites [28]. In another study, Jose *et al.* [29] prepared NR-filled MWCNT nanocomposites by compounding ingredients in a two-roll mixing mill. Figure 4.3 schematically displays the distribution of MWCNTs within the NR matrix. Their findings showed good dispersion of MWCNTs observed at 3.5 phr in NR, which accounted for the enhanced mechanical attributes of the corresponding nanocomposite. Peng *et al.* [32] created NR/CNT composites by latex compounding method following the self-assembly method. In this study, CNT were modified using a

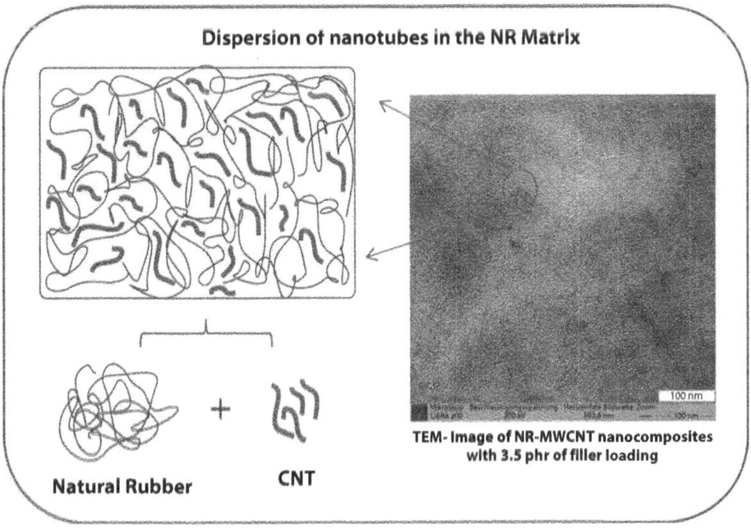

Figure 4.3 Schematic representation showing the dispersion of MWCNT in the NR [29]. Reproduced with permission from Elsevier.

sodium dodecyl sulfate (SDS) surfactant and poly (diallyldimethylammonium chloride) (PDDA). The PDDA acts as a bridging link between the CNTs and the NR matrix and prevents the aggregation of the filler, thereby enhancing the interfacial interaction between the CNT and NR matrix. Figure 4.4 shows schematically the self-assembly process for formulating NR/CNT nanocomposites through the latex compounding method. Figure 4.5 shows the TEM micrographic images of (A) the NR/pristine-MWCNT (1 wt.%), (B) PDDA-modified MWCNT/NR, and (C) a partly enlarged view of part B. The presence of pristine MWCNTs in the NR matrix is clear, where they exist in an aggregated and twisted state. In contrast, MWCNT–PDDA showed good dispersion within the NR matrix.

Mensah *et al.* [38] observed the important role of morphology, dispersion, and functionalization of CNTs in preparing NR/CNT nanocomposites to achieve the desired mechanical properties. Thomas *et al.* [39] observed the effect of 1-octadecanol functionalization on acid-treated CNTs. Subsequently, NR/CNT composites were prepared by adding sonicated CNT (in toluene) to the rubber already dissolved in toluene followed by compounding in a two-roll mill. The findings showed better dispersion of functionalized CNT in the NR compared to pristine CNTs in the NR matrix, as observed in the respective TEM images. In another study, the solution-casting method was used to prepare NR-filled MWCNTs

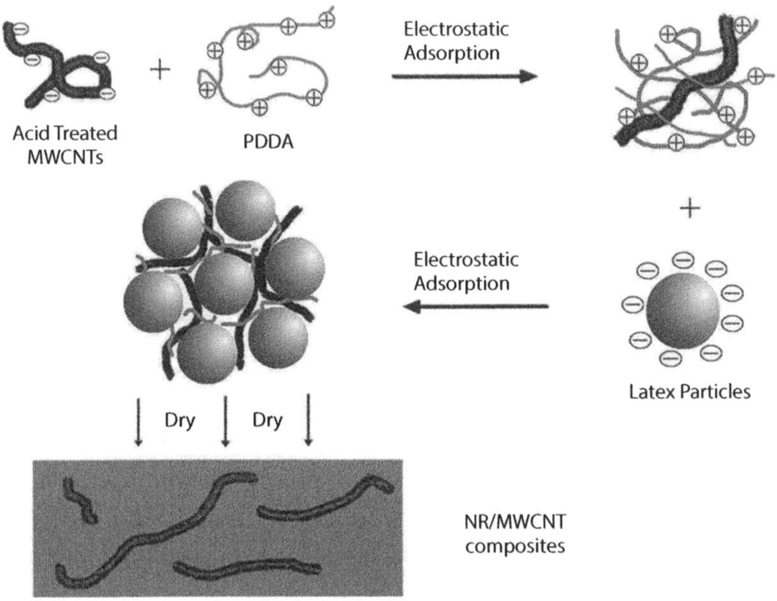

Figure 4.4 Self-assembly process for formulating NR/CNT nanocomposites through latex compounding method [32]. Reproduced with permission from Wiley.

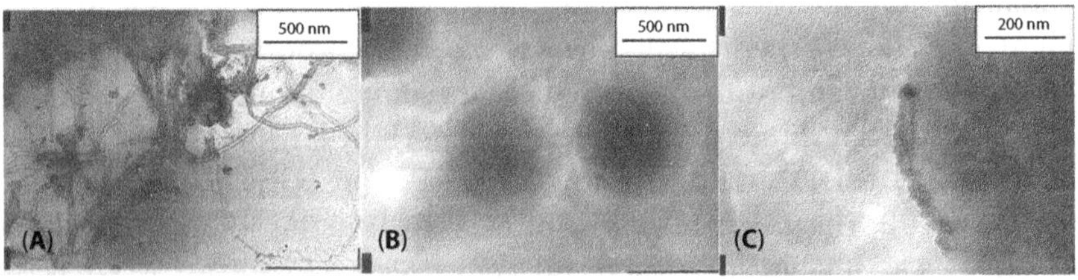

Figure 4.5 TEM micrographs of the NR/MWCNT composites: (A) p-MWCNT (1 wt.%) dispersed in natural rubber latex, (B) MWCNT–PDDA assembled with NR particles, and (C) partly enlarged view of part B [32] (modified). Reproduced with permission from Wiley.

nanocomposites [40]. These findings reveal the uniform dispersion of MWCNT corresponding to 3 wt.% loadings. Abdullateef *et al.* [49] used CNT, modified by nitric acid treatment, and used in the preparation of NR nanocomposites by solution mixing. These findings indicate a better dispersion of 1 wt.% loaded CNT in NR as evident from TEM. Mechanical and solution mixing methods have been used to incorporate different loadings of MWCNT into NR [50]. Subsequent SEM studies of the resulting nanocomposites indicated better dispersion of MWCNT (0.5 phr) in NR compared to the mechanical mixing method. Esmaeili *et al.* [51] applied a method comprising combined compounding approach and observed an optimum CNTs dispersion (3 phr) in NR.

Conductive nanocomposites of NR/CNT by solvent-cast suspensions followed by vulcanization [52] and solution mixing using an ultrasonicator [53] also produced a homogeneous dispersion of CNT. Investigations have also been conducted to study the effects of *in situ* functionalization of MWCNT with silane as coupling agents comprising bis(triethoxysilylpropyl) tetrasulfide (TESPT) and 3-aminopropyltriethoxysilane (APTES) on the properties of ENR composites [54]. TEM images in Figure 4.6 corresponding to 5 phr TESPT modified CNT in ENR (ENR–CNT–TESPT-5) and 5 phr APTES modified CNT in ENR (ENR–CNT–APTES-5) composites suggested homogenous and strong agglomerations of CNTs in ENR matrix, respectively. Anand and his co-workers [55] used SDS to enhance the dispersion of SWCNTs in water and this stabilized aqueous solution of SWCNTs to subsequently fabricate was used its NR nanocomposites through the latex compounding method.

Mohamed *et al.* [56] investigated the dispersion ability of MWCNT in NR latex using various anionic surfactants. These surfactants are composed of single, double, or triple sulfosuccinate molecules with phenyl ring components. Interestingly, it was found that the surfactant sodium 1,5-dioxo-1,5-bis(3-phenylpropoxy)-3-((3-phenylpropoxy)carbonyl) pentane-2-sulfonate (TCPh), containing triple-chain and phenyl groups, has a better stabilizing capacity of MWCNTs compared to easily available single-chain surfactant sodium dodecylbenzene sulfonate (SDBS).

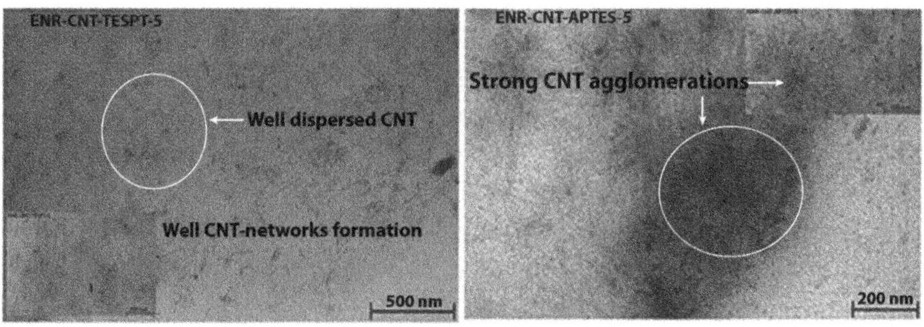

(a) (b)

Figure 4.6 TEM images of (a) ENR-CNT-TESPT-5, and (b) ENR-CNT-APTES-5 composites [54]. Reproduced with permission from Wiley.

Nanocellulose (NCC)/CNT/NR exhibited improved dispersion of the hybrid filler, forming a 3D network in the presence of NCC [57].

4.4 Properties of NR/CNT Nanocomposites

4.4.1 Mechanical Properties

The reinforcing abilities of different filler types in rubber have been extensively studied [14, 37, 58–68]. The presence of these rigid reinforcing fillers enhanced the stiffness of the rubber, which could be ascribed to the increased polymer–filler interactions, filler–filler interactions, and enhanced cross-link density. Accordingly, a continuous filler network occurs once a specific concentration of filler is reached within the polymer matrix, and both are held by weak van der Waals forces. This is also evident in the improved mechanical characteristics of the rubber nanocomposites. However, the mechanical properties of the composites deteriorate after a certain concentration of filler because of their agglomeration in the polymer matrix. Recent studies have established CNT as an exceptional fillers in polymer matrices, owing to their outstanding mechanical and other physical characteristics. CNT are the toughest materials and possess more tensile strength than steel, although their weight is minimal compared to steel [69]. CNT are very effective as nanofillers and are more effective than various traditional fillers, such as carbon black [70]. When dispersed in a polymer matrix, it can provide significant strengthening effects even at lower concentrations [37].

Guo *et al.* [28] enhanced the dispersion of CNT in NR by utilizing water-jet-produced rubber powder (WPRPP) as a carrier. The resulting composite comprising 5 phr of WPRP exhibited about 3% increment in tensile strength. In addition, it resulted in a 4% enhancement in the 300% modulus and a 5% decrease in the Akron abrasion loss compared with the NR/CNT composites. The detected enhancement in the mechanical attributes could be due to the distribution of CNT within the NR and the connection between the WPRP and the NR matrix. According to a study by Jose *et al.* [29], MWCNT (3.5 phr) filled NR showed a 55% enhancement in the tensile strength and improved Young's modulus compared to neat NR. In addition, the NR nanocomposite filled with MWCNT (5 phr) showed maximum improvement in modulus (at 100% MPa), elongation at break, and hardness (Shore A) corresponding to 1.04, 466, and 46, respectively, compared to 0.82. 438 and 34, respectively, for the NR, respectively. In another study, PDDA acted as a bridging link between the MWCNTs (1–2 wt.%)

and NR matrix in PDDA-modified MWCNT/NR nanocomposites [32]. Furthermore, the modulus and tensile strength of the NR filled with MWCNT (1–2 wt.%) exhibited significant improvements. The uniform distribution of the MWCNT in the NR matrix and their strong attachment to the NR are the reasons for this.

Thomas *et al.* [39] investigated the mechanical responses of NR nanocomposites containing 1-octadecanol-modified CNT. According to this, both the tensile strength (TS) and elongation at break (EB) increased at all loadings. Probably, the formation of a modified CNT-reinforced NR matrix could be ascribed to a more effective transfer of stress from the NR to the modified CNT. The NR/MWCNT (1–10 wt.%) nanocomposites, prepared by solution mixing showed a moderate increase in the stress level on adding CNTs in NR [40]. At 10% CNT loading in NR, the stress reached 2.55 MPa and was nine times higher than that of the neat NR. Similar variations were observed in Young's modulus for CNT loading in NR. The Young's modulus of the NR/MWCNT (10 wt.%) increased by 850.375% compared to neat NR as the CNT content increased.

The effect of sodium dodecyl sulfate (SDS) and its comparison with sodium alginate (Alg) surfactants on ENR latex filled with CNT has been evaluated in terms of their mechanical properties [46]. Based on stress-strain plots of gum ENR, ENR/CNT, ENR/CNT-SDS, ENR/CNT-$Alg_{0.5}$, and ENR/CNT-Alg_1 (0.5 and 1 refer 0.5 and 1 phr loading) nanocomposite films, it is noted that 100 and 300% moduli increase on adding CNT in ENR. This is caused by the existence of a non-stretchable filler (CNT), which leads to the strain amplification effect. The 100 and 300% moduli increased further in the CNT-SDS- and CNT-Alg-filled ENR, with maximum values for ENR/CNT-Alg_1 [46]. In addition, decreased TS and EB were also noted in the ENR/CNT-SDS nanocomposites and attributed to the defects in the nanocomposite film and ENR matrix.

The carboxylic acid groups (–COOH) were deposited on the CNT surface through acid treatment and subsequently incorporated into the NR matrix [49]. Nanocomposites of NR/MWCNTs prepared using the solution mixing method showed improved tensile properties compared to composites prepared using the mechanical mixing method [50]. These results are attributed to the improved dispersion of MWCNT in the NR matrix and their improved interactions. The NR/MWCNT nanocomposites also exhibited higher elongation at break (M100 and M300) than those prepared by the mechanical mixing technique.

The mechanical characteristics of ENR have been studied by incorporating *in situ* functionalized CNT using TESPT and APTES as the silane coupling agents [54]. Figure 4.7 displays the tensile test plots of pristine

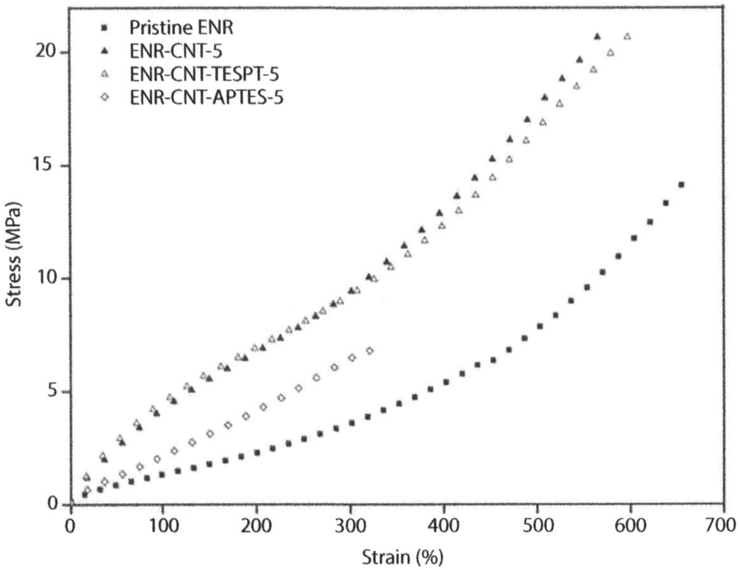

Figure 4.7 Stress–strain curves of pristine ENR, ENR–CNT-5, ENR–CNT–TESPT-5, and ENR–CNT–APTES-5 composites prepared by *in situ* functionalization [54]. Reproduced with permission from Wiley.

ENR, and its nanocomposites with 5 phr of unmodified CNTs and 5 phr of modified CNTs (TESPT-CNT and APTES-CNT) in ENR. According to these findings, improvements in the tensile strength and 100% modulus were noted in the ENR nanocomposites with 5 phr CNT and 5 phr TESPT-modified CNT respectively, while no significant effect was observed in the case of APTES-modified CNT in ENR compared to pure ENR. In contrast, the APTES-modified CNT (5 phr)/ENR composite exhibits inferior tensile properties. These findings also revealed that despite having enhanced mechanical strength and modulus (at 100%), the ENR nanocomposites with pure CNT and TESPT-modified CNT exhibited reduced EB, while the amount of CNT increased in the ENR nanocomposites in comparison with the neat polymer. The inferior results of the ENR nanocomposites containing APTES-modified CNT are ascribed to the poor NR-CNT interaction and agglomeration of CNTs. The tensile strength and tensile modulus of NR/SWNT (2 phr) nanocomposites prepared by latex compounding through surfactant-assisted sonication were significantly enhanced compared to those of NR [55]. Mohamed *et al.* [56] measured the mechanical behaviors of MWCNTs in NR latex nanocomposites following the enhanced dispersion of surfactants bearing phenyl groups.

In another study, Medupin *et al.* [71] utilized the two-roll mill method to prepare NR nanocomposites loaded with MWCNT and studied their (3–15 phr) effect on the mechanical properties. These findings indicate an increasing trend in the tensile modulus with increasing filler loading in the NR. The modulus (at different elongations, such as 100%, 200%, and 300%) of NR/MWCNT also improved with increasing filler loading. These findings are aligned with the decrease in the elastic properties of the NR nanocomposites. Bakošová *et al.* [72] noticed slightly improved TS (9.5%) and Young's modulus (3.60 MPa) in NR/CNT (2 phr) nanocomposites compared to neat NR (3.03 MPa). The NR/SWCNT nanocomposites exhibited improvements in tensile strength (75%), tear strength (27%), moduli (83%), and hardness (10%) [73]. Saravari *et al.* [74] prepared NR nanocomposites filled with different loadings of CNT (0.1 to 0.4 phr) by compounding of latex and studied their mechanical response. These findings revealed enhancements (at 300% strain) in the tensile strength modulus and elongation at break for the 0.2 phr optimum loading of CNT. Barbosa *et al.* [75] studied how sonication enhances the mechanical behaviors of NR latex/MWCNT (0.5 to 2.5 wt.%) nanocomposites. Sui *et al.* [76] incorporated acid-treated CNTs and CB into the NR in a mixer followed by placing the contents in two-roll mill for their dispersion. The mechanical properties of pure NR and NR nanocomposites with purified CNTs, ball-milled CNTs, and CB are listed in Table 4.1 [76]. The hardness, tensile strength, and modulus of all NR composites were significantly higher than those of neat NR. In contrast, the EB decreased within a tolerable range. The observed varying reinforcing effects of CNTs with the same loading can be ascribed to the dispersion, aspect ratio, and filler–polymer interaction. Wang *et al.* [77] fabricated highly conductive NR-CNT composite films through pickering emulsion interfacial assembly and observed a 57.4% enhancement in tensile strength compared to pure NR.

Kueseng *et al.* [78] synthesized NR nanocomposite using SWNTs and SiC nanoparticles separately as fillers *via* the solution intercalation technique and compared their mechanical behaviors. The outcome revealed an improved tensile strength even at a low filler content (1.5 wt.%). However, NR composites containing SiC exhibited better mechanical properties than NR/SWCNT nanocomposites. This improved modulus in NR/SiC is attributed to the reinforcing effect of SiC on the polymer matrix, which was more noticeable at larger levels of strain. In another report, a considerable increase in the mechanical properties was observed in CNT-reinforced NR compared to the virgin NR and traditional NR/CB nanocomposite [79]. Atieh *et al.* [80] also conducted a study on how MWCNT impacts the mechanical properties of NR.

Table 4.1 The properties of neat NR and NR nanocomposites. Reproduced with permission from Wiley [76].

	NR	P-CNTs-NR	B-CNTs-NR	CB-NR
Hardness/Shore A	41 ± 2	54 ± 2	2 62 ± 2	2 54 ± 2
Modulus at 300%/ MPa	1.8 ± 0.2	7.5 ± 0.3	12.3 ± 0.2	9.6 ± 0.3
Tensile strength/ MPa	7.1 ± 0.6	19.4 ± 2.2	25.5 ± 1.8	23.5 ± 1.5
Elongation at break/%	690 ± 20	505 ± 5	490 ± 16	480 ± 20
rebound degree/%		74 ± 3	72 ± 2	2.54 ± 3
Goodrich compress permanent set/%		12.5 ± 0.3	12.2 ± 0.3	13.5 ± 0.5
Goodrich compress heat accumulation (ΔT)/°C		3.5 ± 0.2	3.2 ± 0.2	4.4 ± 0.3

The studies are reported using nanocellulose (NCC) in the fabrication of hybrid filler with CNTs (NCC/CNT) and using it in reinforcing NR to compare further mechanical properties of NR nanocomposites with NCC, CNTs, and NCC/CNT fillers [52]. These results showed superior mechanical properties for the NCC/CNT hybrid filler compared to the individual materials (CNT and CB) as fillers in NR. Thus, NR comprising a hybrid filler of 10 phr enhanced the tensile strength and modulus (at 300% elongation) by 57% and 137%, respectively, owing to the improved dispersion and pronounced filler–NR interactions.

NR nanocomposites were prepared by adding CNT and CB to the matrix, which substantially enhanced the tensile strength, tear strength, and hardness of the polymer until the CNT content reached 2 phr [81]. The homogeneous dispersion and interaction between CNT and NR could be the possible reasons for this. Wang *et al.* [82] fabricated CNT@CNC (cellulose nanocrystal)/NR nanocomposites with a conductive network which has a 3D hierarchical structure, created using a latex assembly method. The choice of CNC is guided by their function in customizing the interconnected network of conductive composites made of NR/CNTs.

Figures 4.8a–d show the variations in the electrical conductivity of CNTs/NR and CNT@CNC/NR nanocomposites with CNT volume fractions and the stress–strain curves, comprising the tensile strength and Young's modulus of NR/CNTs and NR/CNT@CNC nanocomposites with different CNT contents, respectively. It is also noted that the CNT@CNC/NR samples showed significantly enhanced tensile strength and Young's modulus compared to the CNT/NR composites because of the comparatively high mechanical rigidity of the CNT@CNC/NR samples.

The effect of CB in hybrid fillers comprising MWNTs and CB on the mechanical characteristics of NR composites has been explored by several researchers. Natarajan *et al.* [83] studied the mechanical behaviors of NR composites filled with 11 phr of CB, 3 phr of CNT, various ratios of CNT:CB hybrid filler (2:3, 1.5:5.5, 0.5:9 phr). The NR composites exhibited excellent EB values (550–650%). The composites have tensile strengths in the range of 27–31 MPa, which indicates the effectiveness of the fillers in

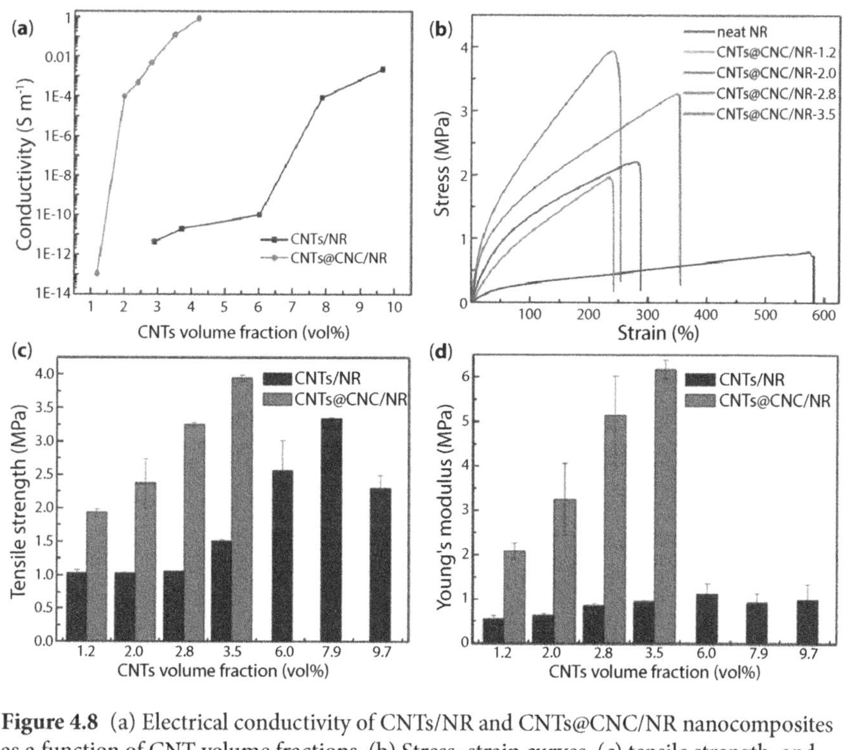

Figure 4.8 (a) Electrical conductivity of CNTs/NR and CNTs@CNC/NR nanocomposites as a function of CNT volume fractions. (b) Stress–strain curves, (c) tensile strength, and (d) Young's modulus of CNT/NR and CNT@CNC/NR nanocomposites with different CNT content [82]. Reproduced with permission from RSC.

reinforcing the NR matrix. In another work, the exchange of 5 phr CB with MWCNT led to enhancements in the mechanical properties, which is the result of the inherent characteristics of MWCNT and their collaboration with CB [84]. According to these investigations, the modulus at 100% and 300% strain increased by 72% and 54%, respectively, with the elongation at break. NR's mechanical properties were improved in the surface-modified MWCNT/CB hybrid filler owing to the symbiotic phenomena between MWCNT and CB [85]. Dong and Lie [86] prepared CNTB/CB/NR composites with a two-roll mill to study their mechanical properties are summarized in Table 4.2. These findings showed an increase in tensile strength, tensile modulus (at 100 and 300% elongation), tear strength, hardness, and a decrease in EB in CNTB/CB/NR composites compared with neat NR.

Zhang *et al.* [87] documented how partial replacement of CB with MWNTs affected the mechanical characteristics of NR/CB composites. NR foam reinforced with nanographene/CNT-filled hybrid fillers showed improved mechanical properties, including hardness, resilience, and compression [88]. In addition, investigations of the mechanical properties of NR/MWCNTs have been reported. NR nanocomposites were formed by combining the volume exclusion of SiO_2 microspheres [89], nano-silica-decorated CNT hybrids filled with NR [90], SiC nanowires/CNT/NR [91], ENR/graphite/CNTs [92], and SiO_2/CNT/NR [93].

The combination (blending) of various rubber types is considered an efficient approach to attaining the necessary performance characteristics [94, 95]. Furthermore, the field of polymeric blend nanocomposites has been receiving considerable interest due to the collective properties of polymeric blend and the advantages of polymer nanocomposites [96, 97]. In view of this, several studies have been conducted on the development of

Table 4.2 Mechanical properties of CNTB (carbon nanotube bundles) reinforced natural rubber composites (modified). Reproduced with permission from Wiley [86].

Sample	Stress at 100% (MPa)	Stress at 300% (MPa)	Tensile strength (MPa)	Elongation at break (%)	Tear strength (kN/m)	Shore A hardness
CNTB-0	2.2	8.7	24.9	659	91.8	71
CNTB-1	2.5	9.9	25.9	640	99.8	73
CNTB-3	2.7	11.2	26.5	620	103.0	74
CNTB-5	3.6	13.7	28.2	596	110.8	75

CNT-filled NR/SBR [98–101], NR/PLA [102–104], NR/EPDM [105, 106], and NR/BR [114] blend nanocomposites. The impact of ionic liquids as dispersing and coupling agents was studied by Le *et al.* [98] in the case of SBR/NR blends with CNT as a filler, and they noted enhanced dispersion. They concluded that the nonuniform distribution of crosslinks in the elastomer network accounts for the improvements in the material's mechanical properties. In another study, a blended latex of NR and SBR (80/20) was compounded with CNT and nano-SiO_2 [100]. The mechanical properties of the obtained nanocomposite latex revealed improved tensile strength and modulus at 300% strain but a reduced elongation at break. These findings also showed that CNT-filled nanocomposites had higher tensile strength than nano-SiO_2-filled nanocomposites at the same filler loading. The entanglement of CNTs and the transfer of stress between CNT and the rubber matrix are responsible for this.

Boonmahitthisud *et al.* studied the mechanical behavior of NR/SBR (80/20) [100] and NR/carboxylated SBR (XSBR) (80/20) [101] composite blends, which were developed through the mixing of blends with other chemical ingredients and were influenced by the addition of CNTs at levels ranging from 0.1 to 0.4 phr. These findings showed the corresponding tensile strength attained an optimal value of 21.1 and 21.4 MPa at the filler loading of 0.2 phr and was found to be approximately 13.4% and 14.4% greater than the respective pure blends. In addition, the modulus at 300% strain (MPa) improved with increasing filler loading, while an inverse trend was observed in the EB (%) for NR/SBR and NR/XSBR blends. The observed increase in modulus has been explained by the higher modulus of CNT, which effectively restricts chain mobility and deformation.

Jahed *et al.* [105] prepared NR/ethylene propylene diene monomer (EPDM)/MWCNT composites in a two-step process using a two-roll mill mixer and analyzed the resulting mechanical properties. They observed an enhancement in mechanical strength and modulus (at 300%), though there was a decline in elongation at break. In another study, an NR/EPDM blend containing CNTs was produced using a conventional mixing process involving a two-roll mixer, followed by the usual vulcanization for curing [106]. Subsequently, they studied how the quantity of CNTs influences the mechanical characteristics of the blend. Notably, the NR/EPDM blend with a 75:25 weight ratio of the polymers and 0.5 wt.% of filler enhanced the mechanical properties more than the pure blend. The moduli (at 100% and 300% elongation) of the NR/EPDM/CNT nanocomposites increased, while no significant effect was observed on the EB. The observed improvement in the mechanical characteristics of CNT/NR/EPDM nanocomposites occurred without any substantial decrease in elongation at break.

4.4.2 Dynamic Mechanical Properties

The elastic or storage modulus (E′) of rubber materials quantifies the impact of strain energy that can be recovered in a distorted sample, whereas the loss modulus (E″) is linked to the energy that is dissipated as heat due to damping. In view of this, several investigations have been conducted on the dynamic mechanical property measurement of natural rubber containing CNTs [2]. There are not many investigations dealing with the dynamic mechanical (DMA) measurement of CNT-filled NR nanocomposites. Jose *et al.* investigated the DMA of NR/MWCNT nanocomposites. Figures 4.9(a,b) illustrate how the storage modulus (log E′) varies with temperature at different loadings of MWCNT [29]. A notable rise in E′ is observed in the rubbery segment with the inclusion of MWCNT, which is attributed to the enhanced polymer-filler interaction. Figures 4.9(a,b) also show a

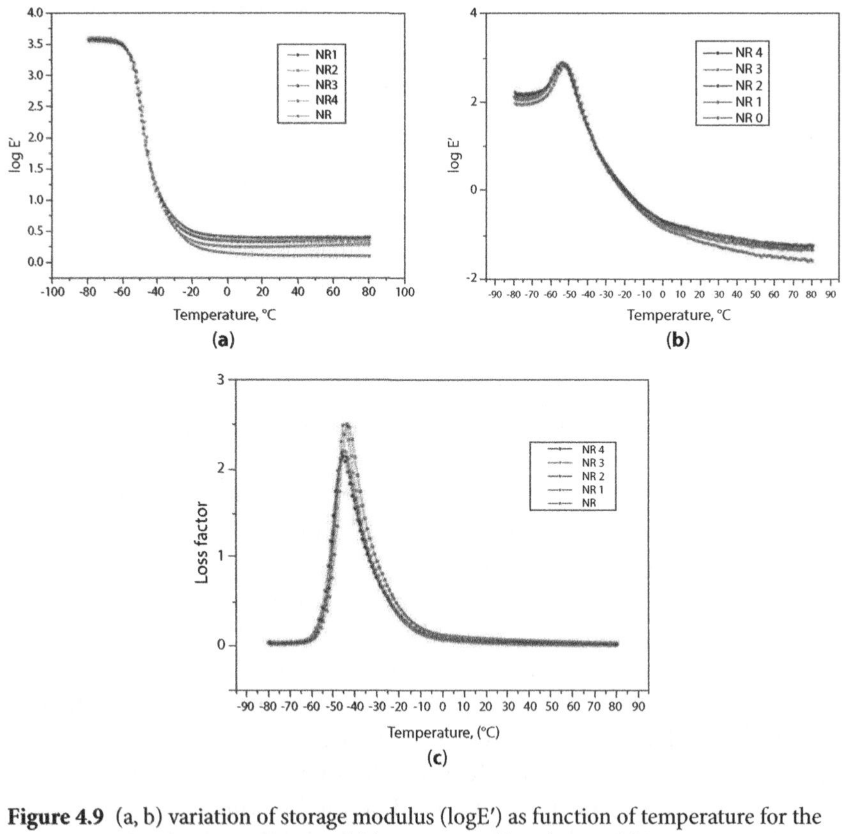

Figure 4.9 (a, b) variation of storage modulus (logE′) as function of temperature for the various MWCNT concentrations and (c) Variation of loss factor of the NR/MWCNT nano composites with temperature [29]. Reproduced with permission from Elsevier.

higher storage modulus (in the glassy region) for 5 phr loaded MWCNT in NR. However, beyond the T_g, E′ significantly improved in the rubbery portion. The NR/MWCNT nanocomposites exhibit a temperature-dependent variation in the loss factor, resulting in a higher loss factor compared to pure NR.

Thomas and his coworkers [47] investigated the DMA properties of NR composites filled with CNT and phenol-functionalized CNT (Figure 4.10). The findings illustrate how surface-modified CNTs impact the E′ and tan δ of NR and its composites at diverse temperatures. These findings at 25°C indicated significant increases in the storage modulus of phenol-functionalized CNT (1, 5, and 10 phr)/NR composites compared to pure NR and CNT-filled NR. The variation of the ratio of E'_c/E'_0 (where E'_c is the composite's E′ and E'_0 is pure NR's E′) with filler loading at 25°C indicated a sharp increase for the functionalized CNT, owing to the enhanced interaction between NR and CNT. The variation in tan δ showed a similar decrease in NR-filled CNT and phenol-functionalized CNT nanocomposites, especially at 5 and 10 phr loadings. The nanocomposites formed by Abdullateef and his team showed a significant enhancement in storage and loss modulus as the CNT loading increased [49]. This outcome suggested improved interaction between the polymer and filler, as well as improved dispersion of modified CNT in NR, which appears to play an important role.

Esmaeili *et al.* [51] observed that the insertion of 10 phr CNTs into the NR matrix enhanced the storage modulus (201%), loss modulus (1232%), and loss angles (339%) of the pure NR. This increment is attributed to the homogeneous CNT dispersion, which is obtained through a combined compounding approach. Medupin *et al.* [71], in one such report, studied the DMA of NR/CNT nanocomposites, and the corresponding findings on E′, E″, and tan δ are shown in Figure 4.11. It showed improved storage modulus (2,000–7,500 MPa) and loss modulus (500–1,413 MPa) with the addition of MWCNT in NR. In addition, NR loaded with 3 phr MWCNT showed the highest capability to disperse energy *via* segmental motion. According to Bakošová and Bakošová [72], NR nanocomposites with 2 phr CNT exhibited 19% higher storage at −80°C and a reduced tan δ value compared to the pristine rubber.

Dynamic mechanical analysis in CB/MWCNT/NR showed substantial improvements performed on the E′, E′, and loss factor (tan δ) by the substitution of 5 phr CB with MWCNT [84]. In another study, DMA had a positive effect on surface-modified MWCNT (P-MWCNT) and oxidized MWCNT (O-MWCNT)-filled NR after vulcanization showed a positive effect [85]. Zhang *et al.* [87] explored the impact of partially substituting

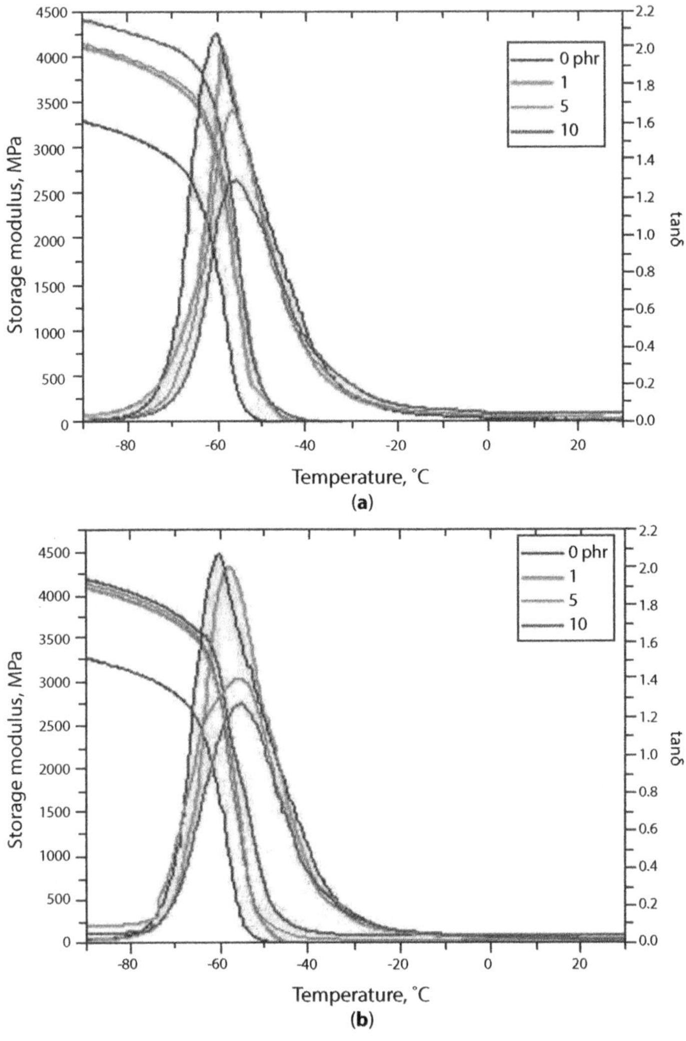

Figure 4.10 Storage modulus and tan δ plots of NR composites with (a) CNT and (b) phenol functionlized CNT [47]. Reproduced with permission from Wiley.

CB with MWCNTs on the dynamic mechanical characteristics of NR composites that contained CB. These findings showed better dynamic properties of NR/MWNT/CB for MWCNT up to a definite concentration compared to NR/CB composites.

Boonmahitthisud *et al.* [100] noticed improved dynamic mechanical behaviour in NR/SBR blend nanocomposites, reinforced with a very small quantity of CNT (0.1 to 0.4 phr). In another study [101], NR/XSBR

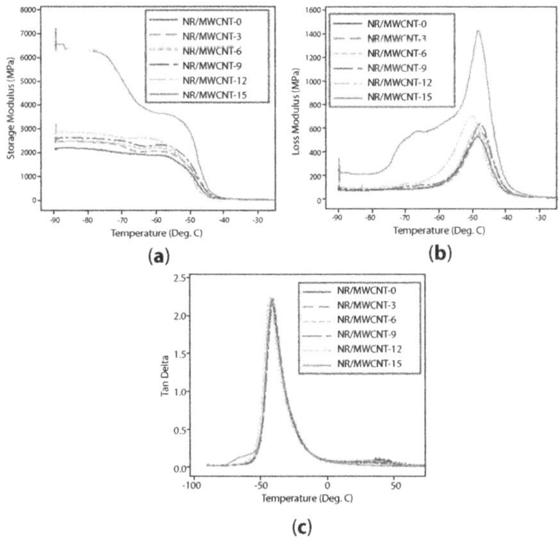

Figure 4.11 DMA Thermographs of NR/CNT nanocomposites (a) Storage modulus, (b) Loss modulus, and (c) Tan delta [71]. Reproduced with permission from Nature Publications.

blend nanocomposites loaded with a very low amount of CNT also showed an enhanced storage modulus in a dose-dependent manner. Dynamic mechanical thermal analysis of MWCNT-reinforced ethylene propylene diene monomer (NR/EPDM) nanocomposites demonstrated slow enhancement of T_g with increasing MWCNT amount [105]. NR/MWCNT (3 phr) exhibited maximum energy storage and dissipation capacities [108]. Tarawneh *et al.* [109] observed an increase in the glass transition temperature (T_g) with the increase in MWNT content.

4.4.3 Electrical Properties

Incorporating conducting fillers into the matrix can boost the low electrical conductivity of highly insulating NR by creating a conductive network within the rubber matrix [110–116]. Notably, improved filler dispersion typically leads to a higher number of conducting pathways in the rubber matrix. The percolation threshold is the specific volume fraction at which a significant decrease in electrical resistance occurs (Figure 4.12) [38]. Jose *et al.* [29] examined the electrical properties of NR/MWCNT and observed percolation transition at 3.5 phr CNT loading. The AC conductivity of neat NR (~0 S/m) was enhanced to 6×10^{-6} S/m for the NR/CNT (3.5 phr) nanocomposite. Mensah *et al.* [38] explained in their review that NR/CNT

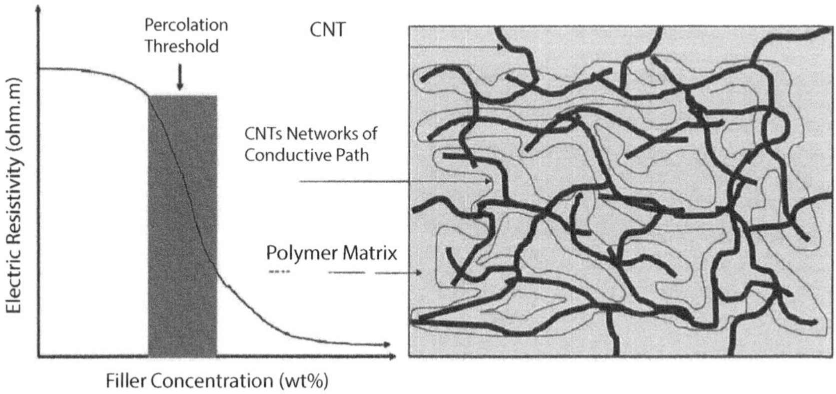

Figure 4.12 An illustration of percolation threshold by network formation of CNTs in a rubber matrix [38]. Reproduced with permission from Taylor and Francis.

nanocomposites showed a significant increment in electrical conductivity owing to the larger conducting paths created by CNT in the rubber matrix. It has been seen that the percolation threshold is determined by the type of CNTs (SWCNT, MWCNT), its aspect ratio, the amorphous carbon content in CNTs, dispersion and arrangement of CNT in rubber matrix, morphology, and type of rubber. It is generally accepted that a percolation threshold can be attained at a lower loading ($\geq$0.1 phr). The electrical properties of NR nanocomposites containing 1-octadecanol functionalized CNTs have also been explored [39]. At room temperature, the log of the DC volume resistivity of the composites decreased as the filler loading increased. Furthermore, the change in volume resistivity with filler loading showed an abrupt decrease after reaching a filler loading above 1 phr for unmodified CNT. Thomas *et al.* [47] observed lower electrical resistivity of the pure NR (~5 Ω·m) improved up to ~15 Ω·m by incorporating phenol-functionalized CNT into the NR matrix owing to the higher dispersion of CNTs into it.

Abdullateef *et al.* [49] observed a decline in the volume resistivity of CNT incorporated in the NR matrix (~5 Ω·m). In contrast, the surface-modified CNT-based NR composite showed an increase in resistivity (~13 Ω·m). In another study, Nakaramontri *et al.* [54] investigated the electrical conductivity of ENR with CNT (1 phr), CNT functionalized with TESPT, and APTES prepared by *in situ* functionalization. These findings, displayed in Figure 4.13, indicate very low percolation thresholds for the TESPT-modified and APTES-modified CNTs. It is also noted that NR/TESPT-CNT showed the highest electrical conductivity (0.01 μS/cm) compared to neat NR (10^{-7} μS/cm) and NR/APTES-CNT nanocomposite (~10^{-3} μS/cm). Such improvements were ascribed to the enhanced chemical linkages between NR and functional groups attached to the CNT's upper face,

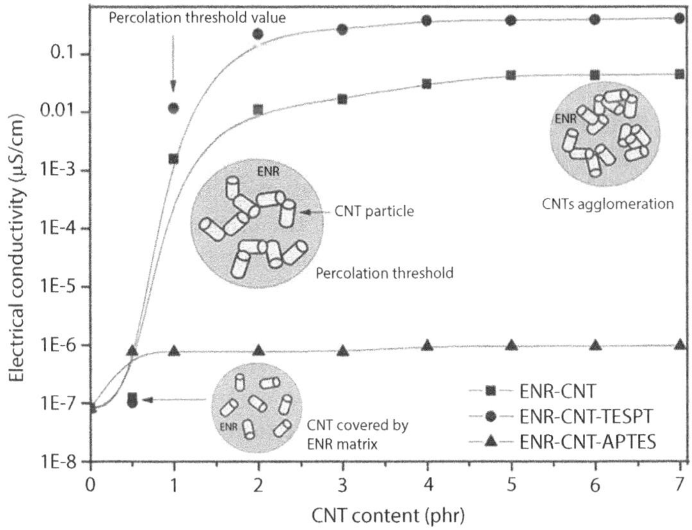

Figure 4.13 Electrical conductivity of ENR–CNT, ENR–CNT–TESPT, and ENR–CNT–APTES composites prepared by *in-situ* functionalization with various CNT contents of 0–7 phr [54]. Reproduced with permission from Wiley.

causing homogenous dispersion of the filler in the polymer matrix. Anand *et al.* [55] in their study also found electrical percolation to the insulating NR ($\sigma = 10^{-12}$ S cm^{-1}) at lower loading (<2 phr) of SWCNT. They also observed electrical conductivity attaining approximately 10^{-5} S cm^{-1} at 1.5 phr SWCNT loading in NR.

Saravari *et al.* [74] noticed that electrical conductivity improved up to 0.3 phr loading of CNT into the NR and beyond that it declined. This increase in electrical conductivity is attributed to the continuous conductive network of CNT in the NR matrix. Wang *et al.* [77] fabricated highly conductive NR-based composite films *via* pickering emulsion interfacial assembly. Bokobza [110] observed a decrease in the volume resistivity with increasing CNT content in NR/CNT nanocomposites with an electrical percolation threshold of >1 phr CNT, which was ascribed to the creation of a conductive network. The effect of substituting CNT (2, 4, and 8 wt.%) in soft NR composites on DC electrical conductivity (σ) and percolation was investigated by Banerjee *et al.* [112]. They discovered that a percolating network was established at a small CNT content of CNT, which was approximately four-fold less than that previously reported. The decrease in the percolation threshold may be ascribed to the outstanding dispersion of the filler.

In a different study, NR/CNT nanocomposites showed an increase in current density from 2.49 to 7.63 Am^{-2} as the concentration of CNTs in the

NR matrix increased [117]. This was attributed to the formation of continuous conductive networks of CNTs and hopping conduction throughout the NR matrix. The electrical percolation threshold in ENR/MWCNT with few-layer graphene (FLG) composites was reduced by 70%, while using the 0.36 vol% of hybrid nanofiller [118]. This is attributed to the bridging between MWCNT and FLG through π–π interactions, filler–rubber interactions, and uniform dispersion of the filler and the matrix. Aguilar-Bolados *et al.* [119] conducted measurements on the electrical properties of hybrid nanocomposites that included a mix of few-layer thermally reduced graphite oxide (TRGO) and MWCNTs combined with NR. They observed an increase in the electrical conductivity of the polymer by approximately 13 times for a combination of TRGOs (2 wt.%) and MWCNTs (2 wt.%) hybrid filler and ascribed it to the construction of the interconnected percolation network.

4.4.4 Thermal Stability Behavior and Differential Scanning Calorimetry (DSC) Analysis

Enhancing the thermal stability of polymer composites continues to be a significant obstacle in expanding their use across various fields. Therefore, several studies have investigated the thermal stability on carbon nanotubes filled natural rubber. Several studies in the literature have been conducted to enhance the thermal stability of NR filled with CNTs. Jose *et al.* [29] examined the thermal stability of the NR/MWCNT nanocomposites from ~25°C to 700°C in N_2 atmosphere and noted maximum enhanced thermal stability in NR3 (48°C) due to the better dispersion in matrix of gum sample. Medupin *et al.* [71] explained the thermal stability behavior of NR and its MWCNT nanocomposites by TGA and DTG (Figure 4.14). According to this, T_{onset} of NR composites with 3 and 12 phr of MWCNT exhibited maximum thermal stability enhancement of 7.89 and 7.87°C, respectively, compared to neat NR. The explanation was based on the combined impact of nanoconfinement and the barrier effect of the nanofillers. In another study [120], the highest thermal stability was observed in NR nanocomposites with 3 phr MWCNT compared with other MWCNT composites [120].

Sui *et al.* [76] used acid treated CNT to prepare CNT/NR composites followed by ball-milling and subjected it to DSC analyses (Figure 4.15). They observed an increase in the glass transition temperature (T_g) of these NR/CNT nanocomposites compared to that of NR. Such an improvement in T_g was due to the molecular mobilization of the rubber matrix of the composites because of the interactions between the NR matrix and CNTs.

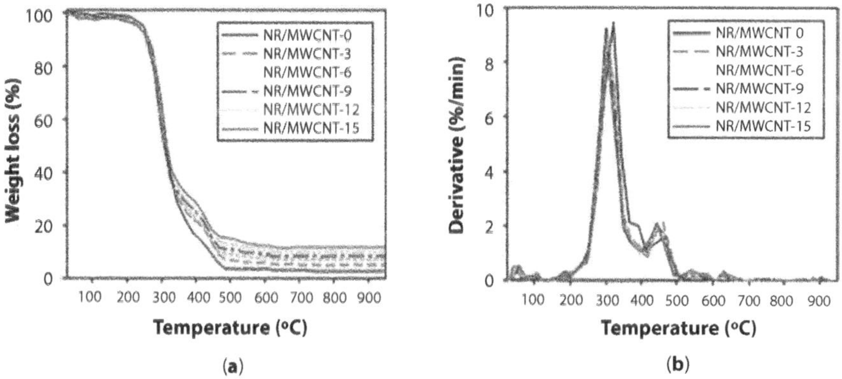

Figure 4.14 Thermogravimetric analysis (a) TGA micrographs of composites and (b) DTG of composites [71]. Reproduced with permission from Nature Publications.

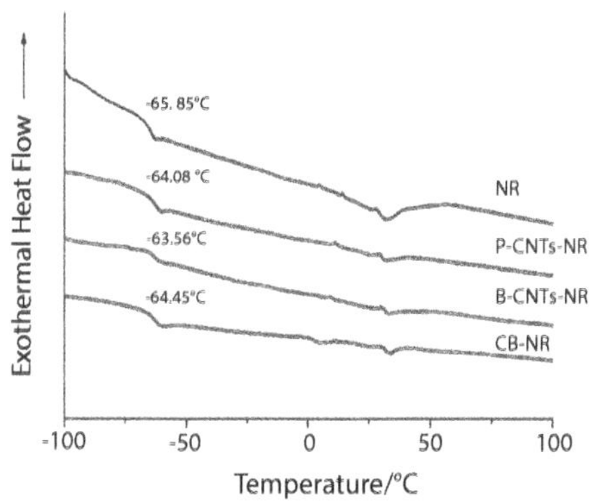

Figure 4.15 DSC curves of neat NR and NR nanocomposites [121]. Reproduced with permission from Wiley.

Another peak emerged at approximately 30°C owing to heat absorption, corresponding to melting.

Rahman *et al.* carried out TG–DTA of NR/CNT composites and observed the weight reduction corresponding to the three-stage degradation [117]. These findings indicate increased thermal stability and delayed degradation of the composites. A composite with 1.0 wt.% of CNTs shows minimal weight loss in TG. These studies also verified that the composites exhibited enhanced thermal stability and delayed degradation. Studies have been conducted on NR foams reinforced with a combination of nanographene and CNTs to assess their thermal stabilities [88]. These findings show that the degradation temperature (T_d) increased by ~68°C, corresponding to 5 wt.% and 50 wt.% weight loss. Damampai *et al.* studied the thermal behavior of ENR (50 mol% epoxy, ENR-50) filled with $FeCl_3$/CNTs. These findings show the shifting of T_d of ENR composites containing $FeCl_3$/CNTs with increasing CNT content to the higher temperature range [121]. Zhang *et al.* [107] while investigating morphology and properties of NR/BR/CNT nanocomposites of CNT reinforced NR/BR 80/20 observed the improvements in its thermal stability.

4.5 Application of NR/CNT Nanocomposites

CNT-filled NR has recently been investigated for its application in several new research fields [117–131]. Rahman *et al.* [117] produced a stretchable electronic material using a low-cost solution casting method, formulating NR/CNT nanocomposite. The nanocomposites have major contributions in the field of flexible and stretchable electronics such as in sensing, aerospace sectors, non-metallic flexible capacitors, robotic arms, etc. It is anticipated that CNT-reinforced natural rubber nanocomposites may have major contributions in these fields. MWCNT-reinforced natural rubber nanocomposites have also been investigated for applications in anthropomorphic prosthetic devices/feet [71].

Esmaeili *et al.* [51] investigated NR nanocomposite doped with CNTs focusing on its balanced dynamic shear properties and energy dissipation for applications in wave energy. Diatomaceous earth/CNT-filled NR composites exhibit improved energy harvesting efficiencies owing to their increased dielectric behavior [123]. These synthetic rubber mixtures have the potential to be used for capturing electrical energy from sustainable mechanical power sources.

Le *et al.* [124] created a self-healing material that consists of CNT-filled butyl imidazole modified bromobutyl rubber (BIIR)/NR blends. According to their studies, the high electrical conductivity of CNT/NR/BIIR blends with imidazole can accelerate self-healing when an electrical voltage is applied. Recently, studies have reported the electric heating behavior of reduced graphene oxide (0.5 wt.%)/CNT (2.5 wt.%)/NR with macro-porous structure and segregated filler in the polymer [125]. The resulting composite exhibited higher electrical conductivity (0.60 S m^{-1}) and demonstrated exceptional electric heating performance.

Natarajan *et al.* [83] studied how NR nanocomposites filled with MWCNT, CB, and combinations of both performed when subjected to strains ranging from 50% to 150%, with the specific strain range dependent on the ratio of the two-fillers. Their findings suggested that these flexible materials could be effectively utilized for health monitoring and as sensors in various dynamic rubbery components, such as tires, gaskets, and engine mounts. In another study, piezoresistive CNT-filled rubber-based composites were used to determine the strain sensing application and main mechanisms underlying the conductivity of NR/CNT composites at different strains [126]. Stretchable conductive polymer composites containing isoprene rubber (IR), CNTs, and CB exhibit strain-sensing properties with high sensitivity to strain and good recoverability at a low percolation threshold [127]. Furthermore, experimental demonstrations have demonstrated their successful use in detecting human motions and emotional expressions. The mechanical properties and sensitivity to the resistance–strain response of MWCNT/NR composites prepared using a solution method (with an electrical percolation threshold of approximately 3.5 wt.%) have been investigated to determine their suitability for monitoring strain load in rubber isolation bearings [128]. Banerjee *et al.* [112] fabricated an extremely sensitive, ultra-stretchable piezoresistive sensor utilizing NR matrix with CNT (percolation threshold: around 2 vol%, tensile strength: ~10.5 MPa, elastic modulus: ~350 kPa at 100% EB, Shore A harness~19), could detect human motion and extended it for soft robotic applications.

Investigations have also been reported on two-way shape memory effects in sulfur-vulcanized NR/paraffin wax (PW)-CNT nanocomposites [129]. It appears that research on solvent vapor-triggered reversible shape memory systems has been conducted for the first time, without the application of any external stress or pre-soaking treatment. Di Silva *et al.* [130] reported the application of piezoresistive materials comprising MWCNT incorporated in NR and synthetic rubber composite for their tensile and pressure sensing applications. Computational studies have investigated the tribological properties of CNT-reinforced NR matrices as a function of sliding velocity [131]. Teng *et al.* [120] developed molecular models of CNT reinforced vulcanized natural rubber (VNR) composites and observed enhanced tribological properties and also revealed the reinforced mechanism.

4.6 Conclusion

In the last few years, significant focus has been placed on the advancement of CNT-natural rubber nanocomposites. Considering this, this chapter aims to thoroughly examine and bring together the research conducted on NR/CNT nanocomposites in recent years owing to their significantly improved properties. It is inevitable that the enhancement of these properties is primarily influenced by the dispersion of CNT in NR and has been the primary focus of this review, which encompasses the mechanical, DMA, electrical, and thermal properties. The enhancements in properties are notably superior in CNT-filled NR compared to traditional fillers (CB and silica). Moreover, the conductivity, flexibility, stretchability, and durability of rubber-CNT nanocomposites have garnered considerable attention for their various advanced applications beyond their traditional uses.

Acknowledgement

Prof. Suneel Kumar Srivastava is a former Professor in Department of Chemistry and thankful to Indian Institute of Technology Kharagpur for this contribution.

References

1. https://en.wikipedia.org/wiki/Isoprene
2. Srivastava, S.K., *Natural Rubber Materials, in Composites and Nanocomposites*, S. Thomas, C. Han Chan, L. Pothen, J. Joy, H. Maria, S. Thomas, H. Maria (Eds.), vol. 2, pp. 504–549, The Royal Society of Chemistry, Burlington House, Piccadilly, London, 2013.
3. Le, H.H., Abhijeet, S., Ilisch, S., Klehm, J., Henning, S., Beiner, M., Sarkawi, S.S., Dierkes, W., Das, A., Fischer, D., Stöckelhuber, K.W., Wiessner, S., Khatiwada, S.P., Adhikari, R., Pham, T., Heinrich, G., Radusch, H.J., The role of linked phospholipids in the rubber-filler interaction in carbon nanotube (CNT) filled natural rubber (NR) composites. *Polymer*, 55, 4738–4747, 2014.
4. Lagashetty, A. and Venkataraman, A., Polymer nanocomposites. *Resonance*, 10, 49–57, 2005.
5. Srivastava, S.K. and Mishra, Y.K., Nanocarbon Reinforced Rubber Nanocomposites: Detailed Insights about Mechanical, Dynamical Mechanical Properties, Payne, and Mullin Effects. *Nanomaterials*, 8, 945, 2018.
6. Bokobza, L., Natural rubber nanocomposites: A review. *Nanomaterials*, 9, 12, 2019.
7. Candau, N., Laghmach, R., Chazeau, L., Chenal, J.-M., Gauthier, C., Biben, T., Munch, E., Strain-induced crystallization of natural rubber and cross-link densities heterogeneities. *Macromolecules*, 47, 5815–5824, 2014.
8. Bala, P., Samantaray, B.K., Srivastava, S.K., Nando, G.B., Organomodified montmorillonite as filler in natural and synthetic rubber. *J. Appl. Polym. Sci.*, 92, 3583–3592, 2004.

9. Pramanik, M., Srivastava, S.K., Samantaray, B.K., Bhowmick, A.K., Synthesis and characterization of organosoluble, thermoplastic elastomer/clay nanocomposites. *J. Polym. Sci., Part B: Polym. Phys.*, 40, 2065–2072, 2002.
10. Srivastava, S.K. and Pionteck, J., Recent advances in preparation, structure, properties and applications of graphite oxide. *J. Nanosci. Nanotechnol.*, 15, 1984–2000, 2015.
11. Kuila, C., Maji, A., Murmu, N.C., Kuila, T., Srivastava, S.K., Recent advancements in carbonaceous nanomaterials for multifunctional broadband electromagnetic interference shielding and wearable devices. *Carbon*, 210, 118075, 2023.
12. Roy, B.C., Kundu, A., Biswas, P., Roy, S., Singha Mahapatra, T., Recent Advances in Stimuli-Responsive Luminescent Supramolecular Lanthanide-Based Metallogels. *ChemistrySelect*, 9, e202304755, 2024.
13. Yu, M.-F., Lourie, O., Dyer, M.J., Moloni, K., Kelly, T.F., Ruoff, R.S., Strength and breaking mechanism of multiwalled carbon nanotubes under tensile load. *Science*, 287, 637–640, 2000.
14. Feng, Y., Qin, M., Guo, H., Yoshino, K., Feng, W., Infrared-actuated recovery of polyurethane filled by reduced graphene oxide/carbon nanotube hybrids with high energy density. *ACS Appl. Mater. Interfaces*, 5, 10882–10888, 2013.
15. Roy, S., Srivastava, S.K., Pionteck, J., Mittal, V., Assembly of layered double hydroxide on multi-walled carbon nanotubes as reinforcing hybrid nanofiller in thermoplastic polyurethane/nitrile butadiene rubber blends. *Polym. Int., 65*, 93–101, 2016.
16. Pradhan, B. and Srivastava, S.K., Layered double hydroxide/multiwalled carbon nanotube hybrids as reinforcing filler in silicone rubber. *Compos. Part A Appl. Sci. Manuf.*, 56, 290–299, 2014.
17. Roy, S., Bhalani, D. V., Jewrajka, S.K., Surface segregation of segmented amphiphilic copolymer of poly(dimethylsiloxane) and poly(ethylene glycol) on poly(vinylidene fluoride) blend membrane for oil–water emulsion separation. *Sep. Purif. Technol.*, 232, 115940, 2020.
18. Roy, S., Srivastava, S. K., Pionteck, J., Mittal, V., Montmorillonite–multiwalled carbon nanotube nanoarchitecture reinforced thermoplastic polyurethane. *Polym. Compos.,* 37, 1775–1785, 2016.
19. Long, X.-H., Ma, Y.-T., Yue, R., Fan, J., Experimental study on impact behaviors of rubber shock absorbers. *Constr. Build. Mater.*, 173, 718–729, 2018.
20. MMMM, K.L., Ramlow, H., Floriano, J.F., Acosta, E.D., Faita, F.L., Machado, R.A.F., Latex and natural rubber: recent advances for biomedical applications. *Polímeros*, 32, e2022015, 2022.
21. Davies, B., Natural rubber—Its engineering characteristics. *Mater. Des.*, 7, 68–74, 1986.
22. Guerra, N.B., Pegorin, G.S., Boratto, M.H., R de Barros, N., Graeff, C.F.O., Herculano, R.D., Biomedical applications of natural rubber latex from the rubber tree Hevea brasiliensis. *Mater. Sci. Eng., C*, 126, 112126, 2021.

23. Kohjiya, S. and Ikeda, Y., *Chemistry, Manufacture and Applications of Natural Rubber*, 2nd Edition, Elsevier, Amsterdam, Netherlands, 2021.
24. Srivastava, S.K. and Bhuyan, B., Rubber Nanocomposites for Tire Tread Application, in: *Rubber Nanocomposites and Nanotextiles in Automobiles*, B. Banerjee (Ed.), pp. 31–74, Rapra, 2019.
25. Roy, S., Srivastava, S.K., Pionteck, J., Mittal, V., Mechanically and thermally enhanced multiwalled carbon nanotube–graphene hybrid filled thermoplastic polyurethane nanocomposites. *Macromol. Mater. Eng.*, 300, 346–357, 2015.
26. Pradhan, B., Roy, S., Srivastava, S.K., Saxena, A., Synergistic effect of carbon nanotubes and clay platelets in reinforcing properties of silicone rubber nanocomposites. *J. Appl. Polym. Sci.*, 132, 41818, 2015.
27. Gao, J-S., Liu, Z., Yan, Z., He, Y., A novel slurry blending method for a uniform dispersion of carbon nanotubes in natural rubber composites. *Results Phys.*, 15, 102720, 2019.
28. Guo, X., Guo, S., Liu, G., Bai, L., Liu, H., Xu, Y., Zhao, J., Chai, H., Jian, X., Guo, L., Liu, F., Improving dispersion of carbon nanotubes in natural rubber by using waterjet-produced rubber powder as a carrier. *Polymers*, 15, 477, 2023.
29. Jose, T., Moni, G., Salini, S., Raju, A.J., George, J.J., George, S.C., Multifunctional multi-walled carbon nanotube reinforced natural rubber nanocomposites. *Ind. Crops Prod.*, 105, 63–73, 2017.
30. Abraham, E., Deepa, B., Pothan, L.A., John, M., Narine, S.S., Thomas, S., Anandjiwala, R., Physicomechanical properties of nanocomposites based on cellulose nanofibre and natural rubber latex. *Cellulose*, 20, 417–427, 2013.
31. Lorenz, H., Fritzsche, J., Das, A., Stöckelhuber, K.W., Jurk, R., Heinrich, G., Klüppel, M., Advanced elastomer nano-composites based on CNT-hybrid filler systems. *Compos. Sci. Technol.*, 69, 2135–2143, 2009.
32. Peng, Z., Feng, C., Luo, Y., Li, Y., Yi, Z., Kong, L.X., Natural rubber/multiwalled carbon nanotube composites developed with a combined self-assembly and latex compounding technique. *J. Appl. Polym. Sci.*, 125, 3920–3928, 2012.
33. Aziz, A.A., Ismail, N.I.N., Che, S.M.S., Rusop, M., Characterization of functionalized multi-walled carbon nanotubes in pre-vulcanized natural rubber latex. *AIP Conf. Proc.*, 1455, 124–130, 2012, DOI: 10.1063/1.4732480.
34. Atieh, M.A., Nazir, N., Yusof, F., Fettouhi, M., Ratnam, C.T., Alharthi, M., Abu-Ilaiwi, F.A., Mohammed, K., Al-Amer, A., Radiation vulcanization of natural rubber latex loaded with carbon nanotubes. *Fuller. Nanotub. Carbon Nanostruct.*, 18, 56–71, 2010.
35. Aziz, A.A., Assim, D.H.A.I., Suriani, A.B., Mahmood, M.R., *JOSTT*, 8, 131–140, 2012.
36. Bras, J., Hassan, M.L., Bruzesse, C., Bruzesse, C., Hassan, E.A., El-Wakil, N.A., Dufresne, A., Mechanical, Barrier, and Biodegradability Properties of

Bagasse Cellulose Whiskers Reinforced Natural Rubber Nanocomposites. *Ind. Crops Prod.*, 32, 627–633, 2010.
37. Sethulekshmi, A.S., Saritha, A., Joseph, K., A comprehensive review on the recent advancements in natural rubber nanocomposites. *Int. J. Biol. Macromol.*, 194, 819–842, 2022.
38. Mensah, B., Kim, H.G., Lee, J.-H., Arepalli, S., Nah, C., Carbon nanotube-reinforced elastomeric nanocomposites: a review. *Int. J. Smart Nano Mater.*, 6, 211–238, 2015.
39. Selvin Thomas, P., Abdullateef, A.A., Al-Harthi, M.A., Atieh, M.A., De, S.K., Rahaman, M., Chaki, T.K., Khastgir, D., Bandyopadhyay, S., Electrical properties of natural rubber nanocomposites: effect of 1-octadecanol functionalization of carbon nanotubes. *J. Mater. Sci.*, 47, 3344–3349, 2012.
40. Fakhru'l-Razi, A., Atieh, M.A., Girun, N., Chuah, T.G., El-Sadig, M., Biak, D.R.A., Effect of multi-wall carbon nanotubes on the mechanical properties of natural rubber. *Compos. Struct.*, 75, 496–500, 2006.
41. Manikandan, N., Suresh, K., Siva, S., Rathis, G., Vishnu, S., Shabariganesh, T., Carbon nanotubes and their properties-The review. *Mater. Today Proc.*, 47, 4682–4685, 2021.
42. Georgakila, V., Perman, J.A., Tucek, J., Zboril, R., Broad Family of Carbon Nanoallotropes: Classification, Chemistry, and Applications of Fullerenes, Carbon Dots, Nanotubes, Graphene, Nanodiamonds, and Combined Superstructures. *Chem. Rev.*, 115, 11, 4744–4822, 2015.
43. Dubey, R., Dutta, D., Sarkar, A., Chattopadhyay, P., Functionalized carbon nanotubes: synthesis, properties and applications in water purification, drug delivery, and material and biomedical sciences. *Nanoscale Adv.*, 3, 5722–5744, 2021.
44. Zhou, Y., Fang, Y., Ramasamy, R.P., Non-covalent functionalization of carbon nanotubes for electrochemical biosensor development. *Sensors*, 19, 392, 2019.
45. Verma, R., Rathod, M.J., Goyal, R.K., *National Conference on Advanced Materials and Applications (NCAMA 2019)*, IOP Conf. Series: Materials Science and Engineering, vol. 798, IOP Publishing, p. 012031, 2020, doi:10.1088/1757-899X/798/1/012031.
46. Krainoi, A. and Boonkerd, K., Mechanical/electrical properties and strain sensibility of epoxidized natural rubber nanocomposite filled with carbon nanotube: Effect of sodium alginate as a surfactant on latex technology process. *eXPRESS Polym. Lett.*, 17, 850–866, 2023.
47. Thomas, P.S., Abdullateef, A.A., Al-Harthi, M.A., Basfar, A.A., Bandyopadhyay, S., Atieh, M.A., De, S.K., Effect of phenol functionalization of carbon nanotubes on properties of natural rubber nanocomposites. *J. Appl. Polym. Sci.*, 124, 2370–2376, 2012.
48. Norizan, M.N., Moklis, M.H., Demon, S.Z.N., Halim, N.A., Samsuri, A., Mohamad, I.S., Knight, V.F., Abdullah, N., Carbon nanotubes:

Functionalisation and their application in chemical sensors. *RSC Adv.*, 10, 43704–43732, 2020.

49. Abdullateef, A.A., Thomas, S.P., Al-Harthi, M.A., De, S.K., Bandyopadhyay, S., Basfar, A.A., Atieh, M.A., Natural rubber nanocomposites with functionalized carbon nanotubes: mechanical, dynamic mechanical, and morphology studies. *J. Appl. Polym. Sci.*, 125, E76–E84, 2012.
50. Ismail, H., Ramly, F., Othman, N., Multiwall carbon nanotube-filled natural rubber: the effects of filler loading and mixing method. *Polym. Plast. Technol. Eng.*, 49, 260–266, 2010.
51. Esmaeili, A., Masters, I., Hossain, M., A novel carbon nanotubes doped natural rubber nanocomposite with balanced dynamic shear properties and energy dissipation for wave energy applications. *Results Mater.*, 17, 100358, 2023.
52. Capezza, A., Andersson, R.L., Strom, V., Wu, Q., Sacchi, B., Farris, S., Hedenqvist, M.S., Olsson, R.T., Preparation and comparison of reduced graphene oxide and carbon nanotubes as fillers in conductive natural rubber for flexible electronics. *ACS Omega*, *4*, 3458–3468, 2019.
53. Kang, B.-H., Jeong, I.-Y., Park, S.-H., Research Progress of Natural Rubber Wet Mixing Technology. *Polymers*, 14, 2551–2562, 2022.
54. Nakaramontri, Y., Nakason, C., Kummerlowe, C., Vennemann, N., Effects of *in-situ* functionalization of carbon nanotubes with bis(triethoxysilylpropyl) tetrasulfide (TESPT) and 3-aminopropyltriethoxysilane (APTES) on properties of epoxidized natural rubber–carbon nanotube composites. *Polym. Eng. Sci.*, 55, 2500–2510, 2015, doi.org/10.1002/pen.24140.
55. Anand, K.S., Jose, T.R., Alex, R.J., Natural rubber-carbon nanotube composites through latex compounding. *Int. J. Polym. Mater.*, 59, 33–44, 2009.
56. Mohamed, A., Anas, A.K., Bakar, S.A., Ardyani, T., Zin, W.M.W., Ibrahim, S., Sagisaka, M., Brown, P., Eastoe, J., Enhanced dispersion of multiwall carbon nanotubes in natural rubber latex nanocomposites by surfactants bearing phenyl groups. *J. Colloid Interface Sci.*, 455, 179–187, 2015.
57. Kazemi, H., Mighri, F., Park, K.W., Frikha, S., Rodrigue, D., Hybrid nanocellulose/carbon nanotube/natural rubber nanocomposites with a continuous three-dimensional conductive network. *Polym. Compos.*, 43, 2362–2374, 2022.
58. Roy, B. C., Ramlal, V. R., Basak, D., Basak, S., Roy, S., Ghosh, S., Mahapatra, T. S., Light-Emitting Coordination Polymers: Stimuli-Responsive Gels, PMMA-Based Composite Films, and UV-Shielding. *Inor. Chem.*, 63, 23044–23056, 2024.
59. Roy, S., Srivastava, S.K., Mittal, V., Facile noncovalent assembly of MWCNT-LDH and CNF-LDH as reinforcing hybrid fillers in thermoplastic polyurethane/nitrile butadiene rubber blends. *J. Polym. Res.*, *23*, 36, 2016.
60. Pramanik, M., Srivastava, S.K., Samantaray, B.K., Bhowmick, A.K., Rubber–clay nanocomposite by solution blending. *J. Appl. Polym. Sci.*, 87, 2216–2220, 2003.

61. Acharya, H., Pramanik, M., Srivastava, S.K., Bhowmick, A.K., Synthesis and evaluation of high-performance ethylene–propylene–diene terpolymer/organoclay nanoscale composites. *J. Appl. Polym. Sci.*, 93, 2429–2436, 2004.
62. Valadares, L.F., Leite, C.A.P., Galembeck, F., Preparation of natural rubber–montmorillonite nanocomposite in aqueous medium: evidence for polymer–platelet adhesion. *Polymer*, 47, 672–678, 2006.
63. Jacob, A., Kurian, P., Aprem, A.S., Cure characteristics and mechanical properties of natural rubber–layered clay nanocomposites. *Int. J. Polym. Mater.*, 56, 593–604, 2007.
64. Wang, Y., Zhang, H., Wu, Y., Yang, J., Zhang, L., Structure and properties of strain-induced crystallization rubber–clay nanocomposites by co-coagulating the rubber latex and clay aqueous suspension. *J. Appl. Polym. Sci.*, 96, 318–323, 2005.
65. Sharif, J., Yunus, W.M.Z.W., Dahlan, K.Z.H.M., Ahmad, M.H., Preparation and properties of radiation crosslinked natural rubber/clay nanocomposites. *Polym. Test.*, 24, 211–217, 2005.
66. Varghese, S. and Karger-Kocsis, J., Natural rubber-based nanocomposites by latex compounding with layered silicates. *Polymer*, 44, 4921–4927, 2003.
67. Seentrakoon, B., Junhasavasdikul, B., Chavasiri, W., Enhanced UV-protection and antibacterial properties of natural rubber/rutile-TiO2 nanocomposites. *Polym. Degrad. Stab.*, *98*, 566–578, 2013.
68. Roy, S., Srivastava, S.K., Mittal, V., Noncovalent assembly of carbon nanofiber–layered double hydroxide as a reinforcing hybrid filler in thermoplastic polyurethane–nitrile butadiene rubber blends. *J. Appl. Polym. Sci.*, 133, 43470, 2016.
69. Treacy, M.M.J., Ebbesen, T.W., Gibson, J.M., Exceptionally high Young's modulus observed for individual carbon nanotubes. *Nature*, 381, 678–680, 1996.
70. Cataldo, F., Ursini, O., Angelini, G., MWCNTs elastomer nanocomposite, part 1: The addition of MWCNTs to a natural rubber-based carbon black-filled rubber compound. *Fuller. Nanotub. Carbon Nanostruct.*, 17, 38–54, 2009.
71. Medupin, R.O., Abubakre, O.K., Abdulkareem, A.S., Muriana, R.A., Abdulrahman, A.S., Carbon nanotube reinforced natural rubber nanocomposite for anthropomorphic prosthetic foot purpose. *Sci. Rep.*, 9, 20146, 2019.
72. Bakošová, D. and Bakošová, A., Testing of rubber composites reinforced with carbon nanotubes. *Polymers*, 14, 3039–3056, 2022.
73. Vázquez-Martínez, Y., Ramírez-Herrera, C.A., Mondragón, M., Elías-Zúñiga, A., Elizalde, L.E., Effect of single-walled carbon nanotubes on the cross-linking process in natural rubber vulcanization. *Polymers*, *15*, 126–142, 2023.
74. Saravari, O., Boonmahitthisud, A., Satitnaithum, W., Chuayjuljit, S., Mechanical and Electrical Properties of Natural Rubber/Carbon Nanotube

Nanocomposites Prepared by Latex Compounding. *Adv. Mater. Res.*, *664*, 543–546, 2013.

75. Barbosa, R., Gonçalves, R., Tozzi, K.A., Saccardo, M.C., Zuquello, A.G., Scuracchio, C.H., Improving the swelling, mechanical, and electrical properties in natural rubber latex/carbon nanotubes nanocomposites: effect of the sonication method. *J. Appl. Polym. Sci.*, 139, 52325, 2022.
76. Sui, G., Zhong, W., Yang, X., Zhao, S., Processing and material characteristics of a carbon-nanotube-reinforced natural rubber. *Macromol. Mater. Eng.*, 292, 1020–1026, 2007.
77. Wang, Y., Xin, Z., Wang, B., Feng, X., Mao, Z., Sui, X., Fabrication of highly conductive natural rubber-based composite films *via* Pickering emulsion interfacial assembly. *Colloids Surf., A*, 651, 129774, 2022.
78. Kueseng, K. and Jacob, K.I., Natural rubber nanocomposites with SiC nanoparticles and carbon nanotubes. *Eur. Polym. J.*, 42, 220–227, 2006.
79. Sui, G., Zhong, W.H., Yang, X.P., Yu, Y.H., Zhao, S.H., Preparation and properties of natural rubber composites reinforced with pretreated carbon nanotubes. *Polym. Adv. Technol.*, 19, 1543–1549, 2008.
80. Atieh, M.A., Girun, N., Mahdi, E.-S., Tahir, H., Guan, C.T., Alkhatib, M.F., Ahmadun, F.R., Baik, D.R., Reinforced Natural Rubber Nanocomposites: Next Generation Advanced Material. *Fuller. Nanotub. Carbon Nanostruct.*, 14, 641–649, 2006.
81. Jawahar, P., Payak, V., Chandradass, J., Prabhu, P., Optimization of mechanical properties of CNT-rubber nanocomposites. *Mater. Today Proc.*, 45, 7183–7189, 2021.
82. Wang, S., Zhang, X., Wu, X., Lu, C., Tailoring percolating conductive networks of natural rubber composites for flexible strain sensors *via* a cellulose nanocrystal templated assembly. *Soft Matter*, 12, 845–852, 2016.
83. Natarajan, T.S., Bhagavatheswaran, E.S., Stöckelhuber, K.W., Wieβner, S., Potschke, P., Heinrich, G., Das, A., Strong strain sensing performance of natural rubber nanocomposites. *ACS Appl. Mater. Interfaces*, 9, 4860–4872, 2017.
84. Shahamatifard, F., Rodrigue, D., Park, K.W., Frikha, S., Mighri, F., Natural rubber nanocomposites: Effect of carbon black/multi-walled carbon nanotubes hybrid fillers on the mechanical properties and thermal conductivity. *Polym.-Plast. Technol. Mater.*, 60, 1686–1696, 2021.
85. Shahamatifard, F., Rodrigue, D., Park, K.-W., Frikha, S., Mighri, F., Surface modification of MWCNT to improve the mechanical and thermal properties of natural rubber nanocomposites. *Can. J. Chem. Eng.*, 101, 1881–1896, 2023.
86. Dong, B., Liu, C., Lu, Y., Wu, Y., Synergistic effects of carbon nanotubes and carbon black on the fracture and fatigue resistance of natural rubber composites. *J. Appl. Polym. Sci.*, 132, 42075, 2015.
87. Zhang, H., Wei, Y., Kang, Z., Zhao, G., Liu, Y., Influence of partial substitution for CB with MWNTs on performance of CB-filled NR composites. *Micro Nano Lett.*, 12, 117–122, 2017.

88. Shojaie, S., Vahidifar, A., Naderi, G., Shokri, E., Mekonnen, T.H., Esmizadeh, E., Physical hybrid of nanographene/carbon nanotubes as reinforcing agents of NR-based rubber foam. *Polymers*, 13, 2346, 2021.
89. Huang, K., Jiang, L., Du, Y., Wang, J., Zheng, X., Jiang, X., Sui, Y., Che, M., Xu, Q., Li, L., Xin, Z., Zhao, S., Multifunctional NR/MWCNTs nanocomposites constructed *via* combining volume exclusion of SiO_2 microspheres with interface reinforcement of tannic acid. *Eur. Polym. J.*, 151, 110424, 2021.
90. George, N., Venugopal, B., John, H., Mathiazhagan, A., Joseph, R., Nanosilica decorated multiwalled carbon nanotubes (CS hybrids) in natural rubber latex. *Polymer*, 161, 170–180, 2019.
91. Janyakunmongkol, K., Nhuapeng, W., Thamjaree, W., Effect of added silicon carbide nanowires and carbon nanotubes on mechanical properties of 0–3 natural rubber composites. *Jpn. J. Appl. Phys.*, 55, 0AE21, 2016.
92. Pichaiyut, S., Kitisavetjit, W., Nakason, C., Synergistic effects of graphite and carbon nanotube hybrid fillers on key properties of epoxidized natural rubber nanocomposites. *J. Polym. Res.*, 30, 412, 2023.
93. Ismail, H., Ramly, A.F., Othman, N., Effects of silica/multiwall carbon nanotube hybrid fillers on the properties of natural rubber nanocomposites. *J. Appl. Polym. Sci.*, 128, 2433–2438, 2013.
94. Srivastava, S.K., *Conducting Polymer Composites*, V. Mittal (Ed.), pp. 157–194, Central West Publishing, Australia, 2019.
95. Srivastava, S.K., *Layered Double Hydroxide Polymer Nanocomposites*, S. Thomas and S. Daniel (Eds.), pp. 111–137, Elsevier, Amsterdam, Netherlands, 2019.
96. Bhuyan, B., Srivastava, S.K., Pionteck, J., MWCNT/Hectorite hybrid filled acrylonitrile butadiene rubber/ethylene-co-vinyl acetate blend nanocomposites: preparation and properties. *J. Polym. Res.*, 24, 150, 2017.
97. Kuila, T., Srivastava, S.K., Bhowmick, A.K., Saxena, A.K., Thermoplastic polyolefin based polymer–blend-layered double hydroxide nanocomposites. *Compos. Sci. Technol.*, 68, 3234–3239, 2008.
98. Le, H.H., Das, A., Basak, S., Tahir, M., Wießner, S., Fischer, D., Reuter, U., Stöckelhuber, K.W., Bhowmick, A.K., Do, Q.K., Heinrich, G., Radusch, H.-J., Effect of different ionic liquids on the dispersion and phase selective wetting of carbon nanotubes in rubber blends. *Polymer*, 105, 284–297, 2016.
99. Bernal-Ortega, P., Bernal, M.M., Gonz, A., Posadas, P., Navarro, R., Valentín, J.L., New insight into structure-property relationships of natural rubber and styrene-butadiene rubber nanocomposites filled with MWCNT. *Polymer*, 201, 122604, 2020.
100. Boonmahitthisud, A. and Chuayjuljit, S., Use of carbon nanotube and nanosilica as reinforcement nanofillers in NR/SBR blended latex. *Adv. Mater. Res.*, 347–353, 3197–3200, 2012.
101. Boonmahitthisud, A. and Chuayjuljit, S., Effects of carbon nanotube on tensile and dynamic mechanical properties of NR/SBR and NR/XSBR

nanocomposites prepared by latex compounding. *Adv. Mater. Res.*, 488–489, 612–616, 2012.
102. Mat Desa., M.S.Z., Hassan, A., Arsad, A., Mechanical properties of poly (lactic acid)/multiwalled carbon nanotubes nanocomposites. *Adv. Mater. Res.*, 747, 639–642, 2013.
103. Ali, A.M. and Ahmad, S.H., Mechanical characterization and morphology of polylactic acid/liquid natural rubber filled with multi walled carbon nanotubes. *AIP Conf. Proc.*, 1571, 83–89, 2013.
104. Ali, A.M. and Ahmad, S.H., *AIP Conf. Proc.*, 1528, 254–259, 2013, DOI: 10.1063/1.4803604.
105. Jahed, M., Naderi, G., Ghoreishy, M.H.R., Effect of processing parameter and filler content on tensile properties of multi-walled carbon nanotubes reinforced polylactic acid nanocomposite. *Polym. Compos.*, 39, E745–E753, 2018, DOI: 10.1002/pc.24153.
106. Paran, S.M.R., Naderi, G., Babakhani, A., An experimental study of the effect of CNTs on the mechanical properties of CNTs/NR/EPDM nanocomposite. *Polym. Compos.*, 39, 4071–4079, 2018.
107. Zhang, J., Gu, Z., Meng, C., Wang, J., Sui, J., Morphology and selected properties of NR/BR/CNT nanocomposites effect of ethanol-assisted mixing. *Polimery/Polymers*, 68, 249–304, 2023.
108. Medupin, R.O., Abubakre, O.K., Abdulkareem, A.S., Muriana, R.A., Lawal, S.A., Dimensional and thermal reliability of multi-walled carbon nanotube filled natural rubber nanocomposites. *Int. J. Eng. Res. Afr.*, 51, 177–189, 2020.
109. Tarawneh, M.A., Ahmad, S.H., Rasid, R., Yahya, S.Y., Lau, K.-T., Kong, I., S.Y.E., Thermal behavior of MWNT-reinforced thermoplastic natural rubber nanocomposites. *J. Reinf. Plast. Compos.*, 30, 216–221, 2011.
110. Bokobza, L., Multiwall carbon nanotube-filled natural rubber: Electrical and mechanical properties. *eXPRESS Polym. Lett.*, *6*, 213–223, 2012.
111. Damampai, K., Pichaiyut, S., Stöckelhuber, K.W., Das, A., Nakason, C., Ferric ions crosslinked epoxidized natural rubber filled with carbon nanotubes and conductive carbon black hybrid fillers. *Polymers, 14*, 4392, 2022.
112. Banerjee, S.S., Arief, L., Berthold, R., Wiese, M., Bartholdt, M., Ganguli, D., Mitra, S., Mandal, S., Wallaschek, J., Raatz, A., Heinrich, G., Das, A., Superelastic ultrasoft natural rubber-based piezoresistive sensors for active sensing interface embedded on soft robotic actuator. *Appl. Mater. Today*, 25, 101219, 2021.
113. Pojanavaraphan, T., Chirasakulkarun, A., Muksing, N., Magaraphan, R., Electrolytic admicellar polymerization of pyrrole on natural rubber/clay nanocomposites. *J. Appl. Polym. Sci., 112*, 1552–1564, 2009.
114. Zhang, C., Zhai, T., Dan, Y., Turng, L., Reinforced natural rubber nanocomposites using graphene oxide as a reinforcing agent and their *in situ* reduction into highly conductive materials. *Polym. Compos.*, *38*, E199–E207, 2017.

115. Dong, Zhang, L., Wu, Y., Highly conductive natural rubber–graphene hybrid films prepared by solution casting and *in situ* reduction for solvent-sensing application. *J. Mater. Sci.*, *51*, 10561–10573, 2016.
116. Qin, H., Deng, C., Lu, S., Yang, Y., Guan, G., Liu, Z., Yu, Q., Enhanced mechanical property, thermal and electrical conductivity of natural rubber/graphene nanosheets nanocomposites. *Polym. Compos.*, *41*, 1299–1309, 2020.
117. Rahman, M.J., Hossain, M.F., Islam, M.J., Mieno, T., Chowdhury, M.N., Carbon nanotube reinforced natural rubber nanocomposite as a stretchable electronic material. *J. Mater. Eng. Perform.*, 32, 5338–5345, 2023.
118. Salaeh, S. and Kaoian, P., Conductive epoxidized natural rubber nanocomposite with mechanical and electrical performance boosted by hybrid network structures. *Polym. Test.*, 108, 107493, 2022.
119. Aguilar-Bolados, H., Yazdani-Pedram, M., Contreras-Cid, A., Lopez-Manchado, M.A., May-Pat, A., Avilés, F., Influence of the morphology of carbon nanostructures on the piezoresistivity of hybrid natural rubber nanocomposites. *Composites, Part B*, 109, 147–154, 2017.
120. Teng, F., Wu, J., Su, B., Wang, Y., Internal polymerization of epoxy group of epoxidized natural rubber by ferric chloride filled with carbon nanotubes: Mechanical, morphological, thermal and electrical properties of rubber vulcanizates. *Nanomaterials*, 11, 2464, 2021.
121. Damampai, K., Pichaiyut, S., Das, A., Nakason, C., Interfacial heat flow in carbon nanotube suspensions. *eXPRESS Polym. Lett.*, 16, 812–826, 2022.
122. Huxtable, S.T., Cahill, D.G., Shenogin, S., Xue, L.P., Ozisik, R., Barone, P., Usrey, M., Strano, M.S., Siddons, G., Shim, M., Keblinski, P., *Nat. Mater.*, *2*, 731–734, 2003.
123. Alam, M.N., Kumar, V., Jung, H.-S., Park, S.-S., Fabrication of High-Performance Natural Rubber Composites with Enhanced Filler–Rubber Interactions by Stearic Acid-Modified Diatomaceous Earth and Carbon Nanotubes for Mechanical and Energy Harvesting Applications. *Polymers*, 15, 3612, 2023, https://doi.org/10.3390/polym1517361.
124. Le, H.H., Hait, S., Das, A., Wießner, S., Stöckelhuber, K.W., Böhme, F., Reuter, U., Naskar, K., Heinrich, G., Radusch, H.-J., Self-healing properties of carbon nanotube filled natural rubber/bromobutyl rubber blends. *eXPRESS Polym. Lett.*, 11, 230–242, 2017.
125. Zhan, Y., Li, Y., Meng, Y., Xie, Q., Lavorgna, M., Electric heating behavior of reduced oxide graphene/carbon nanotube/natural rubber composites with macro-porous structure and segregated filler network. *Polymers*, 12, 2411, 2020, doi: 10.3390/polym12102411.
126. Selvan, N.T., Eshwaran, S.B., Das, A., Stöckelhuber, K.W., Wießner, S., Pötschke, P., Nando, G.B., Chervanyov, A.I., Heinrich, G., Piezoresistive natural rubber-multiwall carbon nanotube nanocomposite for sensor applications. *Sens. Actuators, A*, 239, 102–113, 2016.

127. Chen, J., Li, H., Yu, Q., Hu, Y., Cui, X., Zhu, Y., Jiang, W., Strain sensing behaviors of stretchable conductive polymer composites loaded with different dimensional conductive fillers. *Compos. Sci. Technol.*, 168, 388–396, 2018.
128. Liu, X., Guo, R., Lin, Z., Yang, Y., Xia, H., Yao, Z., Resistance-strain sensitive rubber composites filled by multiwalled carbon nanotubes for structural deformation monitoring. *Nanomater. Nanotechnol.*, 11, 1–13, 2021.
129. Lai, S.-M. and Guo, G.-L., Two-way shape memory effects of sulfur vulcanized natural rubber (NR) and NR/paraffin wax (PW)/carbon nanotube (CNT) nanocomposites. *Polym. Test.*, 77, 105892, 2019.
130. Silva, D.J.D., Abeysinghe, H., Pamunuwa, P., Nisansala, D., Etampawala, T.N.B., Carbon nanotube/rubber composites for pressure sensing applications. *MRS Adv.*, 7, 1004–1009, 2022.
131. Chawla, R., Sliding tribological behavior of carbon nanotube/natural rubber composites. *Tribol. Ind.*, 40, 263–273, 2018, DOI: 10.24874/ti.2018.40.02.10.

5

Biodegradation of Natural Rubber and their Composites and Nanocomposites

Samuel Eshorame Sanni[1]*, Praise Uma[1], Denen Ashiekaa Vershima[1], Rotimi Emmanuel Sadiku[2], Babalola Aisosa Oni[3] and Emeka Emmanuel Okoro[4]

[1]Department of Chemical Engineering, Covenant University, Ota, Ogun State, Nigeria
[2]Department of Chemical, Metallurgical and Materials Engineering, Tshwane University of Technology, Pretoria, South Africa
[3]Department of Chemical Engineering, North Dakorta University, Grand Forks, United States
[4]Department of Petroleum and Gas, University of Port Harcourt, Choba, Rivers State, Nigeria

Abstract
Biodegradation is the process by which biological materials are reduced to lesser composites by several substances generated by living microbial organisms. The laborious and energy-intensive process of separating polymeric substances from the earth or other hazardous substances sometimes makes it impossible to recycle these materials. Several engineers believe that a radical approach to this concern is the availability of compounds that can undergo biodegradation under determined and established conditions with the formation of carbon dioxide and water, which are non-toxic to the environment. This study highlights various natural rubber (NR) composites and their properties. NRs with unique synthetic methods and important chemical, mechanical, and thermal stabilities were explored. This study recognizes that the biodegradability of NR composites depends on their chemical framework and structure. The physicochemical properties of NR can be influenced by its biodegradability. Novel nanoreinforced NR composites have been explored and utilized in several engineering applications. Thus, research on the movement of solution molecules *via* these materials and their flow characteristics,

**Corresponding author*: samuel.sanni@covenantuniversity.edu.ng

Visakh. P.M. and Shiv Kumari Panda (eds.) Natural Rubber-Based Composites and Nanocomposites, (153–186) © 2025 Scrivener Publishing LLC

accompanied by a thermal breakdown evaluation, is imperative. The study concludes by proposing the use of novel methods to analyze the biodegradation abilities of NR.

***Keywords*:** Rubber, natural rubber, biodegradation, natural rubber fillers, hydrogenated dicyclopentadiene, molding

5.1 Introduction

Rubber (*Hevea brasiliensis*) is a highly useful resource present in most daily activities, ranging from rubber gloves to rubber footwear. Rubber is categorized as an elastomer because it regains its shape after stretching or distortion. Rubber can be produced either artificially or naturally. Natural rubber (NR) is mainly used in countries such as Brazil and India. NR is a coagulated or precipitated substance extracted from the milky product of rubber crops called latex [1–3].

Latex is a clogging material used during wound healing as a result of mechanical damage to crops. It creates non-linked but slightly processable polymer chains with a molecular mass of approximately 106 Da and elastic characteristics. Polymers are materials with a molecular framework composed mostly or totally, of a large number of similar sets. NR is a biopolymer produced by over 2,000 crop specimens, the majority of which are members of the Euphorbiaceae or Compositaceae family, as well as certain fungi [4–7].

Biodegradation is the process by which natural materials are degraded into lower composites by enzymes generated by microbial organisms (MO). MOs alter materials *via* biological processes. Biodegradation can occur in two ways, aerobic and anaerobic. Aerobic biodegradation uses oxygen in chemical processes to reduce or degrade substances [8–11].

The end product of the degradation process is carbon dioxide or methane. Anaerobic digestion is a succession of procedures in which MOs reduce the amount of biodegradable substances in the absence of oxygen (O_2). Anaerobic biodegradation is popularly utilized to process wastewater sludge and biodegradable materials because it reduces the amount of input substances [12–15]. The management of waste substances is one of the many issues humanity tackles in the 21st century, with one of the most prolific of such wastes being plastic. The proliferation of plastic contamination is associated with the low cost and high durability of polymers as well as the fact that they are currently indispensable in certain regions, which dictates the high degree of human use of this material. The management of waste polymers is a major environmental concern because of the indecomposability of polymeric materials [16–20].

The laborious and energy-intensive process of separating polymeric substances from the earth or other contaminants sometimes makes it impossible to recycle these materials. Several professionals believe that a radical answer to this concern is the development of polymers (PO) that can undergo biodegradation under suitable conditions with the formation of carbon dioxide and water, which are non-toxic to crops. The most significant factors influencing the biodegradation of polymers are the physicochemical characteristics of the synthetic materials, molecular mass, molecular framework, extension of the macrochain, and supramolecular framework [21]. Polymers with a formless supramolecular framework are usually less impervious to biodegradation than crystalline species [22]. The impervious nature of polymeric substances to the activities of MOs also depends on the materials and other advanced additives included in

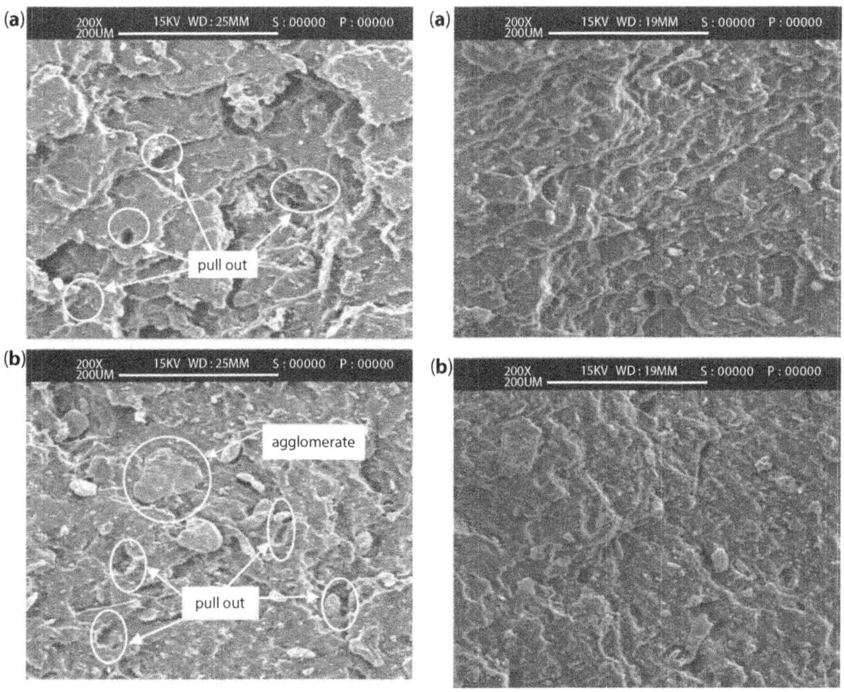

Figure 5.1 Tensile fracture surface of NR doped with (a) 10 phr of paper sludge with/without MANR and (b) tensile fracture surface 40 phr of paper sludge with/without MANR. Adapted from Ismail *et al.* [24].

their composition, as well as the extent to which these substances can be a source of several elements, such as nitrogen, for MOs. Inorganic compounds do not assist in the growth of MOs [23].

In Figure 5.1, Ismail *et al.* [24] highlighted the tensile properties of NR doped with several components.

Since biopolymers are usually reduceable, they could be reduced by environmental MOs into CO_2 and H_2O and they have recently gained a great deal of interest because of their ability to decrease the utilization of fossil sources and nonbiodegradable materials to break down plastic contaminants [25]. Recent studies have relied on enhancing the characteristics of biopolymer compounds, including starch-based substances, cellulose-based synthetic materials, and natural rubber (NR) compounds [26]. Mixing procedures, which include integrating biofillers, can be utilized to modify the characteristics of biological polymer compounds. Biodegradable materials have a wide variety of applications in electronics, including dielectric materials, electronic packaging, insulators, and (semi) conductors [27].

5.2 Biodegradation

Biodegradation, a fundamental process in environmental science, encompasses the reduction of organic substances *via* the biological activities of living organisms. It is characterized by microbial decomposition of organic contaminants, thus treating them as microbial food sources or substrates. The intricate nature of biodegradation unfolds as a sequence of biological degradation steps, culminating in oxidation of the parent compound [28].

In Figure 5.2, Rouilly *et al.* [29] highlighted the characteristics of H_2O on both surfaces of NR films.

Microorganisms, serving as key players, orchestrate microbial decomposition process within the biosphere, limiting their occurrence to specific environments [30]. Diverse biodegradation environments, such as soil, aquatic settings, landfills, and compost, present unique conditions, and host distinct microorganisms. For instance, fungi assume a pivotal role in soil, predominantly facilitating the breakdown of organic materials, including polymeric materials [31]. In aquatic environments, two types of bacteria, such as surface and sediment bacteria, dominate, with their bacterial weight content diminishing with increasing depth [32]. Complete biodegradation or mineralization is at the core of biodegradation. This intricate process involves oxidation of the parent compound, followed by the generation of

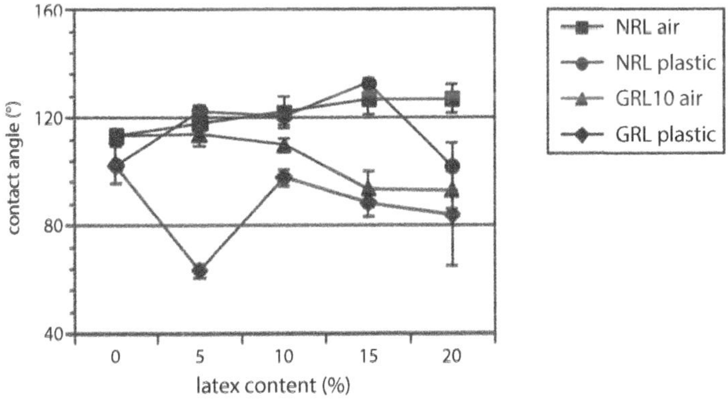

Figure 5.2 Average contact angle and standard deviations for water on both surfaces of films for the NR. Adapted from Rouilly *et al.* [29].

CO_2 and H_2O. Notably, mineralization provides both carbon and energy, which are essential for cellular growth and reproduction [33].

In Figure 5.3, Xu *et al.* [34] highlighted the SEM images of NR/silica composites.

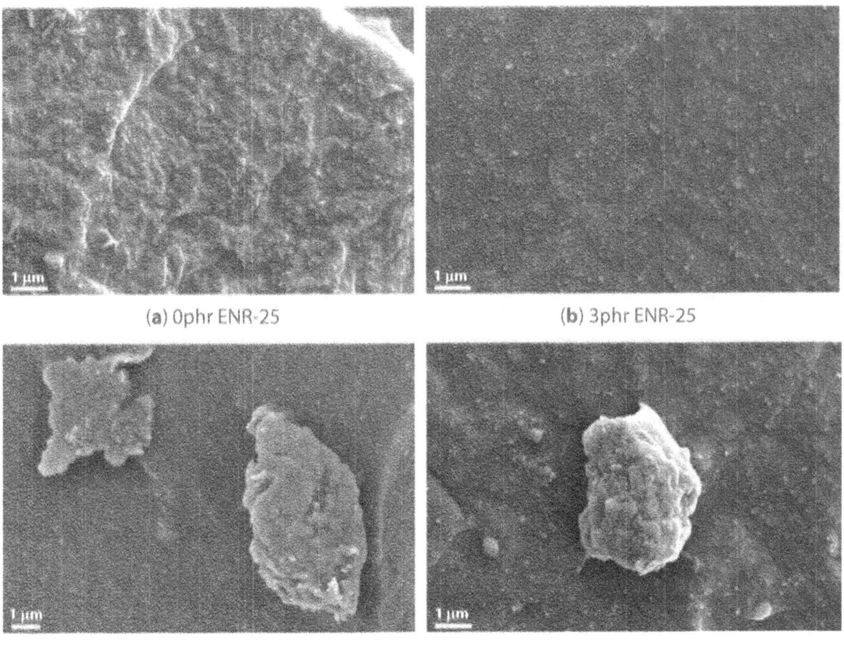

Figure 5.3 SEM of NR/silica composites. Adapted from Xu *et al.* [34].

Regardless of whether the carbon source is a simple sugar, plant polymer, or pollutant molecule, the mineralization steps remain consistent. Each degradation step depends on the specific enzymes produced by the degrading cell. Although enzymes are typically housed within the cell, the significance of extracellular enzymes located outside the cell cannot be overstated, especially in the breakdown of macromolecules, such as cellulose [35]. The occurrence of biotransformation is contingent on the presence of the requisite enzyme, either internally or extracellularly. The absence of suitable biodegrading enzymes is a common contributor to the persistence of organic contaminants, particularly those with unconventional chemical structures [36]. Biodegradation is of paramount significance in the presence of toxic chemicals. The biodegradation rate profoundly influences the concentration and, consequently, the toxic effects of these chemicals [37]. The spectrum of biodegradation rates is vast, ranging from easily biodegradable compounds to highly refractory substances such as Dichlorodiphenyltrichloroethane (DDT) and dioxins, which exhibit excellent biological properties [38], estimating the biodegradability of organic compounds involves considering various rules:

- Higher molecular load shows decreased biodegradability,
- Aliphatic compounds are generally more biodegradable than aromatic ones,
- Substitutions by halogens and nitrogenous compounds decrease biodegradability,
- The introduction of double and triple bonds generally increase biodegradability, whereas the presence of oxygen and nitrogen bridges tends to decrease biodegradability.
- The branches are generally less biodegradable compared to the corresponding primary compounds.

5.3 Natural Rubber Composites and Nanocomposites

According to Rose and Steinbüchel [39], isoprene (C_5H_8) molecules, which comprise natural rubber, have a single double bond arranged in a cis orientation.

Nevertheless, the terminal area of *H. brasiliensis* polyisoprene also contains two trans-isoprene units. While almost two thousand crops can synthesize these compounds, only two types of rubber (RB) are economically produced: guayule rubber from *Parthenium argentatum* (accounting for 1% of the global market) and natural rubber from *H. brasiliensis* (accounting for

Table 5.1 Natural rubber composites, their major additives, general synthesis, and major findings.

Material	Major additive	General synthesis process (method)	Materials	Major findings	Ref.
Rubber composites/ silica	Silica	Wet mixing process (WSR)	-	• The WSR exhibits superior curing properties. • Excellent mechanical properties. • Excellent chemical properties. • Excellent thermal properties. • Enhanced dynamic properties.	[70]

(Continued)

Table 5.1 Natural rubber composites, their major additives, general synthesis, and major findings. (*Continued*)

Material	Major additive	General synthesis process (method)	Materials	Major findings	Ref.
NR-HC	Hydrochar	Hydrothermal carbonization	Natural rubber • N-isopropyl-N0 -phenyl-p-phenylenediamine (IPPD) • Stearic acid • zinc oxide (ZnO) • sulfur • N-cyclohexylbenzothiazol-2-sulfenamide (CBS) • Carbon black N330	• Excellent mechanical properties. • Excellent chemical characteristics. • Excellent thermal characteristics. • Enhanced dynamic properties.	[71]

(*Continued*)

Table 5.1 Natural rubber composites, their major additives, general synthesis, and major findings. (*Continued*)

Material	Major additive	General synthesis process (method)	Materials	Major findings	Ref.
CNT/NR	Carbon nanotubes	• Acid bath • Ball milling with HRH bonding systems	• Natural rubber • 2-Mercaptobenzothiazole • Stearic acid • zinc oxide (ZnO) • sulfur • N-cyclohexylbenzothiazol-2-sulfenamide (CBS) • Carbon black N330 • Ball-milled CNTs	• Excellent mechanical properties. • Excellent chemical properties. • Excellent thermal properties. • Enhanced dynamic properties.	[72]

(*Continued*)

Table 5.1 Natural rubber composites, their major additives, general synthesis, and major findings. (*Continued*)

Material	Major additive	General synthesis process (method)	Materials	Major findings	Ref.
NR-sisal/oil palm hybrid fiber	Sisal/oil palm fiber	• Fiber preparation • Composite preparation	• Sisal fiber • Oil palm fiber • Natural rubber	• Excellent mechanical characteristics. • Excellent chemical properties. • Excellent thermal properties. • Enhanced dynamic properties.	[73]

(*Continued*)

Table 5.1 Natural rubber composites, their major additives, general synthesis, and major findings. (*Continued*)

Material	Major additive	General synthesis process (method)	Materials	Major findings	Ref.
CNT/NR	Multiwalled carbon nanotubes	Novel method	• Natural rubber • Jute yarn • Stearic acid • N-cyclohexyl-2-mercaptobenzothiazole sulfonamide • Sulfur • zinc oxide • Ethanol • toluene • sodium hydroxide • pristine MWCNTs	• Excellent mechanical properties. • Excellent chemical characteristics. • Excellent thermal characteristics. • Enhanced dynamic properties.	[74]

(*Continued*)

Table 5.1 Natural rubber composites, their major additives, general synthesis, and major findings. (*Continued*)

Material	Major additive	General synthesis process (method)	Materials	Major findings	Ref.
Polyethylene-cellulose fibers/NR	Polyethylene-cellulose fibers	–	• Commercial microcrystalline fibers • Technocel-2500 fibers • Alfa fibers • Bleached softwood (pine) fibers • The Avicel fibers • commercial silanes • polyethylene (LDPE)	• Excellent mechanical characteristics. • Excellent chemical properties. • Excellent thermal properties. • Enhanced dynamic properties. • Mechanical characteristics of the compounds rose with raising the average fiber length.	[75]

(*Continued*)

Table 5.1 Natural rubber composites, their major additives, general synthesis, and major findings. (*Continued*)

Material	Major additive	General synthesis process (method)	Materials	Major findings	Ref.
Starch/NR	Starch	Esterification	• Natural rubber latex • Cassava starch	• Excellent mechanical properties. • Excellent chemical properties. • Excellent thermal characteristics. • Enhanced dynamic properties. • The mechanical properties of composite are enhanced with the increase of starch loading.	[76]

(*Continued*)

Table 5.1 Natural rubber composites, their major additives, general synthesis, and major findings. (*Continued*)

Material	Major additive	General synthesis process (method)	Materials	Major findings	Ref.
Nano-lignin/NR	Nano-lignin	Self-assembly technology	• Lignin • Ammonia • Sodium hydroxide, • hydrochloric acid • sulfuric acid • sulfur • stearic acid • zinc oxide • N-tert-butyl-2-benzothiazole sulfonamide (CBS)	• Excellent mechanical properties. • Excellent chemical properties. • Excellent thermal properties. • Enhanced dynamic properties.	[77]

(*Continued*)

Table 5.1 Natural rubber composites, their major additives, general synthesis, and major findings. (*Continued*)

Material	Major additive	General synthesis process (method)	Materials	Major findings	Ref.
Short coir fiber reinforced natural rubber composites	Short coir fiber	Reinforcement	• Coir fiber • Natural rubber	• Frequency increases decrease whereas the frequency values of loss modulus increase in the instance of both gum and the compounds. • Excellent mechanical properties. • Excellent chemical properties. • Excellent thermal properties. • Enhanced dynamic properties.	[78]

(*Continued*)

Table 5.1 Natural rubber composites, their major additives, general synthesis, and major findings. (*Continued*)

Material	Major additive	General synthesis process (method)	Materials	Major findings	Ref.
MWCNTs-NR	MWCNTs	–	• MWCNT • H2SO4 • HNO3 • NR latex of dry rubber content (DRC)	• Excellent AC conductivity. • High dielectric constant. • Excellent mechanical properties. • Excellent chemical characteristics. • Excellent thermal characteristics. • Enhanced dynamic properties.	[79]

(*Continued*)

Table 5.1 Natural rubber composites, their major additives, general synthesis, and major findings. (*Continued*)

Material	Major additive	General synthesis process (method)	Materials	Major findings	Ref.
MST-PBA/NR	Modified starch paste (MST) with polybutylacrylate (PBA)	–	• Natural starch • Butylacrylate • Ammonium persulfate	• Good mechanical properties. • Excellent chemical properties. • Excellent thermal properties. • Enhanced dynamic properties.	[80]

(*Continued*)

Table 5.1 Natural rubber composites, their major additives, general synthesis, and major findings. (*Continued*)

Material	Major additive	General synthesis process (method)	Materials	Major findings	Ref.
Silica/NR	Silica	-	• Concentrated NR latex • Precipitated silica • bis-(3-triethoxysilylpropyl) tetrasulfide (Si-69) • Paraffin wax • 2,6-di-tert-butyl-p-cresol (Yoshinox BHT) • Stearic acid • zinc oxide Elemental sulfur (S8) • diphenyl guanidine (PerkacitDPG) • tetramethylthiuram disulfide (Perkacit TMTD) • N-tert-butyl-2-benzothiazole sulfenamide	• Excellent mechanical properties. • Good chemical characteristics. • Good thermal properties. • Enhanced dynamic properties.	[81]

(*Continued*)

Table 5.1 Natural rubber composites, their major additives, general synthesis, and major findings. *(Continued)*

Material	Major additive	General synthesis process (method)	Materials	Major findings	Ref.
MWNT-NR	MWNT	Surface functionalization	• Multiwalled CNT (MWNT) • Potassium dichromate • 3-aminopropyltriethoxysilane • Natural rubber (SMR 5) • Polymerized trimethyl • quinoline (TMQ) • mercapto benzothiazole sulfenamide • (MBTS) • tetramethyl thiuram disulfide (TMTD) • Zinc oxide (ZnO) • stearic acid • sulfur • concentrated sulfuric acid. • acetone	• Good mechanical properties. • Excellent chemical properties. • Excellent thermal properties. • Enhanced dynamic properties.	[82]

99% of the world market) [40]. Approximately 6% of *H. brasiliensis* dehydrated natural rubber is composed of non-polyisoprene components. The average concentration of latex varies according to the clone, seasonal variations, and soil conditions. Table 5.1 highlights the natural rubber composites, their major additives, general synthesis, and major findings. As in most industrial product applications, rubber is not used in its pure form but is mixed with other materials known as additives to improve its physical characteristics and performance. The inclusion of these additives in rubber gives rise to composites. Several composites of rubber exist and are discussed below.

5.3.1 Carbon Black

Since the years of World War 1, carbon black has been the major popularly utilized reinforcing load in RB manufacturing. The rubber industry, particularly the tire industry, uses more than 92% of the carbon black produced globally for items including conveyor wheels and belts [41]. Currently, the general manufacturing of carbon black is at about 8.1 million tons, making it among the top 50 industrial substances generated regularly [42]. The mechanical characteristics of the specified RB compounds, such as their tensile and tear strengths, can be enhanced by combining them with carbon black (CB). CB is said to be a highly effective additive for all filler ingredients because it makes rubber-based compositions harder and more viscous [43].

5.3.2 Silica

Silica (SiO_2) is a reinforcing weight that is frequently utilized to create colored NR products. Owing to its low rolling resistance and low heat build-up, it is specifically utilized in the treadding of green tires [44]. Using a co-vulcanization technique, scientists have created nano-SiO_2 and SiO_2-integrated NR compounds. They contrasted the matching nano-SiO_2 filled NR with the physical characteristics of NR compounds. According to a previous study, nano-SiO_2 composites spread out well in NR at low filler loadings (0.8 phr), which may have been caused by better NR and nano-SiO_2 contact [45].

In Figures 5.4 a and b, Ruksakulpiwat *et al.* [46] presented the SEM images of NR.

In the case of SiO_2-loaded NR vulcanizates, the mechanical properties first increased somewhat before steadily declining from 5 phr forward owing to the increasing SiO_2 content. This could be caused by both the less uniform filler distribution and poor NR and SiO_2 interactions [47].

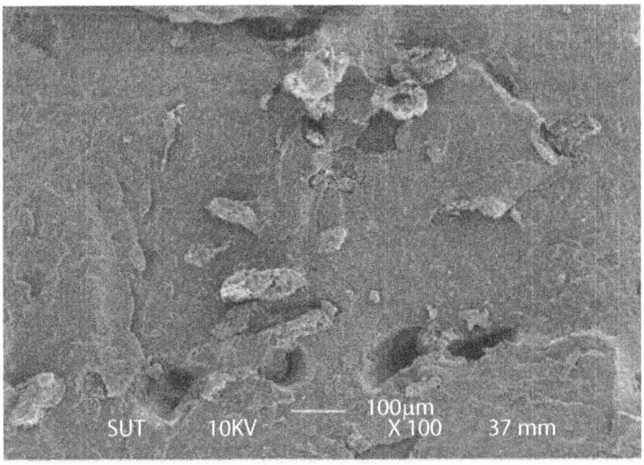

(a)

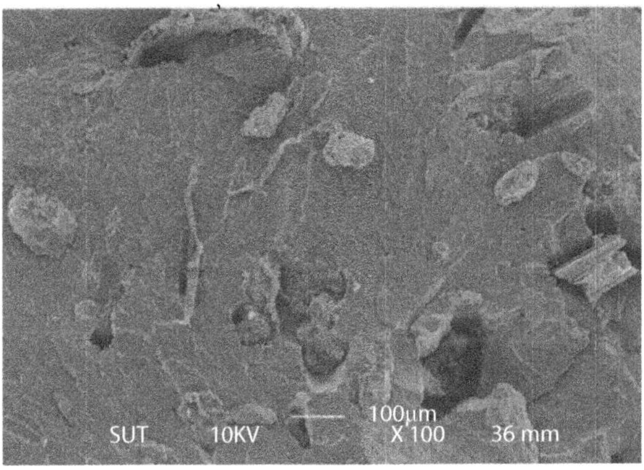

(b)

Figure 5.4 SEM images of NR. Adapted from Ruksakulpiwat *et al.* [46].

In Figure 5.5, Yan *et al.* [48] illustrated the TEM images of rGO/NR compounds.

5.3.3 Calcium Carbonate

In NR-based composites, calcium carbonate ($CaCO_3$), often known as "whiting," is frequently utilized as a load to add color and reduce product costs. Ground limestone (GL) and precipitated $CaCO_3$ are the two main forms of $CaCO_3$. Processed $CaCO_3$ has been produced by chemical processing from a salt mixture, and GL was produced by processing mineral limestone [49]. When ground $CaCO_3$ is added to NR, the product's tear resistance is reduced; however, in certain situations, precipitated $CaCO_3$ with a finer particle size increases the NR product's tear strength.

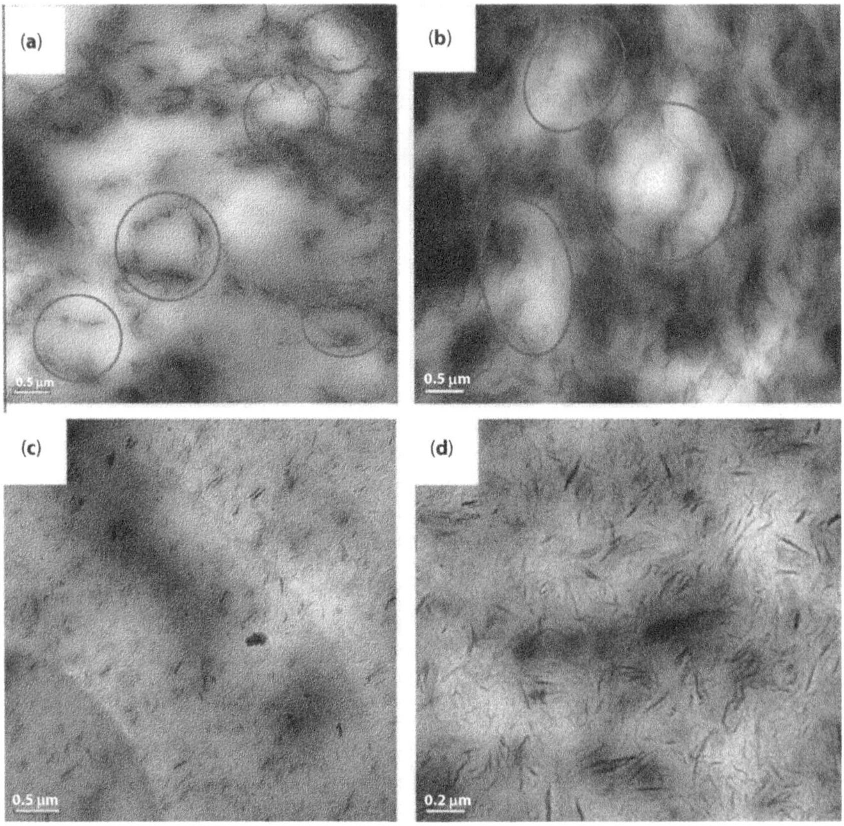

Figure 5.5 TEM images of rGO/NR composites (a) 4 phr, (b) 8 phr (NRLG-8), (c) 4 phr, and (d) 8 phr. Adapted from Yan *et al.*

5.3.4 Bio-Based Fillers

Research on bio-nanofillers made from biomass has accelerated because of the need for biofillers other than petroleum-based fillers. Bio-based nanocomposites occasionally exhibit special benefits over conventional inorganic nanoparticles [50]. NR is frequently used in cellulose whiskers, chitosan, starch nanocrystals, and other bio-nanofillers. The goal of these filler configurations is to produce them at a production cost that is comparable with other petroleum-based fillers while maintaining equivalent qualities [51]. Furthermore, it is anticipated that the bio-based fillers' biocompatibility and biodegradability will be preserved after their dispersion in the NR matrix [52].

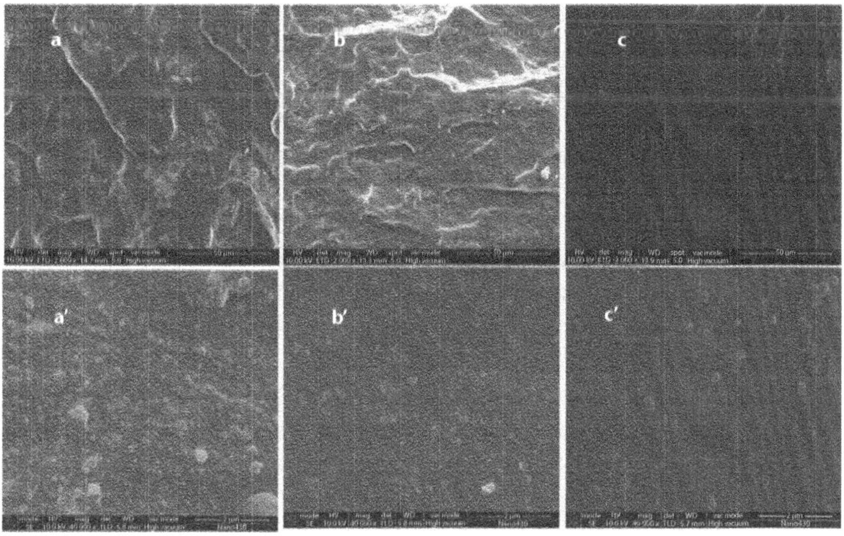

Figure 5.6 SEM images (a, a′: NR/50silica; b, b′: NR/30silica/20lignin; c, c′: NR/50lignin). Adapted from Yu *et al.*

Yu *et al.* [52] displayed the SEM images several NR composites in Figure 5.6.

5.3.5 Carbon Nanotubes

Carbon nanotubes (CNTs) are a unique form of C allotropes characterized by sp^2 hybridization and a cylindrical structure. Typically, CNTs exhibit diameters of several nanometers, whereas their lengths can extend to millimeters or even centimeters. Various synthetic methods, including arc discharge, have been used to produce CNTs. CNTs are generally separated into single-walled carbon nanotubes (SWCNTs) and multi-walled carbon nanotubes (MWCNTs) [53]. However, achieving a homogeneous dispersion of CNTs within a NR composite poses significant challenges because of the increased aspect ratio of CNTs and the improved properties of natural rubber. Strategies such as pretreatment, interfacial alterations, and other techniques have been employed to boost the transportation of CNTs in NR-based CNT nanocomposites [54]. Furthermore, various sample preparation techniques have been utilized to achieve an optimal distribution of CNTs within a NR composite, thereby reducing the CNT combination. These methods include solution mixing, sonication, and the use of surfactants. Ismail *et al.* [55] investigated the characteristics of natural rubber (NR)/multiwalled carbon nanotubes (MCNT) prepared *via* physical and solution mixing (SM) techniques. Their study highlighted the impact of various synthetic techniques on the characteristics of NR/MWCNT nano-compounds. Specifically, the tensile properties of the NR/MWCNT nano-compounds synthesized *via* SM were found to be higher than those synthesized *via* physical mixing. However, it was observed that with the SM approach using five CNTs, although the characteristics improved, the tensile strength decreased from 18 to 12 MPa, accompanied by a decrease in other properties such as fatigue. This reduction in strength was ascribed to CNT integration within the NR composites. Interestingly, NR/MWCNT nano-compounds generated *via* SM with 0.5 CNTs exhibited maximum tensile properties as a result of the improved dispersion of MWCNTs within the NR composite.

5.3.6 Starch Nanocrystals

Starch is insoluble in cold H_2O upon extraction from crops. Native starches contain varying amounts of crystalline material, ranging from 15 to 45%. This crystalline component, referred to as starch crystallite, is obtained through various levels of acidic hydrolysis. While cellulose nanomaterials are 100% crystalline, starch crystal nanomaterials contain approximately 45% crystalline content. To produce starch nanocrystals from waxy maize starch, starch particles were combined with a known concentration of H_2SO_4 solution and mixed at a specific velocity with orbital shaking activity. Following various periods of hydrolysis, the resulting suspension was processed with distilled water until a neutral pH was achieved [56]. Angellier *et al.* [57] prepared nanocrystalline particles from waxy maize starch. These starch nanocrystals, present in aggregated structures exhibited an average size of 4.4 mm. The study also examined the characteristics of unprocessed natural rubber loaded with these nanomaterials. The resulting films were processed at a specific relative humidity. The elongation at break of NR loaded with starch nanocrystals decreased with increasing nanomaterial content, yielding drops from 1,960 to 920% (i.e., leaving the material unfilled up to 30 wt.%). Conversely, the tensile properties experienced a near exponential growth with starch concentration for the unfilled NR compounds in the compound films loaded with starch nanomaterials. Additionally, the storage relaxed modulus under standard conditions for nanocomposite films containing several weight percentages of load was approximately several times higher than that of the unfilled NR membrane [58].

However, at a higher load content, there is a reduction in the tensile modulus when exposed to 43% relative humidity (RH) compared to the equal loading at 0% RH. Furthermore, the effective absorption bandwidth (EAB) was reduced to 1,500% filler loadings of 5–20 wt.% at 43% RH. This decrease in the characteristics of the starch nanomaterial-loaded NR with increasing moisture concentration was attributed to the hydrophilic nature of the starch nanocrystals. The potential creation of water-rich load/NR membrane channels when stored at high RH values may impede stress transfer at the surface during the mechanical tests [59].

5.3.7 Nanocellulose

Cellulose, a homopolysaccharide composed of hybrid composite units, is sourced from natural fibers, wood fibers, and other organic materials. Each cellulose monomer unit contains three hydroxyl groups which facilitate H_2 bonding with other materials and/or micromaterials. This bonding dictates the crystalline compaction of cellulose, ultimately influencing the major mechanical characteristics of the substances derived from it [60]. Nano-cellulose fibers, ranging in diameter from 5 to 60 nm, can be produced from a variety of naturally grown crop fibers, such as hemp, through various chemical processes [61].

5.4 Biodegradation of Natural Rubber and its Composites and Nanocomposites

Beyond the realms of latex processing, it has become increasingly apparent that rubber biodegradation is not confined solely to industrial practices. Both bacteria and fungi exhibit a remarkable capability to degrade rubber, which indicates a gradual and intricate process over time. Bismarck *et al.* [62] discovered that the biodegradability of a compound relies primarily on its framework of the polymeric compound. Petricca *et al.* [63] prepared a biodegradable polymer compound by a physicochemical precipitation method at standard conditions. Mehta *et al.* [64] revealed that biocomposites can be synthesized from a variety of natural fibers (NF) including jute. Hybrid bio-composites were synthesized using this procedure by integrating various NFs with varying mass concentrations. Muniandy *et al.* [65]

investigated the biodegradation of hybrid NR compounds. Different scientists have performed biodegradation studies on NR compounds loaded with various fillers. According to Ismail *et al.* [66], who studied the incorporation of chitosan as a novel type of load in RB composites, one of the major demands for chitosan to be efficiently utilized as a load is ensuring good dynamics with RB at an appropriate loading level, with the loading level being the weight content of chitosan added to the composite.

They studied the influence of chitosan loading on the processing characteristics and structure of chitosan-loaded NR. From their findings, all three rubber compositions showed a decrease in the tensile properties as a result of the dimensions and geometrical factors of the load. Irregularly formed loads tend to reduce the strength of polymeric compounds because of their inability to support stress transferred from the matrix. However, at similar loading levels, the tensile strength of the natural rubber composite was the highest, which was attributed to the ability of the NR chains to undergo strain-induced processes. The hydrophobicity of rubber and the hydrophilicity of chitosan were cited as reasons why the rubber-load dynamics decreased with the integration of chitosan and further reduced with higher chitosan filling. Nanocomposites that are fundamental to polymeric membranes efficiently combine the theories of hybrid substances. It has been well established that "nanocomposites" exhibit a set of multiphase compounds where one of the stages has at minimum one dimension lesser than 100 nm [67].

In Figure 5.7, Dick and dos Santos [68] presents the SEM micrographs of some of these composites.

Nano-reinforced polymeric compounds are popularly utilized in many engineering applications, and thus, research on the movement of solvent particles *via* these matrices and their flow characteristics, along with thermal breakdown evaluation, is imperative [69]. In a study by several researchers, cellulose whiskers extracted from Capim Dourado (CD) were utilized to reinforce NR, and the characteristics of the resulting nanomaterial films were studied. The research revealed that CD whisker-reinforced NR nanomaterials exhibited relatively high characteristics compared to several selected cellulose nanomaterials although with lesser EABs. A potential explanation for this is their stronger inter-nanoparticle interactions than those of other polysaccharide nanocrystals. Their studies also showed that no imperative improvement in the chemical stability of compounds was introduced when integrating starch nanomaterials into natural rubber composites. The composites exhibited a rather high chemical and physical stability, in lieu of the fact that starch starts to break down at similar temperatures where NR starts to disentangle and circulate.

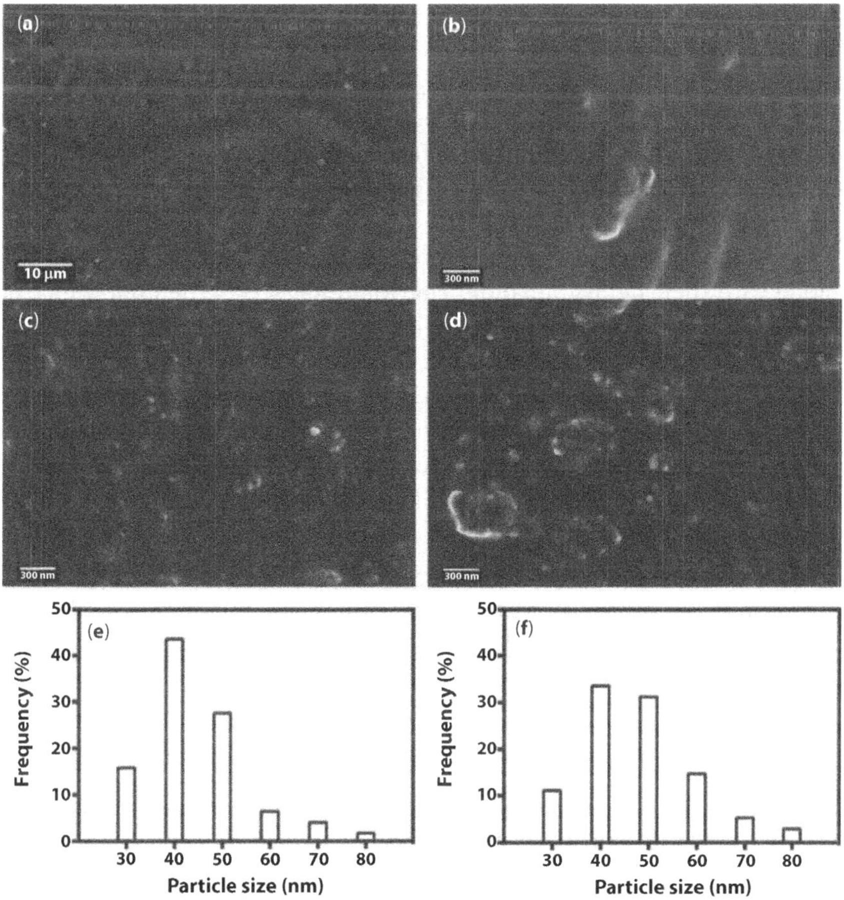

Figure 5.7 SEM micrographs of the composites containing: (a) 10 wt.% hydroxyapatite (HA), (b) 10 wt.% HA, (c) 20 wt.% HA, (d) 30 wt.% HA, (e) 20 wt.% HA, and (f) 30 wt.% HA composites. Adapted from Dick and dos Santos [68].

5.5 Conclusions

In conclusion, the exploration of natural rubber (*H. brasiliensis*) and its composites, as well as the biodegradation of these materials, has provided valuable insights into their properties and environmental impact. Natural rubber, a versatile elastomer, is widely applied in various industries owing to its unique properties including elasticity and resilience. However, the management of rubber waste poses significant challenges, particularly in the context of increasing environmental concerns such as plastic pollution. Biodegradation has emerged as a promising method for addressing the environmental impact of rubber and its composites. Understanding the complex biodegradation process of natural rubber, including the role of microbial organisms and environmental conditions, is crucial for developing sustainable solutions for rubber waste management. Moreover, the incorporation of bio-based fillers and nanocomposites holds promise for enhancing the biodegradability of rubber materials while maintaining their desirable mechanical properties.

The investigation of natural rubber composites and nanocomposites, along with their biodegradation behavior, underscores the importance of interdisciplinary research in materials science and environmental engineering. By integrating knowledge from various fields, including polymer science, microbiology, and environmental chemistry, researchers/scientists can strive towards the development of eco-friendly materials with reduced environmental impact. Continued research and innovation in the field of natural rubber and biodegradable composites are essential for advancing sustainable practices in materials engineering. Collaboration among researchers, industry stakeholders, and policymakers is key to driving meaningful progress towards a more sustainable environmentally conscious approach to rubber production, usage, and disposal.

References

1. Fan, Y., Fowler, G.D., Zhao, M., The past, present and future of carbon black as a rubber reinforcing filler – A review. *J. Clean. Prod.*, *247*, 119115, 2020.
2. Onyeagoro, G.N., Cure Characteristics and Physico-Mechanical Properties of Carbonized Bamboo Fibre Filled Natural Rubber Vulcanizates. *Int. J. Mod. Eng. Res. (IJMER)*, *2*, 4683–4690, 2012.
3. Balan, A.V. and Shivasankaran, N., Noise Control using Waste Materials Reinforced Composites. *London J. Res. Sci.: Nat. Formal*, *19*, 2, 47–53, 2019.
4. Dwivedi, C., Manjare, S., Rajan, S.K., Recycling of waste tire by pyrolysis to recover carbon black: Alternative & environment-friendly reinforcing filler for natural rubber compounds. *Composites, Part B*, *200*, 108346, 2020.
5. Hu, Z., Chen, Z., Meng, F., Zhang, Y., Jia, Y., Fei, H., Li, S., Yuan, X., Novel hydrophobic butyl rubber damping composites modified with bio-based PF/DBA *via* the construction of a three-dimensional network. *Sci. Rep.*, *14*, 1, 5007, 2024.
6. Xia, L., Li, C., Zhang, X., Wang, J., Wu, H., Guo, S., Effect of chain length of polyisobutylene oligomers on the molecular motion modes of butyl rubber: Damping property. *Polymer*, *141*, 70–78, 2018.

7. Tang, D., Zhang, X., Hu, S., Liu, X., Ren, X., Hu, J., Feng, Y., The reuse of red brick powder as a filler in styrene-butadiene rubber. *J. Clean. Prod., 261*, 120966, 2020.
8. Song, M., Yue, X., Wang, X., Huang, M., Ma, M., Pan, W., Qin, Q., Improved high-temperature damping performance of nitrile-butadiene rubber/ phenolic resin composites by introducing different hindered amine molecules. *E-Polymers, 20*, 1, 482–490, 2020.
9. Chen, J., Wang, J., Li, S., Xiang, K., Song, S., Pyridine terminated polyurethane dendrimer/chlorinated butyl rubber nanocomposites with excellent mechanical and damping properties. *Chin. J. Chem. Eng., 53*, 211–221, 2023.
10. Tang, K. and Wang, J., Chlorinated butyl rubber/two-step modified montmorillonite nanocomposites: Mechanical and damping properties. *Chin. J. Chem. Eng., 42*, 437–449, 2022.
11. Park, K.H., Fujiwara, Y., Mazda, T., Kajita, Y., Evaluation of mechanical properties considering hysteresis characteristic of high damping rubber bearing. *J. Phys. Conf. Ser., 1687*, 1, 012019, 2020.
12. Zhang, H., Yang, Y., Liu, Q., Ji, L., Sun, Z., Sun, J., Han, Z., Ma, Q., Yang, H., Ke, Y., Park, C.B.,Jiang, N. Natural coal-derived graphite as rubber filler and the influence of its progressive graphitization on reinforcement performance. *Compos. - A: Appl. Sci. Manuf.*, 184, 108237, 2024. https://doi.org/10.1016/j.compositesa.2024.108237
13. Qu, C., Li, S., Zhang, Y., Wang, T., Wang, Q., Chen, S., Surface modification of Ti3C2-MXene with polydopamine and amino silane for high performance nitrile butadiene rubber composites. *Tribol. Int., 163*, 107150, 2021.
14. Şen, M., Soydaş, O., Aksüt, D., Preparation and characterization of high damping poly(epichlorohydrin) and poly(epichlorohydrin-co-ethylene oxide-co-allyl glycidyl ether) elastomers: I. Effect of curing system and blending on the damping properties. *React. Funct. Polym., 184*, 105520, 2023.
15. Xiaolin, J., Min, X., Minhui, W., Yuanhao, M., Wencong, Z., Yanan, Z., Haoxiang, R., Xun, L., Preparation and molecular dynamics study of polyurethane damping elastomer containing dynamic disulfide bond and multiple hydrogen bond. *Eur. Polym. J., 162*, 110893, 2022.
16. Rathour, R., Kumar, H., Prasad, K., Anerao, P., Kumar, M., Kapley, A., Pandey, A., Kumar Awasthi, M., Singh, L., Multifunctional applications of bamboo crop beyond environmental management: an Indian prospective. *Bioengineered, 13*, 4, 8893–8914, 2022.
17. Azammi, A.M.N., Sapuan, S.M., Ishak, M.R., Sultan, M.T.H., Physical and damping properties of kenaf fibre filled natural rubber/thermoplastic polyurethane composites. *Def. Technol., 16*, 1, 29–34, 2020.
18. Jantachum, P., Khumpaitool, B., Utara, S., Effect of silane coupling agent and cellulose nanocrystals loading on the properties of acrylonitrile butadiene rubber/natural rubber nanocomposites. *Ind. Crops Prod., 195*, 116407, 2023.
19. Unwin, A.P., Hine, P.J., Ward, I.M., Guseva, O.A., Schweizer, T., Fujita, M., Tanaka, E., Gusev, A.A., Predicting the visco-elastic properties of

polystyrene/SIS composite blends using simple analytical micromechanics models. *Compos. Sci. Technol.*, *142*, 302–310, 2017.
20. Vural, K.B., Furtana, Y.B., Yilmaz, M., Yalcin, E., Performance evaluation of bitumen modified with styrene–isoprene-styrene and crumb rubber compound. *Constr. Build. Mater.*, *344*, 128304, 2022.
21. Luo, Y., Wu, Y., Luo, K., Cai, F., Zhai, T., Wu, S., Structures and properties of alkanethiol-modified graphene oxide/solution-polymerized styrene butadiene rubber composites: Click chemistry and molecular dynamics simulation. *Compos. Sci. Technol.*, *161*, 32–38, 2018.
22. Han, H., Tian, G., Gao, Q., Hu, H., Zhao, J., Li, J., Wall slip of styrene-isoprene-styrene (SIS) triblock copolymer induced by micro elastic phase. *Polymer*, *209*, 122990, 2020.
23. Hien, T.T.T., Tsubota, T., Taniguchi, T., Shinogi, Y., Enhancing soil water holding capacity and provision of a potassium source *via* optimization of the pyrolysis of bamboo biochar. *Biochar*, *3*, 1, 51–61, 2021.
24. Ismail, H., Rusli, A., Rashid, A.A., Maleated natural rubber as a coupling agent for paper sludge filled natural rubber composites. *Polym. Test.*, *24*, 7, 856–862, 2005.
25. Dong, Q., Li, H., Niu, M., Luo, C., Zhang, J., Qi, B., Li, X., Zhong, W., Microwave pyrolysis of moso bamboo for syngas production and bio-oil upgrading over bamboo-based biochar catalyst. *Bioresour. Technol.*, *266*, 284–290, 2018.
26. Zhu, J., Zhao, X., Liu, L., Yang, R., Song, M., Wu, S., Thermodynamic analyses of the hydrogen bond dissociation reaction and their effects on damping and compatibility capacities of polar small molecule/nitrile-butadiene rubber systems: Molecular simulation and experimental study. *Polymer*, *155*, 152–167, 2018.
27. Gu, X., Wang, Y., Lai, C., Qiu, J., Li, S., Hou, Y., Martens, W., Mahmood, N., Zhang, S., Microporous bamboo biochar for lithium-sulfur batteries. *Nano Res.*, *8*, 1, 129–139, 2015.
28. Sunali, Mago, J., Negi, A., Pant, K.K., Fatima, S., Development of natural rubber-bamboo biochar composites for vibration and noise control applications. *J. Clean. Prod.*, *373*, 133760, 2022.
29. Rouilly, A., Rigal, L., Gilbert, R.G., Synthesis and properties of composites of starch and chemically modified natural rubber. *Polymer*, *45*, 23, 7813–7820, 2004.
30. Li, C., Zhang, W., Sun, G., Li, Y., Zhang, Y., Research on damping properties and microscopic mechanism of polyurethane cement-based composites. *Constr. Build. Mater.*, *365*, 130137, 2023.
31. Zainathul, A.S.A.S., Nurul, A.M.Z., Mohd, M.A., Siti, S.J., Nor, H.Z., The cure characteristics and mechanical behaviour of bamboo fibre filled natural rubber composite. *Adv. Mater. Res.*, *812*, 53–59, 2013.
32. Hodzic, A., Re-use, recycling and degradation of composites, in: *Green Composites*, pp. 252–271, Elsevier, UK, 2004.

33. Oboh, J.O., Okafor, J.O., Kovo, A.S., Abdulrahman, A.S., Dynamic mechanical properties of crosslinked natural rubber composites reinforced with cellulosic nanoparticles. *Niger. J. Technol.*, *37*, 3, 668, 2018a.
34. Xu, T., Jia, Z., Luo, Y., Jia, D., Peng, Z., Interfacial interaction between the epoxidized natural rubber and silica in natural rubber/silica composites. *Appl. Surf. Sci.*, *328*, 306–313, 2015.
35. Oboh, J.O., Okafor, J.O., Kovo, A.S., Abdulrahman, A.S., Investigation of eco-friendly cellulosic nanoparticles potential as reinforcement agent in the production of natural rubber composites. *Niger. J. Technol.*, *36*, 4, 1078, 2018b.
36. Colom, X., Marín-Genescà, M., Mujal, R., Formela, K., Cañavate, J., Structural and physico-mechanical properties of natural rubber/GTR composites devulcanized by microwaves: Influence of GTR source and irradiation time. *J. Compos. Mater.*, *52*, 22, 3099–3108, 2018.
37. Lazim, N.M. and Abdullah, B., The Emergence of New Inflammatory Markers of Head and Neck Cancer and their Potentials, in: *Head and neck cancer: Hallmarks of the inflamation ecosystem (Frontiers in Inflamarion)*, vol. 2, pp. 185–209, Bentham Science Publishers, UAE, 2012. DOI: 10.2174/97898118032461210200l0.
38. Jørgensen, S.E., Biodegradation, in: *Encyclopedia of Ecology*, pp. 366–367, Elsevier, UK, 2008.
39. Rose, K. and Steinbüchel, A., Biodegradation of Natural Rubber and Related Compounds: Recent Insights into a Hardly Understood Catabolic Capability of Microorganisms. *Appl. Environ. Microbiol.*, *71*, 6, 2803–2812, 2005.
40. Ooi, Z.X., Ismail, H., Abu Bakar, A., Study on the ageing characteristics of oil palm ash reinforced natural rubber composites by introducing a liquid epoxidized natural rubber coating technique. *Polym. Test.*, *37*, 156–162, 2014.
41. Lu, X. and Li, X., Broad temperature and frequency range damping materials based on epoxidized natural rubber. *J. Elastomers Plast.*, *46*, 1, 84–95, 2014.
42. Fan, Y., Fowler, G.D., Norris, C., Potential of a Pyrolytic Coconut Shell as a Sustainable Biofiller for Styrene–Butadiene Rubber. *Ind. Eng. Chem. Res.*, *56*, 16, 4779–4791, 2017.
43. Chandrasekaran, Rubbers and Their Relevant Properties for the Chemical and Mineral Processing Industries, in: *Rubber as a Construction Material for Corrosion Protection*, pp. 79–107, Wiley, US, 2010.
44. Zhang, K. and Xiao, F., Effect of SIS block copolymers on damping properties of natural rubber/AO-80 and the performance enhancement mechanism: experimental study and molecular dynamics simulation. *Colloids Surf., A*, *672*, 131705, 2023.
45. Ayar, M., Dalkiran, A., Kale, U., Nagy, A., Karakoc, T.H., Investigation of the substitutability of rubber compounds with environmentally friendly materials. *Sustainability (Switzerland)*, *13*, 9, 5251, 2021.
46. Ruksakulpiwat, Y., Sridee, J., Suppakarn, N., Sutapun, W., Improvement of impact property of natural fiber-polypropylene composite by using natural rubber and EPDM rubber. *Composites, Part B*, *40*, 7, 619–622, 2009.

47. Chan, C.H., Joy, J., Maria, H.J., Thomas, S., Natural Rubber-Based Composites and Nanocomposites: State of the Art, New Challenges and Opportunities, in: *RSC polymer chemistry series*, pp. 1–33, 2013.
48. Yan, N., Buonocore, G., Lavorgna, M., Kaciulis, S., Balijepalli, S.K., Zhan, Y., Xia, H., Ambrosio, L., The role of reduced graphene oxide on chemical, mechanical and barrier properties of natural rubber composites. *Compos. Sci. Technol.*, *102*, 1, 74–81, 2014.
49. Visakh, P.M., Thomas, S., Oksman, K., Mathew, A.P., Crosslinked natural rubber nanocomposites reinforced with cellulose whiskers isolated from bamboo waste: Processing and mechanical/thermal properties. *Composites, Part A*, *43*, 4, 735–741, 2012.
50. Yin, C., Zhao, X., Zhu, J., Hu, H., Song, M., Wu, S., Experimental and molecular dynamics simulation study on the damping mechanism of C5 petroleum resin/chlorinated butyl rubber composites. *J. Mater. Sci.*, *54*, 5, 3960–3974, 2019.
51. Li, J., Han, X., Huyan, C., Liu, D., Li, X., Li, H., Chen, F., Hindered phenolic antioxidant modified waterborne polyurethane films with excellent thermal stability and anti-aging performance at high temperature. *Prog. Org. Coat.*, *176*, 107398, 2023.
52. Yu, P., He, H., Jia, Y., Tian, S., Chen, J., Jia, D., Luo, Y., A comprehensive study on lignin as a green alternative of silica in natural rubber composites. *Polym. Test.*, *54*, 176–185, 2016.
53. Sun, C., Li, L., Ji, H., Yang, H., Jin, G., Jiang, C., Guo, P., Zhang, L., Yu, P., Wang, R., The use of crude carbon dots as green, low-cost and multifunctional additives to improve the curing, mechanical, antioxidative and fluorescence properties of epoxy natural rubber/silica composites. *Composites, Part A*, *182*, 108177, 2024.
54. Kurian, T. and Mathew, N.M., Natural Rubber: Production, Properties and Applications, in: *Biopolymers: Biomedical and Environmental Applications*, vol. 14, 2011.
55. Ismail, H., Shaari, S.M., Othman, N., The effect of chitosan loading on the curing characteristics, mechanical and morphological properties of chitosan-filled natural rubber (NR), epoxidised natural rubber (ENR) and styrene-butadiene rubber (SBR) compounds. *Polym. Test.*, *30*, 7, 784–790, 2011.
56. Noguchi, T., Bamba, Y., Odaka, I., Iwamoto, R., Endo, M., Isogai, A., Natural rubber composites with high strength, modulus, water-resistance, and thermal stability, prepared with cellulose nanofibrils and sodium methacrylate. *Composites, Part A*, *173*, 107665, 2023.
57. Angellier, H., Molina-Boisseau, S., Dufresne, A., Mechanical Properties of Waxy Maize Starch Nanocrystal Reinforced Natural Rubber. *Macromolecules*, *38*, 22, 9161–9170, 2005.
58. Li, Q., Liu, B., Hu, Z., Jiang, X., Yang, L., Meng, H., Song, Y., Zheng, Q., Preparation of deep eutectic solvents based on metal ions and their influences

on reinforcement and strain softening behaviors of silica filled natural rubber nanocomposites. *Composites, Part A, 181*, 108119, 2024.
59. Ye, Z., Zhao, D., Liu, F., Luo, J., Liu, X., Zhao, W., Flexible and highly conductive Ti3C2Tx/natural rubber composites with interconnected networks for high-performance electromagnetic interference shielding. *Composites, Part A, 180*, 108067, 2024.
60. Duan, X., Cheng, S., Li, Z., Liang, C., Zhang, Z., Zhao, G., Liu, Y., An, D., Sun, Z., Wong, C., Flexible and environmentally friendly graphene natural rubber composites with high thermal conductivity for thermal management. *Composites, Part A, 163*, 107223, 2022.
61. Ismail, H., Edyham, M.R., Wirjosentono, B., Bamboo fibre filled natural rubber composites: the effects of filler loading and bonding agent. *Polym. Test.*, 21, 2, 139–144, 2002.
62. Bismarck, A., Baltazar-Y-Jimenez, A., Sarikakis, K., Green Composites as Panacea? Socio-Economic Aspects of Green Materials. *Environ. Dev. Sustain.*, *8*, 3, 445–463, 2006.
63. Petricca, S.E., Marra, K.G., Kumta, P.N., Chemical synthesis of poly(lactic-co-glycolic acid)/hydroxyapatite composites for orthopaedic applications. *Acta Biomater.*, *2*, 3, 277–286, 2006.
64. Mehta, G., Mohanty, A.K., Thayer, K., Misra, M., Drzal, L.T., Novel Biocomposites Sheet Molding Compounds for Low Cost Housing Panel Applications. *J. Polym. Environ.*, *13*, 2, 169–175, 2005.
65. Muniandy, K., Ismail, H., Othman, N., Biodegradation, morphological, and FTIR study of rattan powder-filled natural rubber composites as a function of filler loading and silane coupling agent. *BioResources*, *7*, 1, 957–971, 2012.
66. Ismail, H., Shuhelmy, S., Edyham, M.R., The effects of a silane coupling agent on curing characteristics and mechanical properties of bamboo fibre filled natural rubber composites. *Eur. Polym. J.*, *38*, 1, 39–47, 2002.
67. Zhao, W., He, J., Yu, P., Jiang, X., Zhang, L., Recent progress in the rubber antioxidants: A review. *Polym. Degrad. Stab.*, *207*, 110223, 2023.
68. Dick, T.A. and dos Santos, L.A., *In situ* synthesis and characterization of hydroxyapatite/natural rubber composites for biomedical applications. *Mater. Sci. Eng., C, 77*, 874–882, 2017.
69. Abraham, E., Thomas, M.S., John, C., Pothen, L.A., Shoseyov, O., Thomas, S., Green nanocomposites of natural rubber/nanocellulose: Membrane transport, rheological and thermal degradation characterisations. *Ind. Crops Prod.*, *51*, 415–424, 2013.
70. Phumnok E, Khongprom P, Ratanawilai S. Preparation of Natural Rubber Composites with High Silica Contents Using a Wet Mixing Process. *ACS Omega*, 7, 10, 8364–8376, 2022. doi: 10.1021/acsomega.1c05848. PMID: 35309431; PMCID: PMC8928548.
71. Lubura, J., Kobera L., Abbrent, S., Pavlova, E., Strachota, B., Bera, O., Pavličević, J. Ikonić, B., Kojić, P., Strachota, A., Natural Rubber Composites Using Hydrothermally Carbonized Hardwood Waste Biomass as a Partial

Reinforcing Filler- Part I: Structure, Morphology, and Rheological Effects during Vulcanization. *Polymers*, 15, 5, 2023. DOI: 10.3390/polym15051176.
72. Sui, G., Zhong, W.H., Yang, X.P., Yu, Y.H., Zhao, S.H. Preparation and properties of natural rubber composites reinforced with pretreated carbon nanotubes. *Polym. Adv. Technol.*, 19, 11, 1543–1549. https://doi.org/10.1002/pat.1163.
73. Jacob, M., Thomas, S., Varughese, K.T. Natural rubber composites reinforced with sisal/oil palm hybrid fibers: Tensile and cure characteristics. *J. Appl. Polym. Sci.*, 93, 5, 2305–2312. https://doi.org/10.1002/app.20696.
74. Tzounis, L., Debnath, S., Rooj, S., Fischer, D., Mäder, E., Das, A., Stamm, M., Heinrich, G., High performance natural rubber composites with a hierarchical reinforcement structure of carbon nanotube modified natural fibers. *Mater. Des.*, 58, 1–11, 2014. https://doi.org/10.1016/j.matdes.2014.01.071.
75. Abdelmouleh, M., Boufis, S., Belgacem, M.N. and Dufrense, A. Short Natural-Fibre Reinforced Polyethylene and Natural Rubber Composite: Effect of Silane Coupling Agents and Fibre Loading. *Compos. Sci. Technol.*, 67, 1627–1639, 2007. http://dx.doi.org/10.1016/j.compscitech.2006.07.003.
76. Wang, Z.F., Peng, Z., Li, S-D., Lin, H., Zhang, K-X, She, X-D., Fu, X. The impact of esterification on the properties of starch/natural rubber composite. *Compos. Sci. Technol.*, 69, 11–12, 1797–1803, 2009. https://doi.org/10.1016/j.compscitech.2009.04.018.
77. Jiang, C., He, H., Jiang, H., Ma, L., Jia, D-M. Nano-lignin filled natural rubber composites; Preparation and characterization. *eXPRESS Polym. Lett.*, 7, 5, 480–493, 2013. https://doi.org/10.3144/expresspolymlett.2013.44.
78. Geethamma, V.G., Kalaprasad, G., Groeninckx, G., Thomas, S. Dynamic mechanical behavior of short coir fiber reinforced natural rubber composites. *Compos. - A: Appl. Sci. Manuf.*, 36, 11, 1499–1506, 2005. https://doi.org/10.1016/j.compositesa.2005.03.004.
79. George, N., Julie Chandra C.S., Mathiazhagan, A., Joseph, R. High performance natural rubber composites with conductive segregated network of multiwalled carbon nanotubes. *Compos. Sci. Technol.*, 116, 8, 33–40, 2015. https://doi.org/10.1016/j.compscitech.2015.05.008.
80. Liu, C., Shao, Y., Jia, D. Chemically modified starch reinforced natural rubber composites. *Polymer*, 49, 8, 2176–2181, 2008. https://doi.org/10.1016/j.polymer.2008.03.005.
81. Prasertsri, S., Rattanasom, N. Mechanical and damping properties of silica/natural rubber composites prepared from latex system. *Polym. Test.*, 30, 5, 515–526, 2011. https://doi.org/10.1016/j.polymertesting.2011.04.001.
82. Nair, A.B., Joseph, R., Eco-Friendly Bio-Composites Using Natural Rubber (NR) Matrices and Natural Fiber Reinforcements, in: *Chemistry, Manufacture and Applications of Natural Rubber*, pp. 249–283, Woodhead Publishing, Sawston, 2014. https://doi.org/10.1533/9780857096913.2.249.

6

Radiative Protective Qualities of Natural Rubber Based Polymer Composites

M. R. Ambika[1*], N. Nagaiah[2†] and M. B. Karthik Kumar[3]

[1]*Department of Physics, M.S. Ramaiah Institute of Technology, M.S.R Nagar, M.S.R.I.T Post, Bengaluru, Karnataka, India*
[2]*Department of Physics, Bangalore University, Jnana Bharathi Campus, Bengaluru, Karnataka, India*
[3]*Department of Physics, Don Bosco Institute of Technology, Kumbalgodu, Bengaluru, Karnataka, India*

Abstract

Natural rubber is a versatile and sustainable material has been widely used in various industries, from automotive to healthcare, known to us since 1700. Charles Goodyear's discovery of vulcanization in 1841was the catalyst for the development of the contemporary rubber industry. Rubber has conquered almost all fields of technology, wherein natural rubber is modified in accordance with requirements. Several researchers have also contributed to the development of synthetic rubbers such as silicones, styrene-butadiene, neoprene, and butyl rubber. These elastomers also have applications in radiation environment in specific.

Owing to the significant studies of several physicists, it is worth appreciating their efforts in discovering radiation. Radiation is the process of transmission or emission of energetic particles or waves that travel through matter-containing media. The natural form of radiation is light and heat from the sun, which are essential for life. Radiations can be classified as ionizing or non-ionizing, based on their effects produced by them. Ionizing radiations are α, β, n, γ, X-rays, etc., whereas non-ionizing radiations are UV, Visible light, IR, radio waves, all of which together constitute the electromagnetic spectrum, also called Maxwell's rainbow. Radiations can also be classified based on their origin as natural or artificial. Cosmic rays, gamma rays from the Earth's crust, radon and its decay products in air, and various radio nuclides in food and drinks are natural sources of radiation,

**Corresponding author*: mr.ambika01@gmail.com
†*Corresponding author*: nagaiahn@redfiffmail.com

Visakh. P.M. and Shiv Kumari Panda (eds.) Natural Rubber-Based Composites and Nanocomposites, (187–224) © 2025 Scrivener Publishing LLC

whereas medical X-rays, radioactive waste, and fallouts from nuclear weapon tests are artificial sources. The entire world is flooded with radiation, which is a part of our lives. It is a double-edged sword; controlled use is beneficial to society, whereas uncontrolled use results in harmful effects.

Two important characteristics of any source of radiation are the dose delivered to the receptor and the effect of such doses. At the outset, we cannot escape from exposure to natural radiation as they are background radiations of cosmic rays, natural radioactivity within the body, and gamma rays. However, it is possible to minimize the radiation dose received from artificial sources, as we can alter or terminate the process of reaching these radiations in living and non-living systems using an appropriate shielding material.

Keywords: Radiation, natural rubber, interaction mechanism, shielding efficiency

6.1 Introduction

6.1.1 Interaction of Nuclear Radiations with Matter

Radiations when travelled through a material medium transfer complete or partial energy to the medium, resulting in the ionization and excitation of the atoms. Ionizing radiation has its effect only when a person receives a sufficient radiation dose, wherein a very high dose received over a period of weeks can cause death. For instance, if an absorbed dose received by a person is 5 Gy or higher, it leads to a fatal state unless treatment is administered, as it damages the bone marrow and gastrointestinal tract. Instead, if the same dose is delivered to a limited area of the body, other early sign effects are observed, but are not fatal. On the other hand, if the same dose is received spontaneously by the skin, erythema may occur within a few days. Furthermore, if the same dose were taken over an extended period of time (long time), the body's cells would have time to repair, and no symptoms of harm would be observed. Even despite the absence of its effects, the probability of tissue damage could also occur to a greater extent, which can be noticed only later (late effects), of which the most important is cancer. Thus, the higher the dose, the greater the damage [1–3].

Ionization/excitation, if measured, serves as the quantitative detection of the presence of radiation. These nuclear radiations can be detected only by understanding the basic mechanisms through which they interact with the material medium.

6.1.2 Interaction of Heavy Charged Particles with Matter

Alpha particles, fission products, protons, and products of many nuclear reactions fall into the category of heavy charged particles whose mass is ≥1 amu and characteristic path length is on the order of 10^{-5} m. When a heavily charged particle interacts with a material medium, there is a transfer of momentum, and hence, the kinetic energy to the electrons of the target. The two mechanisms involved in the transfer of kinetic energy are ionization and excitation. Ionization is easily measurable and often used for the detection of positively charged particulate radiation. When a heavily charged particle is unable to impart sufficient energy to an atomic electron to be ejected from the atom, the excitation process takes place. Rather, a portion of the particle's energy is absorbed by the atoms of the target medium, which then excites them to higher-energy states. Furthermore, it moves to the ground state or lower energy state by emitting energy in the form of electromagnetic radiation.

The high mass and charge of the alpha particles are responsible for imparting high ionizing power and low penetrating power to matter. Because these particles are of short range, they do not travel much distance in the medium; hence, there are no such external effects of these radiations. However, it has severe effects when ingested, inhaled, or absorbed. Lung cancer, kidney damage, DNA damage, and many more are some of their effects owing to their high ionizing ability when compared to other radiations [2, 3].

6.1.3 Interaction of Fast Electrons with Matter

β-particles and energetic electrons are fast and high-speed electrons whose energy is dependent on their velocity. Like heavily charged particles, fast electrons also interact with matter through the process of ionization, and another prominent mechanism that distinguishes fast electrons is radiation loss *via* the production of bremsstrahlung radiation [4, 5]. The fast electrons transfer their energy by ionizing the material medium and ejecting electrons from the K, L, or M shells; in turn, characteristic X-rays are produced during the filling of vacancies in the atomic orbit. These ejected atomic electrons possess so much energy that they can cause ionization upon interacting with other target atoms, resulting in secondary ionization. Suppose that a high-Z material is the absorbing medium and because fast electrons are high-speed electrons, when they pass in close proximity to the atomic nucleus, they experience acceleration due to the deflecting force, and hence, produce braking radiation called as bremsstrahlung radiation. The transfer of kinetic energy (part or all) by beta particles to the medium mainly depends on the path traversed in the medium and the amount of deflection it experiences. Thus, the deflecting force is directly proportional to the nuclear charge of the medium. In addition, the electrons can reach the target at different angles and hence produce a spectrum of bremsstrahlung with a continuous band of energies. The amount of bremsstrahlung radiation produced is proportional to the energy of the electrons and the atomic number of the absorbing medium.

The sum of ionization loss and radiation loss gives the total stopping power for the electrons and is given by Eq. (6.1).

$$-\left(\frac{dE}{dX}\right)_{Tot} = \left(\frac{dE}{dX}\right)_{Ion} + \left(\frac{dE}{dX}\right)_{Rad} \tag{6.1}$$

Radiation loss becomes dominant at higher energies when $E >> 2m_o c^2$. The radiation loss is equal to the ionization loss at the critical energy E_{cr}. This is a characteristic property of the absorbing medium and has shown that for electrons, E_{cr} is 10 and 100 MeV for lead and air, respectively.

6.1.4 Interaction of Neutrons with Matter

Neutrons are neutral particles with no charge. The lack of charge of neutrons is the main reason why neutrons do not undergo ionization with the matter. Hence, when a neutron passes through a medium and collides with atomic nuclei, the probable ways through which it may interact are inelastic scattering, elastic scattering, and neutron capture [7, 8]. Neutrons can be effectively slowed by the elastic scattering process, which is significant in low-Z materials, particularly hydrogen. In elastic scattering, the collision of neutrons of mass m_n with the nuclei of mass M results in scattering of neutrons and recoil nuclei at angles β and α, respectively to conserve momentum. During this process, part of the kinetic energy of the neutron is transferred to the nucleus; thus, the sum of the kinetic energies of the scattered neutrons and recoil nuclei is equal to that of the colliding neutron. The kinetic energy lost by neutrons in this collision [5] is given by Eq. (6.2).

$$E_k = \frac{4Mm_n}{\left(M + m_n\right)^2} Cos^2\beta \tag{6.2}$$

On the other hand, inelastic scattering mostly occurs when a fast neutron collides with a nucleus of a large atomic number. During the process, the neutrons are scattered along with the recoil nuclei by transferring part of their kinetic energy to the recoil nucleus, which is in an excited metastable state, which loses this energy as gamma radiation or remains for a period in the excited metastable state. In the neutron capture process, slow neutrons in close proximity to the nucleus may enter into or be captured by the nuclei, resulting in nuclear reactions. Neutron capture by the nucleus mainly depends on the type of nucleus and energy of the neutron. Thus, neutron capture interaction results in new products that may eject a charged particle from the compound nucleus, emit gamma rays, or produce radioactive atoms that emit different forms of radiation as they undergo radioactive transformation. Hence, it is necessary to understand the mechanism of interaction of neutrons with the medium, as it considers the determination of neutron dose, neutron cross-section, shield plan, and method of detection.

6.1.5 Interaction of Gamma Photons with Matter

Gamma and X-ray photons are the high-energy ionizing radiations with no mass or charge. They transfer their energy and momentum upon interaction with the material medium through several modes, such as Compton scattering, pair production, photoelectric effects, photonuclear effects, Rayleigh scattering, photo fission, and Thomson scattering. The three principal modes of interaction are the photoelectric effect, Compton scattering, and pair production, as shown in Figure 6.1 [6, 9]. During the photoelectric process, a gamma photon interacts with the innermost orbital electron of an atom; the latter gains kinetic energy from the photon and is ejected from the orbit. The atomic number of the atom, electron binding energy,

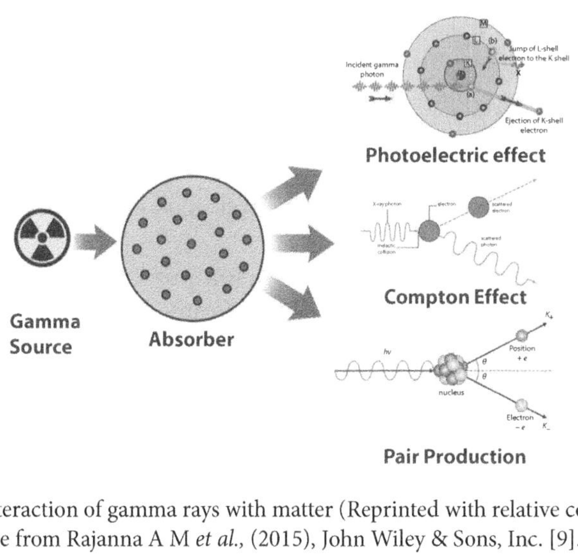

Figure 6.1 Interaction of gamma rays with matter (Reprinted with relative commons CCC by licence from Rajanna A M *et al.*, (2015), John Wiley & Sons, Inc. [9].)

and gamma-ray energy play significant roles in determining the probability of photoelectric absorption, as expressed in Eq. (6.3).

$$\sigma_{PE} \, \alpha \frac{Z^4}{E_\gamma^3} \tag{6.3}$$

The kinetic energy of the photoelectron is given by Eq. (6.4)

$$E = E_\Upsilon - E_b \tag{6.4}$$

where E_Υ and E_b are the energy of the incident gamma photons and binding energy of the electrons, respectively.

A gamma photon interacts with a free or unbound electron in the target medium, resulting in a scattered gamma photon (with lower energy) and a recoil electron, which is known as Compton scattering. Part of the incident photon energy is transferred to the ejected electron because the binding energy of the electron is small compared to the incident photon energy. The kinetic energy gained by the electron is just the energy lost by the incident photon and is given by the following equation:

$$E = E_\gamma - E' \tag{6.5}$$

Where, E_γ is the energy of incident gamma photon and E' is the energy of the scattered one and is given by the following equation

$$E' = \frac{E_\gamma}{1+\alpha(1-cos\theta)} \tag{6.6}$$

Where $\alpha = E_\gamma/m_0c^2$ and m_0c^2 is the rest mass energy of electron and θ is the scattering angle.

Because the interaction in the Compton process is with the free electron, the effect of the nucleus is not significant, and hence the interaction probability is independent of Z; however, it depends on the electron density, which is proportional to the ratio Z/A, which is approximately constant for all the materials.

A pair of electron–positron is created by gamma photon under the influence of a strong electromagnetic field in the vicinity of the nucleus and is called as pair production. In this process, the nucleus receives a very small amount of energy, but the nucleus appears to remain unchanged,

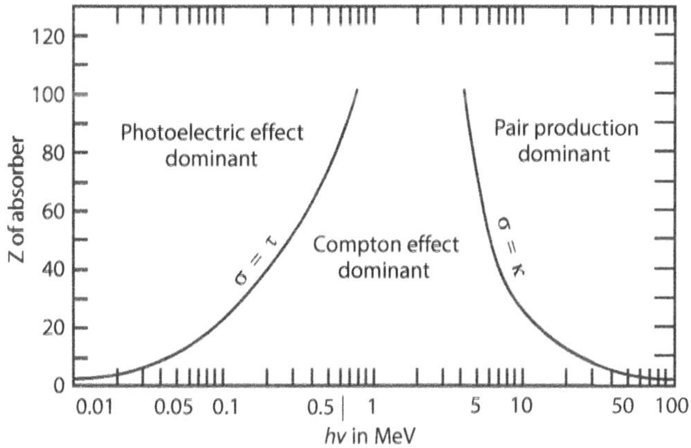

Figure 6.2 Energy dependence of principal modes of gamma photon interaction (Reprinted with relative commons CCC by licence from Evans RD, (1955), McGraw Hill Book Company, Inc. [4]).

while the gamma ray disappears. The minimum energy required to create a positron–electron pair is 1.02 MeV. If the gamma photon energy exceeds the threshold, then the excess energy is shared as the kinetic energy between the electron and the positron. This interaction is not as significant for nuclear material assays, as most of the gamma ray signatures are below 1.02 MeV.

The photoelectric effect predominates at low energies, whereas the Compton interaction occurs at intermediate energies, and the pair is produced at quite high energies. For a given energy, the atomic cross-section for each process depends only on the atomic number Z of the absorber; thus, the dividing lines between the processes can be given on a diagram of Z vs. photon energy, as shown in Figure 6.2. The separating lines are drawn such that the atomic cross sections are equal for neighboring processes along the lines. In almost all interaction processes, secondary electrons are produced, which further undergo ionization to dissipate their energy [4, 6, 10].

6.2 Effects of Ionizing Radiations

Radiations are widely used in several fields, including agriculture, medicine, industry, nuclear science and technology, and spacecraft applications. Irrespective of their field of application, ionizing radiation has its own degrading effects on both living and non-living systems in terms of physical, chemical, and biological effects. Radiations can also affect solids, liquids, and gases, sometimes degrading or strengthening the material properties. Upon the interaction of radiation in a solid medium, the transfer of energy to the lattice atom results in the displacement of atoms from the lattice, formation of interstitial which are some of the physical effects of radiation, resulting in physical and structural changes such as embrittlement, hardening, and thermal creep. In gases, radiation effects result in chemical changes. Noble gases were the least susceptible to radiation effects. In contrast, liquids do not have a fixed internal structure, such as gases, when compared to solids. Only the radiolysis process is involved in altering the composition of gases, resulting in the formation of free radicals.

The biological effects of ionizing radiation on living systems are mainly due to the ionization of atoms. Radiation damage is also caused by reactive oxygen species and free radicals produced by water. Water subjected to ionizing radiation forms hydrogen and hydroxyl free radicals, which can recombine to form gaseous hydrogen, oxygen, hydrogen peroxide, hydroxyl radicals, and peroxide radicals. Free radicals attack the biomolecule-forming structures within the cells, causing oxidative stress (a cumulative damage that may be significant enough to cause cell death, or may cause DNA damage, possibly leading to cancer). The outcome of radiation damage is that cells get damaged, damaged cells can also be repaired and operate normally or abnormally, and there is also a probability of the damaged cells dying. Furthermore, the effects of radiation can be acute or chronic. If a small amount of radiation is received for a long period, it leads to chronic effects, while a large amount of radiation dose received in a short period of time results in acute radiation hazards that are observable within a few days or weeks of exposure. However, the late effects of radiation without any early signs results in cancer, which is always a serious and fatal [11, 12].

Induction of Cancer: The main cause of cancer remains unexplored, and its understanding remains poor. Cancer is induced by exposure to tobacco smoke, asbestos, UV radiation, and ionizing radiation. It is not so easy to determine the exact cause for cancer. Cancer development is a complex cellular process that occurs in several stages over many years. Radiation appears to act principally at the initial stage by inducing certain mutations in the DNA of normal cells in the tissues. These mutations allow a cell to enter the abnormal growth pathway, which can sometimes lead to the development of malignancy. The development of science and technology has resulted in an understanding of the process by which radiation exposure leads to DNA damage and cellular systems that can repair or misrepairing such damage. This knowledge also provides adequate evidence that the risk of cancer may be very small at low doses. Thus, cancer risk increases proportionally with an increase in radiation dose [13, 14].

Hereditary Diseases: Exposure to radiation may also result in several hereditary diseases apart from cancer. The probability of hereditary disease mainly depends on the radiation dose, whereas the severity of the effect does not depend on the latter. Irradiation of the testis and ovaries, which produce sperm cells and egg cells in males and females, respectively, results in genetic damage. Ionizing radiation can induce mutations in these cells, i.e., structural changes in the DNA that subsequently carry hereditary information to future generations. The severity of hereditary diseases may vary, from early death and mental defects to severe abnormalities and minor mental illnesses. In addition, natural radiation and other environmental factors may also contribute to existing hereditary diseases [14].

Irradiation in Pregnancy: If embryo or fetus is exposed to radiation during which the development of several organs takes place, there are defects in the development with reduced size of different organs or certain mental abnormalities. Additionally, a high radiation dose results in death or gross malformations [14].

Hence, there is serious concern to protect both living and non-living systems from radiation exposure. In fact, none of us can escape radiation exposure owing to natural radiation. However, radiation dose can be minimized by following the principles of radiation protection.

6.3 Radiation Protection

The three main principles of radiation protection are distance, time, and shielding, as shown in Figure 6.3. By minimizing the exposure time, increasing the distance and introducing a shielding material between the source emitting radiation and the receptor will certainly help to reduce the radiation dose, which in turn reduces the probability of occurrence of radiation hazards. Of these, shielding plays a vital role in minimizing the radiation dose and hence can be a form of radiation control. When placed between the source and receptor, suitable materials can drastically affect the intensity transmitted from the radioactive source to the receptor. This is mainly due to the absorption or attenuation of radiation in the source itself or in the material used for encapsulating the source or shielding material, which in turn depends on the shield type, feature, design,

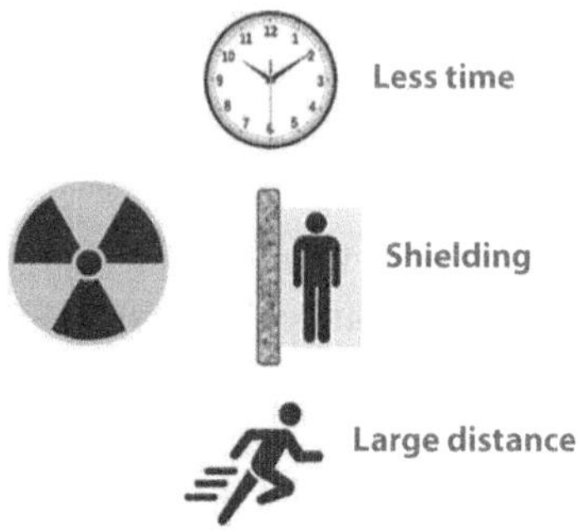

Figure 6.3 Principles of radiation protection.

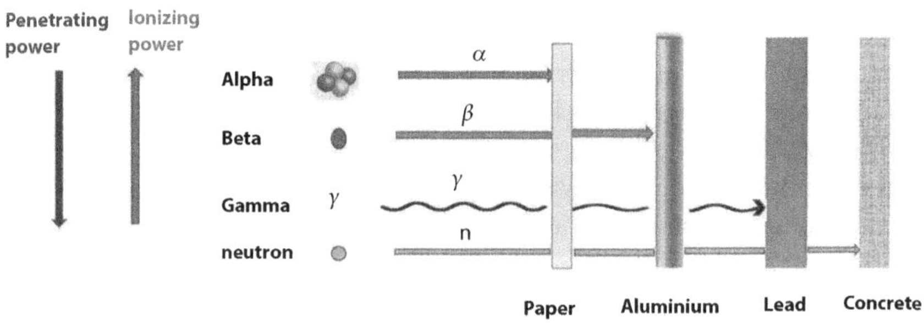

Figure 6.4 Penetration diagram of different radiations.

and geometry setup. Therefore, radiation shielding is a complex progress. Herein, a discussion is presented with respect to the shielding of different types of radiation, as shown in Figure 6.4.

6.3.1 Shielding of Alpha Particles

Alpha particles are known for their high mass and charge. Because the stopping power of most of the materials is very high for alpha particles, they possess a very short range on the order of a few tens of micrometers to a few centimeters in air. A thin piece of paper is sufficient to stop these particles. Because alpha particles have low penetration, dead cells in the outer layer of human skin act as a shield. Thus, shielding of α particles is not a significant issue. However, they cause considerable damage to the body owing to their high ionizing power when inhaled or ingested and undergo alpha decay significantly affecting the living tissue accompanied by gamma, for which shielding is a serious concern [7].

Most of the time, care must be taken while setting up the shielding of alpha particles so that they do not become a source of contamination that enters the body by contact or as airborne particles. Instead, highly contaminated areas can be layered with paint or fixative to reduce this effect. Additionally, regular monitoring and alertness in terms of inspection must be monitored to ensure that the fixative remains undamaged. On the other hand, these short-range alpha particles are not easily detected as they travel much less distance and may not reach the detector.

6.3.2 Shielding of Beta Particles

Beta particles (β) are charged particles that lose their energy upon passing through a medium, mainly by ionization and the bremsstrahlung process and have a limited range. The path traversed by them may be tortuous and long owing to the small number of electrons. Most beta particles were found to possess an average range of less than a few meters in air. At the outset, other low-density materials can also attenuate beta particles better than air, within a range of only a few millimeters. Thus, beta particles can be stopped completely by choosing an appropriate medium thickness and determining the maximum range of the highest-energy beta particles.

When beta particles of intensity I_0 pass through a material medium of thickness x (cm) and density ρ (g/cc), the transmitted intensity is given by $I(x)$ with an attenuation coefficient μ (cm^{-1}), which is a function of the maximum energy of the beta particle. The attenuation of β particles follows an exponential decrease in the intensity of the particles for absorber thicknesses less than the maximum range R_{max}. The transmitted intensity is given by Eq. (6.7).

$$I(x) = I_0 e^{-\mu x} \tag{6.7}$$

6.3.3 Shielding of Neutrons

Neutrons are neutral, and chargeless particles can be shielded by a mixture of materials that can slow down the fast neutrons with those of the materials, which enhances neutron capture. The best material that is effective in shielding neutrons is hydrogen, as half of the incident energy is reduced upon collision between hydrogen and neutrons with a higher cross section. Several materials, such as lithium or boron, are added to neutron shields to minimize the capture yield of gamma rays due to their absorption. Polyethylene, paraffin, water, lithium hydride, and cadmium are frequently used materials for the attenuation of neutrons. Neutron attenuation follows the same good geometry condition as photon attenuation [15]. The neutron attenuation equation is given by Eq. (6.8)

$$I(x) = I_o e^{-\sum n_r x} \tag{6.8}$$

Where, $\sum n_r$ is the neutron removal cross section

Since, it is very difficult to establish the neutron flux at a distance r from the source, dose equivalent rate D(x) outside the shield at a distance x must be determined with respect to D_0 which is the dose equivalent rate per unit neutron flux rate for neutrons and thus the shielded value of D(x) takes the form

$$D(x) = D_o B\, e^{-\sum n_r x} \tag{6.9}$$

Similar to photon attenuation, the role of the build-up factor is significant in neutron attenuation because it considers scattered neutrons resulting in elastic or inelastic collisions in the target medium.

6.3.4 Shielding of Protons and Heavy Ions

Protons, helium ions, deuterons, and tritium are charged particles generally found around accelerators. Charged particles transfer their energy to the target medium through ionization. They possess considerable mass and ionize the atoms in a target medium, giving up their energy by travelling through a short distance. Therefore, shielding of these charged particles is achieved by determining the maximum range of the particles, as they follow the same principle as alpha and beta particles. In addition to ionization, there is also a probability of the emission of bremsstrahlung radiation as they are charged particles. However, this process is limited unless the particles are highly energetic particles, that is, larger than a few tens of MeV. The shielding design for high-energy protons also requires a range of matter. Light materials, which are low-Z materials, are highly effective in attenuating protons. However, care must be taken when selecting the reference thickness for the penetration of protons with a given energy. Furthermore, the different masses and charges for deuterons, tritons, and helium ions are considered by introducing adjustments to this reference value [5].

6.3.5 Shielding of Gamma Photons

The interaction of gamma photons with matter is quite different from that of charged particles such as alpha and beta, as discussed in *Sections 6.1.1* and *6.1.2*. When gamma photons interact with the material medium, some are absorbed, some are transmitted without undergoing any interaction, and a few are scattered in a direction different from that of the actual incident photon. The gamma-ray shielding mechanism is shown schematically in Figure 6.5. The photon interaction and attenuation process are mainly characterized by the experimental setup, which includes "good geometry" or "poor geometry" [8].

Gamma rays are either absorbed or scattered when they are incident on a material medium *via* various processes. The attenuation of gamma rays by a medium under ideal narrow beam conditions, i.e., good geometry set up satisfies the relation

$$I = I_O \, e^{-\mu x} \tag{6.10}$$

where I_O and I represent the initial and final intensities of gamma photons, respectively, 'x' is the thickness (cm) of the absorbing medium, and μ is defined as the total linear attenuation coefficient (cm^{-1}), which accounts for all the interaction processes. In a good geometry setup, every photon that interacts with the target is either absorbed or transmitted and reaches the detector. The attenuation coefficient depends on the absorber medium and energy of the incident photons. μ generally increases with an increase in Z of the absorber medium because photoelectric absorption is high in materials with high Z, especially for low-energy gamma photons. Hence, materials with high Z are in practice being used as shielding materials for the attenuation of X/gamma rays.

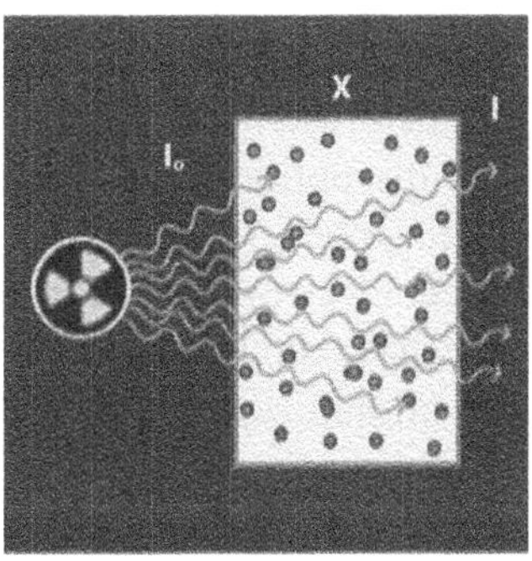

Figure 6.5 Attenuation of gamma photons (Reprinted with relative commons CCC by licence from Rajanna A M *et al.*, (2015), John Wiley & Sons, Inc. [9]).

In the poor geometry setup, the photon flux changes significantly owing to Compton scattered photons, many of which reach the detector with a complicated function of beam size and energy spectrum. This poor geometry represents the most practical situation for radiation protection. The transmitted intensity determined by the good geometry condition implies that the absorption is greater than what actually occurs, and a shield designed on this basis will not be thick enough. So, the effect of the scattered photons in addition to the unscattered primary photon must be considered by introducing 'Build up factor' B which is greater than 1 to account for the photons scattered towards the detector from regions outside the primary beam. Upon introducing B, the shielding equation takes the form [3, 8].

$$I(x) = I_o B\, e^{-\mu x} \tag{6.11}$$

The build-up factor can be obtained by estimating the photon fields using good geometric conditions and then adjusting the results to account for the build-up of scattered photons. Initially, the change in photon intensity upon passing through an absorber in a good geometry setup was determined, which provided the total attenuation of the photon beam. Furthermore, the unscattered intensity is multiplied by the build-up factor for the specific photon energy and absorber [8].

6.3.6 Electromagnetic Interference (EMI) Shielding

The world is full of technologies. Vacuum tubes, or valves, were employed as charge carriers in the early 19th century, but the development of the transistor, a semiconductor, drastically altered the world of technology. As technology is improved there are many advantages and disadvantages to the mankind as well as to the electronic devices. The term "Electromagnetic Interference (EMI)" is one such instance.

External source-induced disruption in an electrical circuit or channel is known as Electromagnetic Interference (EMI). This is also known as radio-frequency interference. Electronic devices, machinery, and systems are being used everywhere. These can be negatively affected by the electromagnetic interference. The miniaturization of electronic devices is one of the main causes of signal interruption in the device. Several studies have demonstrated the effects of EMI on the human body. It is observed to have alarming effects on human health, such as miscarriages, heart disease, cancer, and brain tumors [16, 17].

6.3.6.1 *Electromagnetic Interference Shielding Mechanism*

To avoid the impact of electromagnetic interference, it is better to protect the device using shielding materials such as metal, mesh screening, and conductive polymers. EMI shielding is the process of using a shielding material to reflect and absorb electromagnetic radiation. The substance acts as a shield by stopping low-frequency radiation, such as radio waves, from penetrating. However, in the conventional method, a metallic screen is used, which encloses the electronic used inside.

The three mechanisms governing the operation of electromagnetic interference (EMI) shielding are absorption, multiple reflections, and reflection, as shown in Figure 6.6 [18]. The substance absorbs the incident electromagnetic radiation during absorption. This results from the radiations' interaction with the material's composition. Electromagnetic radiation bounces off a material's surface and is known as reflection. The primary cause of this mechanism is the nature of the material surface. Some of the factors that cause reflection mechanisms are impedance mismatch, metallic surfaces, dielectric surfaces, etc. Multiple reflections of electromagnetic radiation (EM radiation) occur when incident radiations in a material or structure comes into contact with several surfaces or interfaces. The interior geometry and composition of the material are frequently implicated as reasons for multiple reflections.

To understand the shielding performance of the material, one should know about shielding effectiveness (SE). It is the ratio of the intensity of the electric or magnetic field before and after attenuation [19]. In general, we can say that it is the ability of a material to protect itself from

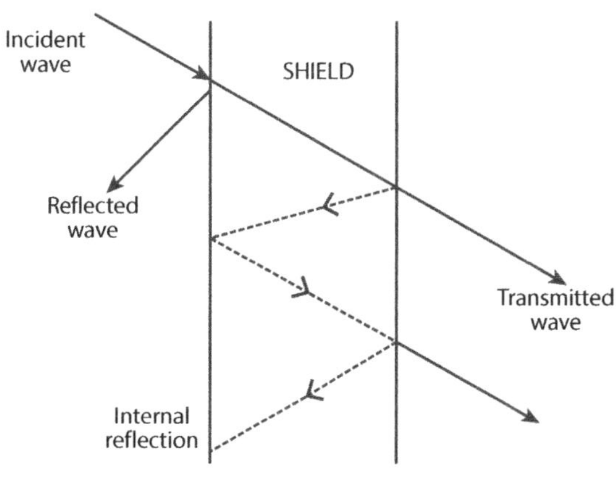

Figure 6.6 Mechanism involved in EMI shielding.

electromagnetic radiation. The total shielding effectiveness is the sum of the absorption, reflection, and multiple reflection effectiveness.

$$SE_{Total} = SE_{Absorbtion} + SE_{reflection} + SE_{multiple\ reflection} \tag{6.12}$$

$$SE(dB) = 20\log\left(\frac{E_i}{E_t}\right) \tag{6.13}$$

$$SE(dB) = 20\log\left(\frac{H_i}{H_t}\right) \tag{6.14}$$

where SE is Shielding Effectiveness in dB, E_t and H_t are the intensities of the transmitted electric and magnetic fields, respectively, and E_i and H_i are the intensities of the incident electric and magnetic fields, respectively.

Electrical conductivity and magnetic permeability are the two key characteristics that provide EMI shielding. Therefore, several studies have been conducted to enhance the EMI shielding effectiveness by enhancing the electrical conductivity of the shielding material.

6.4 EMI Shielding Materials

Natural Rubber (NR) is known for its flexibility and good resistance to extreme temperatures. As such, NR is not recommended as EMI shielding because of its poor conductivity and SE value. However, composites of natural rubber with some metals as fillers can enhance the conductivity and SE of the NR composite materials. Many researchers worldwide are conducting experiments to develop a better EMI shielding materials using NR as part of their composite matrix.

Zhan *et al.* [20] attempted to enhance the EMI shielding properties of natural rubber-based supercritical CO_2 foams. The authors developed an EMI shielding material comprising NR/CNT composites. It is observed that the 1.33 mm thickness of F-NR/$CNT_{6.40}$ exhibits an EMI SE value of 33.74 dB which is significantly higher than that of the commercially available EMI shielding of 20 dB, which is required for the telecom industry. In addition, the authors concluded that the addition of CNT to the NR matrix enhanced the EMI shielding effectiveness and electrical conductivity of the NR/CNT composites.

Wang *et al.* [21] developed $Ti_3C_2T_x$ MXene/NR composite to have a better EMI shielding material. In the study, authors modified the structure of the matrix from brick-mortar to honeycomb structure by the help of vulcanization reaction. It is observed that the SE value of the composite matrix with a honeycomb structure is 63.5 dB when compared to that of 57.1 dB. Similarly, the tensile strength was enhanced to 39.5 MPa when compared to that of 28.5 M Pa after the structural change. In this study, the authors proved that by changing the material structure from brick mortar to honeycomb, the SE and conductivity of the material can be enhanced.

Jeevakumari *et al.* [22] investigated the role of conductive graphene nano platelet and cobalt nano wire in the natural rubber matrix in enhancing the mechanical property of the composite material. In this study, the composite NR_4 (100 phr of NR + 3 phr of graphene + 8 phr of cobalt) showed an SE value of 38.1 dB at a frequency of 18 GHz. In addition, it was found that mechanical properties such as tensile strength, tear strength, and microhardness increased as the concentration of graphene and cobalt in the matrix increased. This proves that the addition of graphene nano platelet and cobalt nano wire to NR matrix enhances the SE and mechanical properties of shielding material.

Rashid *et al.* [23] studied the EMI shielding effect by adding PANI (polyaniline) and NiFe (nickel ferrites nanoparticles) to the natural rubber. The authors attempted to prepare three combinations (NBR/PANI, NBR/NiFe, and NBR/PANI/NiFe) of the composite to understand the effect of additives on the NBR matrix. The composite NBR/PANI30% shows a total SE value of 30 dB in the 2–3 GHz frequency range, whereas NBR/NiFe30% shows a total SE value of 8 dB at 5 GHz, and NBR/PANI15%/NiFe15% shows a total SE value of 52 dB in the frequency range of 2 GHz. In addition, the electrical conductivities of the NBR/NiFe30%, NBR/PANI30%, and NBR/PANI15%/NiFe15% composite matrices were 1×10^{-7} S/cm, 286 S/cm, and 167 S/cm, respectively.

The electromagnetic interference shielding effectiveness of Natural Rubber (NR) and Ethylene Vinyl Acetate (EVA) with conductive short carbon fibers (SCFs) and carbon black was studied by Das *et al.* [24]. In this study, an attempt was made to compare the shielding effectiveness (SE) of two different composites. In addition, the SE was analyzed for two different frequency ranges, i.e., in the X-band region (8–12 GHz) and microwave region (100–2,000 MHz). From the observed results, the authors found that the EMI shielding ability increased in the composite consisting of SCF as a filler compared to that of carbon black for both frequency ranges. This was attributed to the fact that the carbon fiber acted as a conducting mesh inside the polymer matrix. From the comparison study between the NR matrix and EVA matrix, the authors noticed a better SE value for the composite containing the EVA matrix rather than NR, which is mainly due to the conductivity of EVA compared to that of natural rubber.

Yan *et al.* [25] have attempted to develop a unique material of natural rubber latex foam coated with reduced graphene oxide with high density Polyethylene [rGO@HDPE coated NRLF] and observed a good electrical conductivity and EMI Shielding ability. The authors observed that the electrical conductivity of the prepared composite material increased with increasing rGO content and was found to be 0.48 S/m without any compression for the composite with 2Wt% rGO. This is six times higher than that of the unmodified NRLF. However, when the material was compressed by 90%, it showed an electrical conductivity of 1.01 S/m. As shown in Figure 6.7, reducing the percentage of HDPE in the matrix also played an important role in improving the electrical conductivity of the material. Upon reducing the HDPE content in the matrix to 2 wt.% rGO composites, the electrical conductivity changed from 0.48 to 1.6 S/m without any compression. In this study, the EMI shielding effectiveness of a matrix containing 2 wt.% rGO was also studied and observed to have 24 dB for the frequency range of 8.2 to 12.4 GHz.

Jia *et al.* [26] attempted to develop a flexible and strong shielding material with good shielding effectiveness by incorporating the carbon nano tubes in the natural rubber matrix (sCNT/NR). In this study, we developed a polymer composite with a thickness of 600 μm which provides better shielding effectiveness. It is observed that sCNT/NR composite with 10wt% CNT content shows EMI SE of 59.8 dB for the frequency range of 11.2–12.4 GHz along with electrical conductivity of 131 S/m. To understand the toughness of the material, the mechanical strength was also measured and

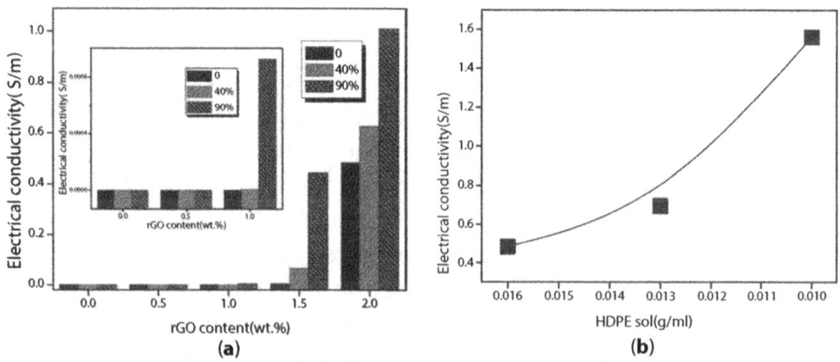

Figure 6.7 Variation of electrical conductivity with respect to (a) rGO content and (b) HDPE content in the composite (Reprinted with relative commons CCC by licence from Sun *et al.*, (2021), Elsevier [25].)

found to be 7.3 ± 0.39 and 8.2 ± 0.27 MPa for Young's modulus and tensile strength, respectively which are found to be higher than those of the pure NR matrix (0.7 ± 0.08 and 3.7 ± 0.28 MPa for Young's modulus and tensile strength, respectively).

The influence of conducting carbon black blended with the NR/IIR [mixture of natural rubber and synthetic rubber] to enhance the mechanical and EMI shielding parameters of the polymer matrix has been reported by Madani [27]. It was observed that there was an improvement in the mechanical properties, such as Young's modulus (MPa) and tensile strength (MPa), with increasing filler content, as depicted in Figure 6.8. This was because of the interlocking of the fillers within the polymer chain. In addition, the shielding effectiveness (SE) was estimated for a frequency range of 0.5 to 5 GHz. It is observed that SE was found to be 30.5 dB for the 40 wt.% filler blended in NR/IIR polymer composite.

Kanking *et al.* [28] have developed the NR polymer composite in which bagasse fiber ash (BFA) was used as filler along with carbon black or precipitated silica. The authors attempted to replace carbon black or precipitated silica in polymer composites. However, the study showed that the mechanical properties of both the BFA:CB/NR and BFA:Psi/NR composites were reduced by the addition of BFA to the polymer matrix. Based on experimental observations, the authors have concluded that bagasse fiber ash cannot be used as a secondary filler along with NR to develop a shielding material.

From Table 6.1, it can be seen that the natural rubber composite, along with the conductive fillers, serves as a flexible shielding material for EMI in different frequency ranges. Depending on the type of filler, it can be used in different applications by considering many other properties, such as conductivity and mechanical strength. However, from the detailed survey of

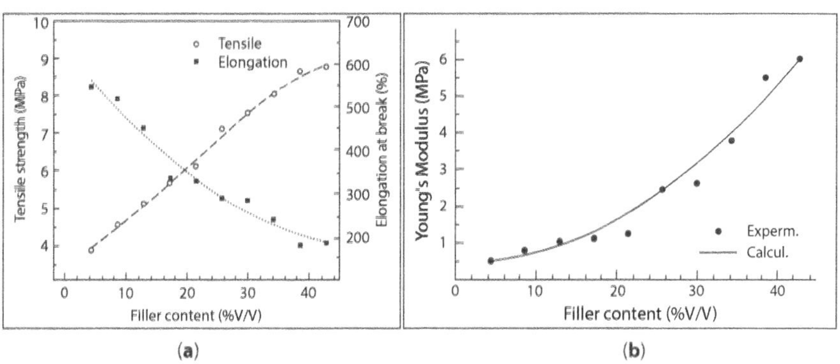

Figure 6.8 Variation of (a) tensile strength and elongation at break and (b) Young's modulus, for various filler wt.% (Madani, 2010).

Table 6.1 Comparison of EMI shielding effectiveness of natural rubber-based polymer composites.

Composite	SE (dB)	Frequency (Hz)	References
NR/CNT	33.74	12.4 G	[20]
$Ti_3C_2T_x$ MXene/NR	57.1	10.4 G	[21]
NR/Graphene/Cobalt	38.1	18 K	[22]
NR/PANI/NiFe	52	2 G	[23]
rGO@HDPE coated NRLF	24	8–12.4 G	[25]
sCNT/NR	59.8	11.2–12.4 G	[26]
NR/IIR + Carbon Black	30.5	0.5–5 G	[27]

literature one can also notice that fiber fillers provide better performance when compared to that of metal fillers within the polymer matrix. It is also noticed that when compared to other polymers, the natural rubber matrix shows poor performance in EMI shielding ability, whereas it acts as a better flexible material.

6.5 Gamma Ray Shielding Materials

Elastomers, especially shielding materials, are light weight and flexible in nature. Likewise, researchers have attempted to innovate and fabricate new materials that can replace conventional shielding materials used for shielding different radiations in radiation environments. Natural rubber-based composites have attracted the interest of researchers. But, before one makes a choice of matrix and filler, the following should be taken into consideration

1. The matrix should be radiation resistant, as it must withstand a high radiation dose without affecting the crosslink density, as it would otherwise result in polymer scission effects.
2. The matrix should be thermally stable, depending on the specific field of application. For example, for radiographers in the medical field, the material should be stable at room temperature. However, the material must withstand very high temperatures when used in nuclear reactor applications.

3. Mechanical strength also plays a vital role if the material is used as a wall, block, or flexible protective garment.
4. In addition, chemical stability is as important as the shielding material used in different chemical environments.
5. The filler particles should be appropriate, based on the type of radiation to be detected. Example, High density high-Z materials are preferable for X/gamma rays, whereas low-Z materials are preferred for beta particles. For neutrons, low-Z materials such as boron and its compounds are considered. Conducting materials are a good choice for EMI shielding.

6.5.1 NR Composites with Metals as Fillers

i. El-Khatib *et al.* [29] fabricated natural and waste rubber composites with different lead compositions. Gamma-ray shielding parameters were estimated for different gamma photon energies using Cs-137, Ba-133, Co-60, and Eu-152 gamma sources. In addition, the mechanical and thermal stability of the composites were investigated. The authors reported an increase in the crosslink density with an increase in the Pb concentration, and the same trend was observed upon adding waste rubber. The tensile strength, modulus, and elongation at break decreased with increasing Pb content, indicating poor matrix–filler bonding/interaction. The authors reported three stages of degradation for all samples. Samples W50 and 100W50 proved to be thermally stable compared to the Pb/NR composites. W50 and 100W50 also exhibit excellent AC and DC conductivity values, which may be attributed to the presence of waste rubber containing carbon black, which is a conductive material. The authors also claimed that the addition of waste rubber increased the attenuation coefficient of NR composites loaded with Pb. The relative increase rate δ% is found to range from 0.7 to 6.9% at energy ranges from 121.8 to 1,408 keV. The half-value layer (HVL) thicknesses of Pb100 and 100W50 were found to be lower than those of the other NR composites. In addition, the % of heaviness was found to increase with the waste content, as shown in Figure 6.9. Despite this, the samples still proved to be lightweight and flexible, i.e., 14.85% and 17.34% lead for Pb100 and 100W50, respectively.

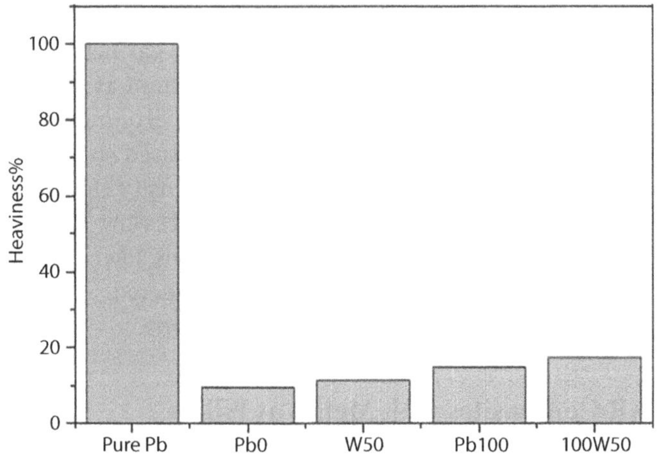

Figure 6.9 % of heaviness of NR/Pb & NR/W/Pb composites (Reprinted with relative Creative Commons Attribution 4.0 licence from El-Khatib *et al.*, (2020), Materials Research Express, IOP Science [29]).

ii. Gwaily *et al.* [30] have developed natural rubber composites with different concentrations of lead (100, 300, 500, 1,000, 1,500, and 2,000 phr) by master batch technique using two roll mill method. The attenuation coefficient and half-value layer thickness of the composites was estimated for 1.25 and 0.662 MeV gamma rays. The authors reported the highest values for the composite with Pb 2,000 phr as 28 and 54 m^{-1}, respectively. In addition, the authors proposed an empirical relation given by Eq. (6.15)

$$\mu = \mu_r + aF^n \tag{6.15}$$

Where, F is the concentration of lead in phr, μ_r is the attenuation coefficient of the rubber matrix at $F = 0$, and a & n are fitting parameters which can be obtained from the plot of $\ln\left(\mu - \mu_r\right)$ versus ln F.

iii. Although several researchers have attempted to find the best substitute for lead, Ninyong *et al.* [31] used lead as a filler particle in natural rubber (NR)-based polymer composites. The study revealed no significant improvements in the curing characteristics upon adding lead to the natural rubber matrix as lead does not interact chemically with the rubber matrix, whereas wood/natural rubber composites were found to be better than natural rubber composites. The mechanical properties increased with the addition of 20 phr of Pb, and a further increase in lead decreased the strength of the material owing to poor interfacial compatibility between the filler and matrix. The attenuation study revealed an increase in the attenuation coefficient with an increase in the lead content in the NR composites compared to that in the wood/NR composites.

6.5.2 NR Composites with Metal Oxides as Fillers

i. Kalkornsuranee *et al.* [32] have fabricated the natural rubber-based composites filled with bismuth oxide and lead oxide as fillers using internal mixer and ball mill mixer. The authors investigated the cure, thermal, mechanical, and gamma-ray shielding properties of these composites. The filler particles were found to disperse effectively, as is evident from the SEM images shown in Figure 6.10. The mass attenuation coefficient decreases with increasing energy. It was found to be the maximum for 223 keV gamma rays for the NR/Bi_2O_3 rubber composite. They compared the half-value layer (HVL) thickness of the prepared composites with commercially available materials, including window, serpentine, ordinary concrete, and hematite–serpentine (Figure 6.11), and were found to be better than these materials. The maximum rate of degradation of NR is found

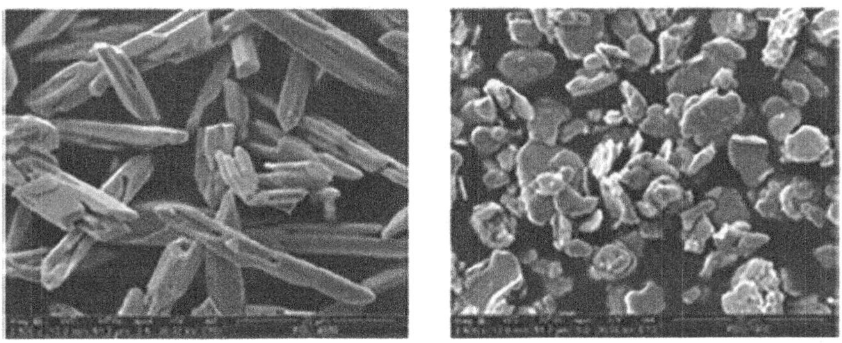

Figure 6.10 SEM images of NR filled with Bi_2O_3 and PbO (Kalkornsuranee *et al.*, 2022).

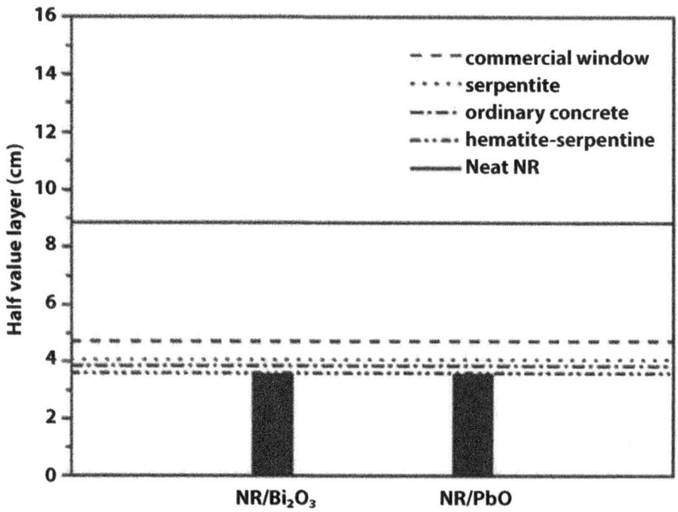

Figure 6.11 HVL thickness of rubber composites (Kalkornsuranee *et al.*, 2022).

increased from 370 to 388°C upon the addition of Bi_2O_3. In addition, the tensile strength of the NR/PbO sample was higher than that of the NR/Bi_2O_3 sample. Because the toxic metal lead is not preferable, the authors have reported that the NR/Bi_2O_3 sample is a good choice for gamma shielding applications.

ii. Natural rubber composites reinforced with Pb_3O_4, WO_3, SnO_2, and Bi_2O_3 were fabricated by Fisli *et al.* [33] using open milling and vulcanization processes. The authors studied the microstructure, physical structure, and attenuation properties of X-rays (50–140 kV) and gamma rays (356, 661, and 1,250 keV). The results of the HVL thickness of these materials for different energies of X-ray and gamma ray energies are shown in Figure 6.12.

It is evident from Figure 6.12 that the HVL value decreases with a decrease in the energy of the X-rays and gamma rays. The lowest values were observed for NR/Bi_2O_3 for gamma rays and SnO_2/NR for X-rays. Thus, these composites have been proposed as novel and flexible materials for use in radiation protection garments.

iii. Intom *et al.* [34] reinforced natural rubber with Bi_2O_3 particles. The mechanical and radiation-shielding properties

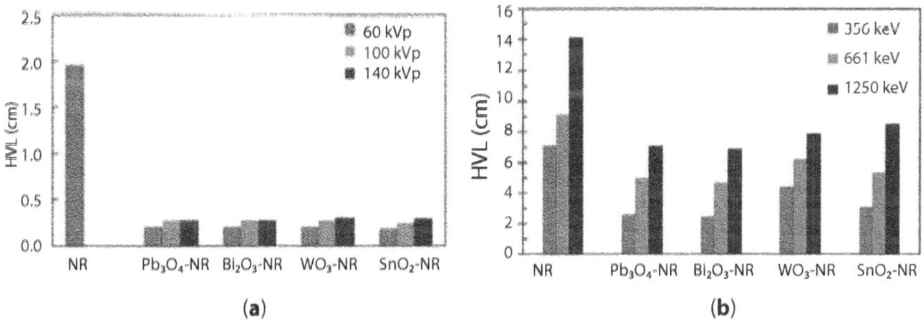

Figure 6.12 HVL values of NR-based composites for (a) X-rays and (b) gamma rays (Reprinted with relative CC BY-NC-SA 4.0 licence from Fisli *et al.*, (2023), Atom Indonesia [33]).

were investigated for different concentrations (10, 20, 40, 80, 120, and 150 phr) of bismuth oxide. The authors have observed the formation of agglomerates in some parts of the samples through the SEM images upon adding a high concentration of filler particles. The attenuation coefficient for X-rays with energies ranging from 50 to 120 kV_p and gamma rays with energies of 662 keV were found to increase with the addition of bismuth oxide. However, the tensile strength of the material decreased owing to the presence of agglomerates. However, the tensile modulus increased owing to the increase in the crosslink density of rubber and metal oxide. The authors also concluded that the rigid nature of bismuth oxide reduced the flexibility of the polymer composite.

iv. Kusiak *et al.* [35] have fabricated rubber-based composites with different metal oxides like Bi_2O_3, Gd_2O_3 and WO_3 as filler particles. The authors have highlighted the efficiency of the smallest size of Bi_2O_3 particles among all the fillers towards the potential interaction between the filler and elastomer matrix. The tensile strength was also good (17 Pa), as reported by the authors. The mass attenuation coefficient was estimated for all samples. The authors claim that although bismuth is efficient in absorbing X-rays, it emits fluorescent radiation upon interaction. Tungsten oxide was added to minimize this, but tungsten serves as a source of additional X-rays with a lower yield, which was further reduced by adding gadolinium oxide. The authors have proposed a composite containing 50 phr Bi_2O_3, 30 phr Gd_2O_3, 20 phr WO_3, and 20 phr Sb_2O_3 as efficient materials for absorbing X-rays.

v. Aroon *et al.* [36] developed lead free, natural rubber filled with bismuth oxide (0, 100, 200, and 500 phr) sponges using Oxybis (benzene sulfonyl) hydrazide (OBSH) (its contents varied from 0, 8, to 16 phr) as blowing agent using two roll mill method. The density of the composite decreased with the addition of OBSH, whereas the X-ray/gamma-ray shielding properties increased with increasing bismuth oxide content. The authors reported an increase in the tensile modulus and hardness owing to the high rigidity of Bi_2O_3, whereas the tensile strength and % of elongation increased initially up to 100 phr and further decreased owing to agglomerations in the high-filler-loaded samples.

vi. Madani *et al.* [37] have used waste newspaper (untreated and treated with magnetite) to fabricate natural rubber-based composites using two roll mill technique. The samples were irradiated using a Co-60 gamma source to investigate any changes in their physicomechanical properties. The conductivity and mechanical properties were found to increase upon mixing natural rubber with the modified wastepaper. Furthermore, gamma and neutron shielding parameters have been studied using a Pu-α-Be source and thermoluminescence detectors. The authors reported excellent attenuation properties in a sample with 54% modified newsprint and 6% paraffin wax.

Several researchers have used different fillers, such as lead, tungsten, bismuth oxide, tungsten oxide, lead oxide, gadolinium oxide, and antimony oxide, to reinforce the natural rubber matrix to enhance the gamma shielding ability, as natural rubber helps only to hold/bind the filler particles and is known to possess poor shielding ability for gamma rays. Natural rubber only helps to achieve flexibility in the composite; the fillers mentioned above are of high Z and are found to be effective in attenuating gamma rays of different energies. Upon comparison as presented in Table 6.2, it is evident that the linear attenuation coefficient is maximum for the composite NR + 100Pb + W50 at 122 keV gamma rays and NR + Pb2,000 at 662 and 1,250 keV gamma rays. Although researchers have attempted to fabricate lead-free shielding materials, the gamma shielding efficiency

Table 6.2 Comparison of shielding ability of natural rubber-based polymer composites.

Material	Energy (keV)	Linear attenuation coefficient (cm^{-1})	References
NR + 100Pb	121.8	2.26	[29]
	244.7	0.502	
	344.3	0.289	
	356.01	0.30	
	661.6	0.156	
	964	0.117	
	1,085	0.110	
	1,173.2	0.102	
	1,332.5	0.095	
	1,408	0.092	
NR + 100Pb + W50	121.8	2.45	
	244.7	0.532	
	344.3	0.320	
	356.01	0.316	
	661.6	0.187	
	964	0.149	
	1,085	0.135	
	1,173.2	0.133	
	1,332.5	0.126	
	1,408	0.112	
NR + Pb2,000	1,250	0.28	[30]
	662	0.54	
NR + 50Pb		21	[31]
NR/Bi_2O_3	662	0.22	[32]
NR/PbO		0.199	
NR + 100 Bi_2O_3	356	0.28	[33]
	662	0.14	
	1,250	0.085	
NR + 100 Pb_3O_4	356	0.27	
	662	0.135	
	1,250	0.082	
NR + 100 WO_3	356	0.22	
	662	0.12	
	1,250	0.072	
NR + 100 SnO_2	356	0.16	
	662	0.125	
	1,250	0.075	

(Continued)

Table 6.2 Comparison of shielding ability of natural rubber-based polymer composites. (*Continued*)

Material	Energy (keV)	Linear attenuation coefficient (cm^{-1})	References
NR + 150 Bi_2O_3	662	0.0966	[34]
NR + 50 Bi_2O_3 + 30Gd_2O_3 + 20 WO_3	122	1.31	[35]
NR + 50 Bi_2O_3 + 30Gd_2O_3 + 20 WO_3 + 20 Sb_2O_3		1.42	
NR + 25 Bi_2O_3 + 25Gd_2O_3 + 10 WO_3		0.93	
NR + 25 Bi_2O_3 + 25Gd_2O_3 + 10 WO_3 + 10 Sb_2O_3		1.01	

is still found to be high in the presence of lead, which is a highly superior, dense material with a high atomic number. However, choosing a matrix with high density and the presence of an element with a high atomic number in the functional group with a reinforcer will certainly help to enhance the gamma shielding efficacy of the polymer composites.

6.6 Applications of Shielding Materials

6.6.1 EMI Shielding Materials

Owing to the increased usage of electronic components in circuits, numerous disruptions in the system due to electromagnetic interference have been observed. This highlights the importance of the fabrication of EMI shielding materials. Many industries, such as military, medical, and electronic communication, require EMI shielding materials to safeguard the instrument, and hence, their proper usage.

6.6.1.1 Electronics and Telecommunication

In electronics, interference of electromagnetic radiation is caused by defective components used during manufacturing, resulting in the malfunctioning of electronic devices such as laptops and smart phones. Similarly, in telecommunication, the electronic components used in the transmitting and receiving stations must be shielded from electromagnetic interference by providing proper shielding to avoid the loss of information. In such cases, EMI shielding materials made up of metals such as copper, silver, or iron will be used as protective materials.

6.6.1.2 Military/Defense and Aerospace

In defense and military services, electronic components are widely used in radio communication. It is necessary to protect these devices from EM attacks not only to prevent the loss of data but also to prevent security threats. Similarly, military-grade EMI shields are preferred in aircraft construction to safeguard devices from energy transmission through electromagnetic interference, such as GPS and communication systems. Hence, research is being conducted to fabricate better shielding materials using different materials such as carbon-based composites and conductive materials [38–40].

6.6.1.3 Medical Applications

In the medical industry, devices such as pacemakers, monitoring devices, and MRI devices are composed of several electronic devices and circuits. The protection of these devices from electromagnetic interference by incorporating the EMI shielding material helps in the proper functioning of the instrument, thus enabling accurate measurements and increasing the durability [41, 42]. Autocatalytic selective plating and conductive paint are also used shielding materials in medical devices [43, 44].

In view of this, the fabrication of novel and highly efficient EMI shielding materials is essential to shield electronic equipment from the interference of electromagnetic radiation and ensure that they function properly by reducing the number of EMI-related attacks.

6.6.2 Gamma Ray Shielding Materials

X/gamma-shielding materials have diverse applications in the medical, agricultural, and space industries, where protection against X/gamma radiation is necessary.

6.6.2.1 Medical Applications

Gamma rays have several important applications in medicine, primarily in diagnosis and treatment. A few to mention are radiography and imaging, radiation therapy, sterilization, gamma knife surgery, thyroid imaging and treatment and quality control in medical imaging devices. Hence, protecting patients, healthcare professionals, and the general public from harmful radiation exposure is important and mandated [45–48].

- Radiation protection for personnel: Healthcare workers who operate or work near radiation-emitting equipment such as X-ray machines, CT scanners, and radiotherapy devices are at risk of exposure to gamma radiation. Shielding materials, such as lead or lead equivalents, are used in the construction of walls, doors, and barriers in radiology departments and radiation therapy facilities to reduce the amount of gamma radiation that can penetrate and reach personnel working in adjacent areas.
- Patient shielding: Patients may be exposed to radiation during medical imaging procedures involving gamma radiation. Shielding materials are used to protect sensitive organs and tissues from exposure. Lead aprons and shields are commonly used to cover reproductive organs during X-ray examinations to minimize radiation exposure in these areas.
- Radioactive material containment: In nuclear medicine facilities, radioactive materials are used for diagnostic imaging and therapy. Gamma shielding materials are used to shield these radioactive sources and prevent radiation exposure to personnel and the environment.
- Transportation of radioactive materials: Gamma-shielding materials are used in the design and construction of containers and packaging for the transportation of radioactive materials, such as radiopharmaceuticals used in nuclear medicine. These materials help minimize radiation exposure to transportation personnel and the public during handling and transportation.

6.6.2.2 Industrial Applications

Gamma-shielding materials are used extensively in the nuclear industry, including nuclear power plants, to protect workers and the surrounding environment from radiation exposure. Concrete, lead, steel and other dense materials are used in the construction of reactor containment structures, reactor shielding and spent fuel storage facilities [49, 50].

- Gamma radiation sources are used in industries to inspect welds, pipelines, and other critical components of flaws and defects. Shielding materials, such as lead, steel, and concrete, are used to contain radiation and protect personnel from exposure during radiographic testing.
- Industries such as aerospace, automotive, construction, and manufacturing utilize radiography and NDT techniques for quality control and inspection. Gamma radiation sources are used to penetrate the materials and detect defects. Shielding materials are used to protect workers and bystanders from radiation exposure during these processes.
- Gamma radiation detection and shielding are crucial in security and defense applications such as border security, nuclear material detection, and radiation monitoring. Shielding materials are used to protect personnel and equipment from radiation hazards associated with handling and detection of radioactive materials.

6.6.2.3 Agriculture

In agriculture, X-rays are not commonly used; hence, gamma rays are used primarily in irradiation processes for food preservation and pest control [51–53].

- Food Irradiation Facilities: Gamma radiation is used to irradiate food products to extend their shelf life, control pests, and eliminate harmful bacteria and pathogens. This process helps kill or inhibit the growth of bacteria, fungi, and insects that cause spoilage and foodborne illnesses. Irradiated foods include fruits, vegetables, grains, spices, meat, and seafood. Shielding materials such as concrete, lead, and steel are used in the construction of an irradiation chamber to prevent radiation leakage and ensure the safety of workers and the surrounding environment from radiation exposure.
- Transportation of Irradiated Food: Shielding materials are used in the design and construction of containers and packaging for the transportation of irradiated food products. Lead or lead equivalents are often incorporated into packaging to shield against gamma radiation and to minimize radiation exposure to transportation personnel and the public.

- Pest control: Gamma radiation is used for pest control in agriculture to eliminate or sterilize pests such as insects, mites, and nematodes. Irradiation can disrupt the reproductive cycles of pests or induce sterility, thereby effectively controlling pest populations without the use of chemical pesticides. This method is used to protect crops, stored grains, and agricultural products from pest damage and infestation. Agricultural workers involved in irradiation processes receive radiation safety training and follow strict procedures to minimize radiation exposure risks. Shielding materials such as lead aprons and protective clothing may be provided to workers as additional safety measures during the handling and processing of irradiated food products.

6.6.2.4 Space Applications

Shielding materials are important in space exploration to protect astronauts, spacecraft, and sensitive equipment from harmful radiation, including X-rays and gamma rays. These high-energy electromagnetic radiations pose significant health risks to humans and can interfere with electronic systems. Several shielding materials are used in space applications [54–56].

- Polyethylene is a lightweight and cost-effective shielding material that is commonly used in space missions and is often used in the form of panels or sheets to line spacecraft walls and modules, providing radiation protection for astronauts and equipment.
- Aluminum is another lightweight material used in space applications. Although aluminum is less effective at shielding against high-energy gamma rays than lead, it is still used in spacecraft construction owing to its low density and structural strength. Aluminum sheets or foils may be incorporated into spacecraft walls or used as part of multilayer shielding systems.
- Lead is a highly effective shielding material for attenuating gamma rays. However, its use in space applications is limited owing to its high density, which increases the mass of spacecraft and impacts the launch costs. Lead is sometimes used in localized shielding applications, such as protective vests for astronauts during extravehicular activities (EVAs) or in critical areas where additional shielding is required.

- Polyethylene-based composite materials consisting of polyethylene and other additives or reinforcing fibers offer improved radiation-shielding performance compared to pure polyethylene. These composite materials can provide enhanced protection against X/gamma rays, while maintaining lightweight and structural properties suitable for space applications.
- Boron-based materials, such as borated polyethylene or boron-loaded polymers, are used for neutron shielding in addition to gamma-ray shielding. Neutrons are another form of radiation encountered in space, and boron has a high neutron capture cross section making it effective for neutron attenuation.

In space missions, a combination of these shielding materials can be used to provide comprehensive protection against various forms of radiation. The design of shielding systems for spacecraft involves careful consideration of factors such as mass constraints, effectiveness against different types of radiation, structural requirements to ensure the safety and well-being of astronauts, and integrity of spacecraft and mission objectives.

Gamma-shielding materials play a critical role in ensuring radiation safety in various medical settings, from diagnostic imaging and radiation therapy to nuclear medicine and research laboratories. By effectively attenuating gamma radiation, these materials help protect individuals from the harmful effects of ionizing radiation. In each of these fields, shielding materials are selected and implemented based on factors, such as the level of radiation exposure, specific application requirements, regulatory standards, and cost-effectiveness. The proper design, installation, and maintenance of shielding systems are essential to ensure effective radiation protection and safety.

6.7 Conclusions

Radiation shielding plays a vital role in the radiation environment by protecting both living and non-living systems from the hazardous effects of radiation. Many researchers have focused on developing materials for shielding electromagnetic radiation, including ionizing radiation such as X-rays, gamma rays, neutrons, and low-energy radiation such as radio waves and microwaves, called EMI interference. In addition, natural rubber was used as the matrix, which was then reinforced with different metal/

metal compounds. As such, rubber alone will not be effective in attenuating radiation; hence, reinforcers will be used. The role of the reinforcer is vital in the shielding mechanism; on the other hand, an appropriate filler/reinforcer is very important. Though several researchers have used different fillers to enhance the gamma shielding and EMI shielding ability of the natural rubber-based composites, it has been observed that

- Increasing the filler content results in agglomeration, which deteriorates the shielding ability and mechanical strength of the material.
- Natural rubber, as such, does not aid the shielding mechanism, but an appropriate filler is very important.
- It is also evident that metals, as reinforcers, are highly efficient in attenuating gamma rays compared to metal oxides. In contrast, highly conducting metal/metal oxides or fibers are best suited for EMI shielding.
- The radiation stability of natural rubber is also important as long-time exposure may affect the crosslink density and hence the interlocking of matrix and filler particles, resulting in the loss of several material properties.
- Limited studies have been conducted on natural rubber-based composite radiation shields owing to the invention of several synthetic rubbers with enhanced properties.
- Conventional materials such as lead and barite can be replaced by rubber-based composites for low-energy gamma shielding applications.

Thus, natural rubber-based composites with suitable fillers can be used in several applications such as electronic gadgets, spacecraft, medical fields, nuclear reactor sites, and in the medical field for radiographers and radiotherapists.

References

1. Holmes-Siedle, A. and Adams, L., *Handbook of radiation effects*, Oxford university press, UK, 2002.
2. Cember, H., *Introduction to health physics,* McGraw-Hill Companies, Inc., 1969.

3. Ambika, M.R., Studies on Radiation Shielding Characteristics of Unsaturated Polyester Based Polymer Composites filled with Bismuth & Tungsten Oxides, Thesis, Bangalore University, 2017.
4. Evans, R.D., *The atomic nucleus*, McGraw-Hill, New York, 1955.
5. L'Annunziata, M.F. (Ed.), *Handbook of radioactivity analysis*, Academic press, London, 2012.
6. Nikjoo, H., Uehara, S., Emfietzoglou, D., *Interaction of radiation with matter*, CRC press, USA, 2012.
7. Turner, J.E., *Atoms, radiation and radiation protection*, Wiley-VCH Verlag GmbH & Co. KGaA, Weinheim, 2008.
8. Martin, J.E., *Physics for radiation protection: a handbook*, Wiley-VCH Verlag GmbH & Co. KGaA, Weinheim, 2006.
9. Rajanna, A.M. and Ningaiah, N., Polyester-Based Polymer Composites for Gamma Shielding Applications-A Substitute for Lead. *ChemistrySelect*, 7, 26, e202201214, 2022.
10. Knoll, G.F., *Radiation detection and measurement*, John Wiley & Sons, USA, 2010.
11. Brannigan, F.L. and Miles, G.S., *Living with radiation*, US Atomic Energy Commission, USA, 1959.
12. Eisenbud, M. and Gesell, T.F., Environmental radioactivity, in: *Elsevier eBooks*, Academic Press, USA, 1997.
13. Chapter 2, Radiation Protection manual, Institute of Nuclear Medicine and Allied Sciences, DRDO, Delhi, India, pp. 16–27, 2010.
14. Clarke, R., *Living with Radiation*, 5th Edn, National Radiological Protection Board, London, 1998.
15. Schaeffer, N.M., *Reactor shielding for nuclear engineers*, Radiation Research Associates, Inc., Fort Worth, Tex.(USA), 1973.
16. Hwang, J.H., Kang, T.W., Kwon, J.H., Park, S.O., Effect of electromagnetic interference on human body communication. *IEEE Trans. Electromagn. Compat.*, 59, 1, 48–57, 2016.
17. Duhaini, I., The effects of electromagnetic fields on human health. *Physica Med.*, 32, 213, 2016.
18. Geetha, S., Satheesh Kumar, K.K., Rao, C.R., Vijayan, M., Trivedi, D.C., EMI shielding: Methods and materials—A review. *J. Appl. Polym. Sci.*, 112, 4, 2073–86, 2009.
19. Gooch, J.W., Daher, J.K., Gooch, J.W., Daher, J.K., Fundamentals of electromagnetic shielding, in: *Electromagnetic shielding and corrosion protection for aerospace vehicles*, pp. 17–24, 2007.
20. Zhan, Y., Oliviero, M., Wang, J., Sorrentino, A., Buonocore, G.G., Sorrentino, L., Lavorgna, M., Xia, H., Iannace, S., Enhancing the EMI shielding of natural rubber-based supercritical CO 2 foams by exploiting their porous morphology and CNT segregated networks. *Nanoscale*, 11, 3, 1011–20, 2019.

21. Wang, Y., Liu, R., Zhang, J., Miao, M., Feng, X., Vulcanization of Ti3C2Tx MXene/natural rubber composite films for enhanced electromagnetic interference shielding. *Appl. Surf. Sci.*, 546, 149143, 2021.
22. Amutha Jeevakumari, S.A., Mothilal, T., Ramesh, G., Arun Prakash, V.R., Mechanically toughened EMI shielding natural rubber composite in microwave frequency bands. *Plast. Rubber Compos.*, 50, 5, 241–8, 2021.
23. Rashid, I.A., Hamza, A., Asim, S., Zubair, K., Shakir, H.F., Afzal, A., Zubair, Z., Achieving enhanced electromagnetic shielding by novel flexible rubber based nanocomposite with the incorporation of nickel spinal ferrites and polyaniline. *Synth. Met.*, 295, 117339, 2023.
24. Das, N.C., Khastgir, D., Chaki, T.K., Chakraborty, A., Electromagnetic interference shielding effectiveness of carbon black and carbon fibre filled EVA and NR based composites. *Composites, Part A*, 31, 10, 1069–81, 2000.
25. Sun, Y., Ma, L., Song, Y., Phule, A.D., Li, L., Zhang, Z.X., Efficient natural rubber latex foam coated by rGO modified high density polyethylene for oil-water separation and electromagnetic shielding performance. *Eur. Polym. J.*, 147, 110288, 2021.
26. Jia, L.C., Yan, D.X., Yang, Y., Zhou, D., Cui, C.H., Bianco, E., Lou, J., Vajtai, R., Li, B., Ajayan, P.M., Li, Z.M., High strain tolerant EMI shielding using carbon nanotube network stabilized rubber composite. *Adv. Mater. Technol.*, 2, 7, 1700078, 2017.
27. Madani, M., Conducting carbon black filled NR/IIR blend vulcanizates: Assessment of the dependence of physical and mechanical properties and electromagnetic interference shielding on variation of filler loading. *J. Polym. Res.*, 17, 1, 53–62, 2010.
28. Kanking, S., Niltui, P., Wimolmala, E., Sombatsompop, N., Use of bagasse fiber ash as secondary filler in silica or carbon black filled natural rubber compound. *Mater. Des.*, 41, 74–82, 2012.
29. El-Khatib, A.M., Doma, A.S., Badawi, M.S., Abu-Rayan, A.E., Aly, N.S., Alzahrani, J.S., Abbas, M.I., Conductive natural and waste rubbers composites-loaded with lead powder as environmental flexible gamma radiation shielding material. *Mater. Res. Express*, 7, 10, 105309, 2020.
30. Gwaily, S.E., Madani, M., Hassan, H.H., Lead—Natural rubber composites as gamma radiation shields. II: High concentration. *Polym. Compos.*, 23, 4, 495–9, 2002.
31. Ninyong, K., Wimolmala, E., Sombatsompop, N., Saenboonruang, K., Properties of natural rubber (NR) and wood/NR composites as gamma shielding materials. *IOP Conf. Ser.: Mater. Sci. Eng.*, 526, 1, 012038, IOP Publishing, 2019 Jun.
32. Kalkornsuranee, E., Intom, S., Lehman, N., Johns, J., Kothan, S., Sengloyluan, K., Chaiphaksa, W., Kaewkhao, J., Mechanical and gamma radiation shielding properties of natural rubber composites: effects of bismuth oxide (Bi2O3) and lead oxide (PbO). *Mater. Res. Innov.*, 26, 1, 8–15, 2022.

33. Fisli, A., Yulianti, E., Hanurajie, B., Sukaryo, S.G., Mashadi, M., Rivai, A.K., Prastanto, H., Fathurrahman, M.I., Some metal oxide-natural rubber composites for gamma-and low-energy X-ray radiation shielding. *At. Indones.*, 1, 1, 45–52, 2023.
34. Intom, S., Kalkornsurapranee, E., Johns, J., Kaewjaeng, S., Kothan, S., Hongtong, W., Chaiphaksa, W., Kaewkhao, J., Mechanical and radiation shielding properties of flexible material based on natural rubber/Bi2O3 composites. *Radiat. Phys. Chem.*, 172, 108772, 2020.
35. Kusiak, E. and Zaborski, M., Characteristic of natural rubber composites absorbing X-radiation. *Compos. Interfaces*, 19, 7, 433–9, 2012.
36. Lim-Aroon, P., Wimolmala, E., Sombatsompop, N., Saenboonruang, K., Manufacturing process and properties of lead-free natural rubber sponge for use in X-ray and gamma ray shielding applications. *IOP Conf. Ser.: Mater. Sci. Eng.*, 526, 1, 012015, IOP Publishing, 2019 Jun.
37. Madani, M., Basta, A.H., Abdo, A.E., El-Saied, H., Utilization of waste paper in the manufacture of natural rubber composite for radiation shielding. *Prog. Rubber Plast. Recycl. Technol.*, 20, 4, 287–310, 2004 Nov.
38. Harussani, M.M., Sapuan, S.M., Nadeem, G., Rafin, T., Kirubaanand, W., Recent applications of carbon-based composites in defence industry: A review. *Def. Technol.*, 18, 8, 1281–300, 2022.
39. Wachowski, M., Kosturek, R., Hutsaylyuk, V., Śnieżek, L., Lightweight multilayer material for military applications with enhanced ballistic properties and ability for electromagnetic interference (emi) shielding, in: *1st international symposium in electronic engineering, information technology, and education*, 2019.
40. Mishra, R.K., Thomas, M.G., Abraham, J., Joseph, K., Thomas, S., Electromagnetic interference shielding materials for aerospace application: A state of the art, in: *Advanced Materials for Electromagnetic Shielding: Fundamentals, Properties, and Applications*, vol. 16, pp. 327–65, 2018.
41. Bonaldi, R., Siores, E., Shah, T., Electromagnetic shielding characterisation of several conductive fabrics for medical applications. *Carbon*, 2, 4, 237–45, 2010.
42. Los, P., Lukomska, A., Jeziorska, R., Metal-polymer composites for electromagnetic interference shielding applications. *Polimery*, 61, 10, 663–9, 2016.
43. Skelly, J., Decorative plating processes for common plastic resins: Resin selection as well as plastic part design is critical to matching the right finishing method with the intended application. *Met. Finish.*, 106, 6, 61–5, 2008.
44. Karmalkar, S. and Banerjee, J., A study of immersion processes of activating polished crystalline silicon for autocatalytic electroless deposition of palladium and other metals. *J. Electrochem. Soc.*, 146, 2, 580, 1999.
45. Parvaresh, R., Jalili, M., Haghparast, A., Khoshgard, K., Eivazi, M.T., Ghorbani, M., Evaluations for determination of optimum shields in nuclear medicine. *J. Biomed. Phys. Eng.*, 10, 5, 651, 2020.

46. Ghassoun, J. and Senhou, N., The evaluation of neutron and gamma ray dose equivalent distributions in patients and the effectiveness of shield materials for high energy photons radiotherapy facilities. *Appl. Radiat. Isot.*, 70, 4, 620–4, 2012.
47. Mansouri, E., Mesbahi, A., Malekzadeh, R., Mansouri, A., Shielding characteristics of nanocomposites for protection against X-and gamma rays in medical applications: effect of particle size, photon energy and nano-particle concentration. *Radiat. Environ. Biophys.*, 59, 583–600, 2020.
48. Erdemir, R.U., Kilic, G., Sen Baykal, D., ALMisned, G., Issa, S.A., Zakaly, H.M., Ene, A., Tekin, H.O., Diagnostic and therapeutic radioisotopes in nuclear medicine: Determination of gamma-ray transmission factors and safety competencies of high-dense and transparent glassy shields. *Open Chem.*, 20, 1, 517–24, 2022.
49. Adrovic, F. (Ed.), *Gamma radiation*, InTech. InTech, Croatia, 2012.
50. Zahran, O., Kasban, H., Abd El-Samie, F.E., Utilization of gamma rays for troubleshooting in distillation columns of petrochemical industry. *J. Electron. Electr. Eng.*, 2, 3, 279–89, 2010.
51. Gautam, S. and Tripathi, J., Food processing by irradiation—An effective technology for food safety and security. *Indian J. Exp. Biol.,* 54, 700–707, 2016.
52. Young, A.L., Food irradiation. *Environ. Sci. Pollut. Res.*, 10, 82–8, 2003.
53. Pillai, S.D. and Shayanfar, S., Electron beam technology and other irradiation technology applications in the food industry, in: *Applications of radiation chemistry in the fields of industry, biotechnology and environment*, pp. 249–68, 2017.
54. Hong, J., Craig, W.W., Hailey, C.J., Laboratory tests on neutron shields for gamma-ray detectors in space. *Nucl. Instrum. Methods Phys. Res., Sect. A*, 452, 1–2, 192–204, 2000.
55. Wilson, J.W., Miller, J., Konradi, A., Cucinotta, F.A., Shielding strategies for human space exploration, 1997 Dec 1.
56. Gohel, A. and Makwana, R., Multi-layered shielding materials for high energy space radiation. *Radiat. Phys. Chem.*, 197, 110131, 2022.

7

The Use of Green Additives in Eco-Friendly Compounds Based on Natural Rubber

Aline Zanchet[1] and Fabiula Danielli Bastos de Sousa[2]*

[1]*Universidade de Caxias do Sul, Rua Francisco Getúlio Vargas, Caxias do Sul, RS, Brazil*
[2]*Technology Development Center, Universidade Federal de Pelotas, Rua Gomes Carneiro, Pelotas, Brazil*

Abstract
Sustainable development is the responsible search for economic development without harming human beings and the environment, using natural resources intelligently so that future generations are not affected.

In the area of elastomeric materials, a strong example of sustainable action, widely discussed in different areas of product development, is the exchange of conventional additives used in the production of elastomeric consumer goods with other options from renewable sources, looking for the production of eco-friendly products. Such an exchange brings many benefits, which will be analyzed in the present work, based on the discussion of some examples available in the literature.

***Keywords*:** Natural rubber, green additive, eco-friendly, renewable source, plasticizer, lubricant, filler, activator

7.1 Introduction

Currently, the high consumption and inadequate disposal of polymeric materials has raised problems related to urban solid residues. Among the most harmful polymeric materials are those that contain heavy-metals, fillers, plasticizers, and vulcanized elastomers. The latter, being reticulated and non-biodegradable, is the most unwanted because of the high time

**Corresponding author*: fabiuladesousa@gmail.com

Visakh. P.M. and Shiv Kumari Panda (eds.) Natural Rubber-Based Composites and Nanocomposites, (225–242) © 2025 Scrivener Publishing LLC

necessary for natural degradation and the generation of serious public health problems, as they may accumulate rainwater when disposed of in landfills where vectors may proliferate, such as the *Aedes aegypti* mosquito, a transmitter of dengue, chikungunya, yellow fever, and zika. Conventional additives used in a large scale in the production of elastomeric consumer goods can be harmful to the environment and human health as well.

According to de Sousa [1], "the vast majority of current additives used in elastomeric compounds are synthetic, non-biodegradable, derived from non-renewable sources, potential environmental contaminants (post-consumer materials), and harmful to human health. There are several reasons for the exchange of conventional vulcanization additives for others from renewable sources: to follow current laws, to change the sources of additives (from fossil fuel) by others from renewable sources, to reduce the final cost of the vulcanized compound, among others."

The increasing use of petroleum additives like paraffinic, aromatic, and naphthenic oils used as processing aids can generate serious environmental problems, especially when end-of-life goods are wrongly disposed of in landfills and dumps. Thus, some studies have been conducted to develop environmentally friend compounds using materials from natural sources. Therefore, non-polluting and biodegradable vegetable oils derived from renewable sources have been used as substitutes for plasticizers/lubricants in elastomeric compounds.

The use of cellulosic fillers are mainly used in the development of eco-friendly nanocomposites. However, despite being quite promising, their applications still require further study, mainly because of the difficulties of such materials to be well dispersed in polymeric matrices.

7.2 The Use of Oils in Elastomeric Compounds

Processing aids are organic substances that improve the flexibility and processability of polymers. In the case of elastomers, the use of these processing aids aims to reduce the viscosity and elasticity of the unvulcanized material, improving its ability to mix with other additives and its flow during molding [2].

This type of additive can be classified into two categories: plasticizers and lubricants. Plasticizers are able to decrease intermolecular interactions and improve molecular mobility, resulting in a decrease in the glass transition temperature (T_g) of the elastomer and lubricating agents, although decreasing the compound viscosity has no significant influence on T_g [2].

Mineral oils, classified as paraffinic, naphthenic, and aromatics oils are among the most widely used processing aids [3, 4].

Paraffinic oil has a predominance of paraffinic hydrocarbons. This type of oil has a high viscosity, high chemical stability, and high fluidity. In naphthenic oils, naphthenic hydrocarbons are predominant, producing lubricating oils with a low carbon residue. In the aromatic oils, there is a predominance of aromatic hydrocarbons. This type of oil is rare, producing excellent quality solvents and petrol with high octane content. This type of oil is not used in the manufacturing of lubricants.

Naphthenic oil has good compatibility with most elastomers in comparison to paraffinic, so it is possible to add higher amounts in elastomeric compounds. Since most of these additives used currently are synthetic and petroleum derivatives, there has been a significant trend in the use of natural materials and renewable sources, making the elastomeric compounds more environmentally friendly [5–7].

Since January 2010, from the European Directive EEC No 1272/2008, it has been set that tire manufacturers should reduce the quantity of oils with a high content of polycyclic aromatic compounds (PCA) to avoid contamination of the environment [8, 9]. In this order, vegetable oils from renewable sources, non-polluting and biodegradable, have been used as substitutes for conventional plasticizers/lubricants for elastomer compounds because of the increasing demand for eco-friendly products.

In the literature, studies have been conducted using different oils, such as soybean oils [8, 10, 11], castor [7], palm [12], coconut [13], peanut, and flaxseed [6], for applications in elastomeric compounds, with satisfactory results.

7.3 Plasticizers

Plasticizers are organic substances added to polymers to improve their flexibility and processability. They increase the flexibility and elongation at low temperatures, decrease the intermolecular forces and glass transition temperature, and affect the mechanical and physical properties as a whole, but they do not change the chemical nature of the polymer macromolecules [5, 14, 15].

According to the International Union of Pure and Applied Chemistry (IUPAC), a plasticizer is defined as a substance or material embedded in a polymeric material to increase its flexibility; however, other definitions for plasticizers are based on molecular weight [16].

Plasticizers include many organic compounds, such as oil, animal fats, and vegetable oils [16, 17]. Currently, plasticizers have been studied in relation to possible risks to both the environment and human health [14].

A good plasticizer should improve the flexibility and processing of polymers by reducing the polymer–polymer interaction and increasing chain mobility, with a consequent reduction in the T_g value [14]. Plastification theory was first developed between 1930 and 1950. Three of them are still used today: the theory of lubrication (a), gel theory (b), and theory of free volume (c), as shown schematically in Figure 7.1 [16].

The theory of lubrication [15] is related to the decrease in the resistance of polymer deformation as a result of the reduction in the intermolecular friction among the polymer segments, in which the plasticizer acts as a lubricant and facilitates the movement of the macromolecules over each other (internal lubrication).

Gel theory proposes that plasticizers act on dipole–dipole interactions and van der Waals forces relaxing and consequently reducing the rigidity of the polymer [15, 16]. The attenuation of these interactions occurs once the plasticizer molecules positionned themselves between the polymer chains, thereby increasing the distance between them.

The strength of electrostatic attraction is inversely proportional to the distance between the electric charges. Therefore, an increase in intermolecular distance reduces the force of attraction among the chains, which makes them more flexible. In other words, the presence of plasticizer molecules, in the midst of the polymeric chains, promotes the breakage of dipole–dipole interactions, creating new dipoles between the polymer and the plasticizer [15, 16].

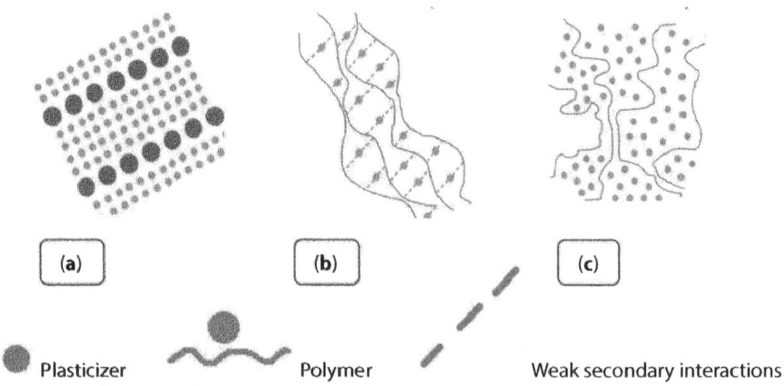

Figure 7.1 Schema showing the theory of lubrication (a), the gel theory (b), and the theory of free volume (c), adapted from Bocqué [16].

In addition, the theory of free volume [15–17] describes the existence of empty spaces in a polymer mass. When a plasticizer is added, the free volume increases, making the polymers less rigid and increasing the movement of polymeric molecules.

Plasticizers are the most common additives used in the polymer industry. The first use of this additive was in 1800 for the plastification of varnish using natural oils. During the last decade, the global production of plasticizers has been approximately 5 million tons per year [16].

Plasticizers have been applied to approximately 60 polymers and over 30 nonpolymeric product groups [17]. Plasticizers are classified as internal or external. For both types of plasticizers, a temperature dependence exists in the properties of the material, and this is more pronounced for internal plasticizers. However, elastomeric goods have the problem of maintaining dimensional stability at high temperatures [14, 16].

In this context, vegetal oils may represent a promising plasticizer because they are readily available. In addition to being biodegradable and of low toxicity, many are extracted from plants and oilseeds [16].

7.4 Lubricants

The polymers have high molecular masses, which make them extremely viscous in the molten state. This viscosity makes processing difficult because they usually need to flow between channels or narrow matrices in relatively short time intervals [17, 18].

In this way, to obtain high production with minimum wear of the machine and minimum energy consumption, it is common to reduce the viscosity. This can be achieved in two ways:

(1) By increasing the temperature, which increases the degradation rate of the molecules, the energy consumption and the cooling time; and
(2) The introduction of lubricant additives assists processing by reducing internal and external friction.

Lubricants facilitate processing and polymer mixing by improving the flow properties and reducing the adhesion of the polymeric mass to the machine components. They can also perform many other functions, such as acting as stabilizers, antistatic agents, pigments, and impact modifiers, in addition to activating vulcanization in the case of elastomers.

The first lubricants used for this purpose were salts derived from carboxylic acids with a low melting point, which not only stabilized the polymer but also improved processing [17, 19].

Lubricants are also important for the surface quality of products because they assist the dispersion of fillers and pigments. The use in polymers containing fillers is also important to provide a better surface quality and to compensate for the reduction in the flowability of the polymer due to the incorporation of fillers (increase in viscosity). In addition, lubricant molecules can be adsorbed on the surface of the filler particles, preventing them from retaining the stabilizers additives, resulting in a greater resistance to degradation [13, 19].

Among the main criteria for evaluating the performance of lubricants, they should facilitate processing (which consequently increases the production) and obtain final goods technically and aesthetically of high quality.

7.5 Lubrication Mechanisms

During mixing and processing, the lubricant coats the polymeric surface, and when this starts to soften, the molten lubricant penetrates into the polymer. The penetration rate depends on its solubility in the molten polymer, which is a function of its structure and molecular polarity in relation to the polymer.

Internal lubricants affect the viscosity and flow characteristics owing to the compatibility of these additives with the polymer, essentially lubricating polymer molecules. External lubricants affect non-stick and sliding characteristics, and their molecules are incompatible with melting; thus, they segregate and migrate to the melting surface, acting as a lubricant between the melting and the metal of the processing equipment [20]. Depending on the polymer–lubricant system, the solubility can range from totally soluble to insoluble [16].

7.6 Plasticizers and/or Alternative Lubricants

Many research groups have investigated the use of plasticizers and/or alternative lubricants from renewable sources.

Vegetable oils have two features that make them potentially good chemical plasticizers:

(1) Fatty acids chains can intercalate between the polymer chains, increasing the intermolecular spacing and, consequently, conferring greater mobility, and
(2) The ester groups can interact with the polymeric chains (e.g., through van der Waals interactions) and generate better compatibility [16].

However, there are still some limitations to the use of alternative products, such as their high cost and mechanical properties, which are different from those of plasticizers and/or conventional lubricants.

It must be noted that the introduction of a new additive in elastomeric formulations may cause some side effects, mainly in relation to the vulcanization reaction of the material, owing to the presence of chemical groups present in these additives and changes in the mobility of the polymer chain.

Therefore, studies are still needed to search for new materials that meet the market and have an environmental appeal [14]. To this end, several papers have been published on this topic, some of which are briefly discussed.

Ismail *et al.* [12] observed that the fatty acids present in palm oil affected the vulcanization properties, slowing the onset of vulcanization and increasing the minimum (ML) and maximum torque (MH) values of the compositions. Regarding the mechanical properties, an increase was observed in the modulus at 100% and 300% elongation, on hardness Shore A, in the tensile strength and tear strength with an increasing concentration of fatty acids present in the compositions. Such changes in the properties were attributed to the consequent increase in the cross-linking density, which generated greater rigidity in the material.

In another study by Dasgupta *et al.* [11], natural rubber (NR) compounds containing various vegetable oils (soybean oil, castor oil, etc.) were characterized in comparison to compositions with aromatic, naphthenic, and paraffinic oils and with low levels of PCA. The addition of these oils affected the action of the vulcanization system, resulting in low ML, MH, and optimum vulcanization time (t_{90}) in relation to the compounds containing conventional petroleum-derived oils. The mechanical properties, especially in the composition containing soybean oil, showed a reduction in the modulus at 100% and 300% elongation, and an increase in the tensile strength and hardness Shore A.

Da Costa *et al.* [7] conducted a study on castor oil used as an activator in the process of NR vulcanization using compositions containing silica. They observed that neat castor oil did not show a significant improvement in vulcanization characteristics. However, when castor oil was combined with stearic acid, better performance was achieved. This effect was attributed mainly to the castor oil lubricant, which promoted an improvement in the vulcanization characteristics.

Gujel *et al.* [4, 5] studied the effect of the incorporation of soybean oil in ethylene-propylene diene monomer rubber (EPDM). The results showed that vulcanization properties were affected and there was a reduction in the processing time due to the reduction in vulcanization. In addition, the incorporation of 10 phr of soybean oil (MD600) reduced the amount of lubricant oil required by approximately 60%. The oil had a lubricating effect on the EPDM matrix because there was no reduction in the T_g values of the compounds.

In a study by Zanchet *et al.* [10], atomic force microscopy (AFM) analysis resulted in phase images, in which it was possible to observe the different materials present in each sample. Garcia *et al.* [21] related that the morphology can be analyzed through the difference of colors in the images, and this color difference represents the hardness scale. As the samples are composed mostly of NR, silica, and oils, depending on the formulation, two or three color scales were observed. The AFM images are shown in Figure 7.2.

As the fillers have a spherical shape, in the images, the lighter areas represent the most rigid materials (silica), areas with intermediate tones represent NR and oil processing, in the case where oil is present between the NR chains, and the darkest areas (fewer rigid materials or "softer" materials) represent areas where the processing oil was secreted from the NR matrix. In sample NO-5, containing 5 phr of naphthenic oil, greater homogeneity and better dispersion of the silanized silica was observed in the NR matrix, which was not verified in the samples containing soybean oil. The higher the amount of soybean oil added, the worse is the filler dispersion in the elastomer, making it possible to observe the segregation of soybean oil from the NR matrix. It is assumed that, despite the silanization treatment performed in the fillers to improve their compatibility with the elastomeric matrix, filler agglomeration was observed in the NR matrix. Another hypothesis is that vulcanization additives are soluble in soybean oil. The sample SO-15 (containing 15 phr of soybean oil) had greater segregation of the oil and was the sample that showed a lower cross-linking

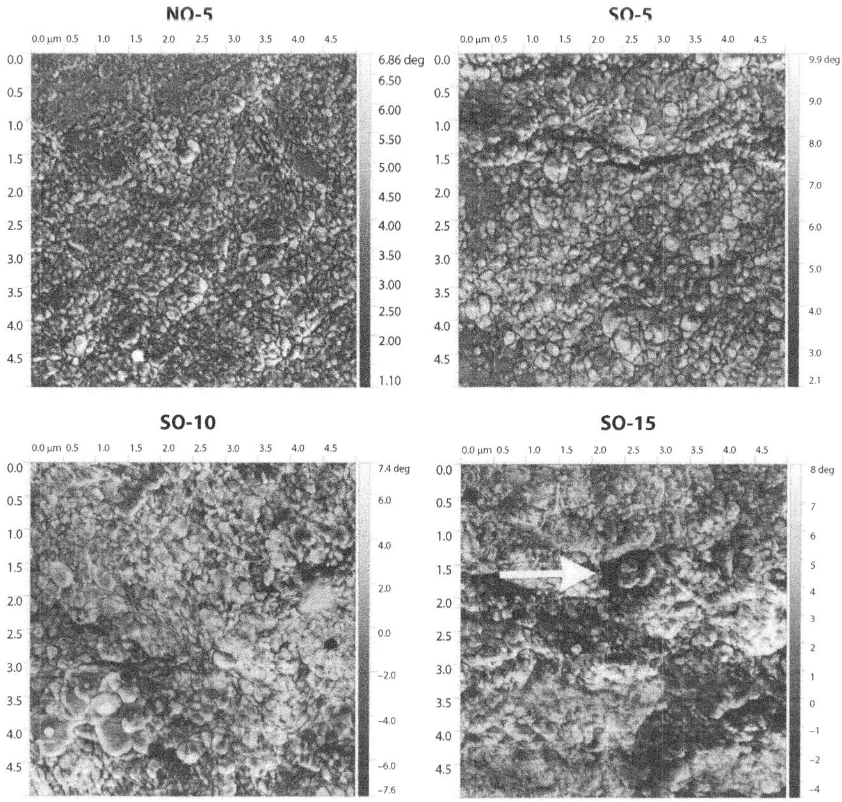

Figure 7.2 Tapping mode AFM phase images of the samples [10].

density value; part of the vulcanization additives possibly was dragged by oil, resulting in smaller values of this property [10].

Some works in the literature [22, 23] have shown that natural oils improve the dispersion and distribution levels of fillers in elastomers when there is an interaction between the oil/filler and oil/elastomeric matrix. Jayewardhana *et al.* [23] analyzed the reason for a better dispersion of fillers such as carbon black on NR matrix containing vegetable oils. In this case, the interactions between the long chains of unsaturated fatty acid esters of vegetable oil with the polar groups present on the surface of carbon black (phenol, carboxyls, and lactones) and the non-polar interaction of these oils with the molecules of NR provided such a result.

7.7 The Use of Fillers in Elastomeric Compounds

Fillers are additives widely used in elastomers to reduce the cost and/or to strengthen the final artifact, change a specific property, and provide certain processing characteristics [24].

A large number of minerals in powder form can be embedded in elastomers, although not all of them have the reinforcement ability. Essentially, two important classes of powdered minerals offer the following features: carbon black and silica, which have a complex three-dimensional structure, large surface area, and active surface that promotes interaction with the matrix.

Reinforcement fillers can increase the hardness, tensile strength, modulus, wear resistance, and tear resistance of elastomeric compounds. They play a role in strengthening elastomers and retaining their characteristics [25].

The ones called filling fillers do not affect or may even decrease the mechanical properties of the artifact and are used mainly to reduce costs and provide good processing characteristics. Some examples include talc, calcium carbonate, kaolin, mica, barite, and alumina [26].

One of the most important parameters affecting the effects of the filler on the elastomeric composition is the average particle size (Figure 7.3). Particles larger than 10^3 nm do not have the capacity to strengthen. They

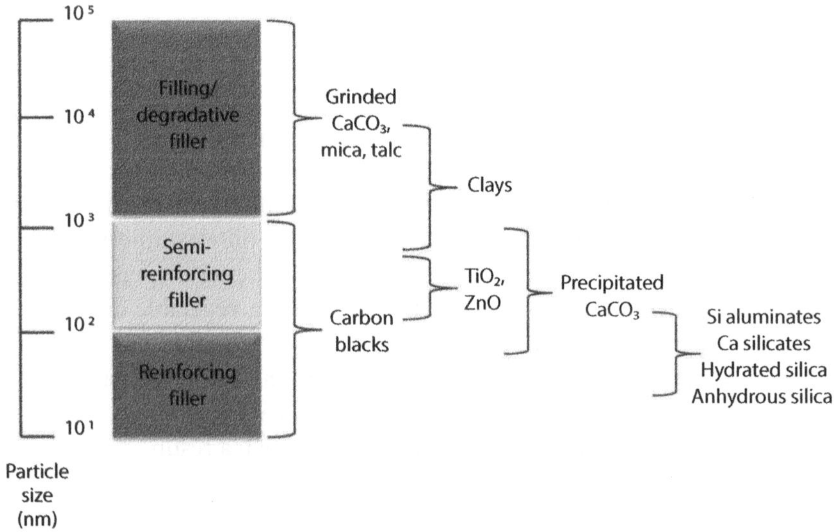

Figure 7.3 Classification of fillers according to particle size. Adapted from Leblanc [26].

only increase the viscosity of the mixture and may harm processing. Reinforcement is obtained with particle sizes of less than 100 nm.

Reinforcement concerns not only influence the final artifact, i.e., the vulcanized elastomer, but can also be noticed during processing. Reinforcement fillers further change the rheological properties of the matrix, which is related to the strong interaction between the elastomer and particles. Therefore, there is a correlation between the rheological properties of a material during processing and its final properties after vulcanization [26].

Among the observed effects of fillers during processing are the disappearance of linear viscoelasticity regions and the reduction of the swelling of the extruded material with an increase in the filler amount. In the vulcanized artifact, an anisotropic effect can be observed, in which the composite offers better mechanical properties in the direction of processing, mainly because of the alignment of the fillers in the direction of the material flow [26].

Generally, the higher the filler amount, the higher the viscosity of the compound, which may cause difficulties in the processing/production of elastomeric goods. Various types of fillers can be combined into the same compound depending on the desired properties and estimated price. The maximum amount of filler that can be incorporated into a compound depends on the elastomer type, type and quantity of plasticizer used, and nature and reinforcing power of the filler.

It is known that the fillers traditionally used in elastomeric formulations (to strengthen and fill) are relatively expensive, and their total or partial replacement for fillers from renewable sources becomes attractive without significant reduction in the properties of the artifact [27].

Global ecological concerns have resulted in great interest in renewable natural materials. Renewable natural organic fibers can offer the possibility of acting as biodegradable reinforcements and the creation of alternative materials, as seen in several studies in the literature [10, 27–36]. The results of some studies that used additives from alternative/renewable sources are briefly discussed.

Gujel *et al.* [4, 5] used the filler derived from cereal (MDCO), with cellulosic structure, in EPDM matrix. The mechanical properties were not influenced by the addition of the filler; therefore, it acted as a filling filler; that is, there was no significant improvement in the mechanical properties of the material.

Da Costa *et al.* [27] used rice husk ash for comparison with two commercial fillers, precipitated silica (Zeosil-175) and carbon black (N762), in the NR matrix. It was observed that the addition of rice husk ash to NR compositions, in comparison to the other fillers, revealed different effects on the vulcanization kinetics, compared to the conventional fillers, such as an increase in rate and a reduction of activation energy of cross-linkings formation.

7.8 The Use of Other Additives in Elastomeric Compounds

Some additives that are used are unsafe to human health and may cause illnesses. NR latex is widely used as medical items like gloves, catheters, among others. Dithiocarbamate accelerators, commonly used in NR latex, are likely to cause allergies [37], problems that can be avoided by changing sulfur vulcanization by radiation vulcanization [37–39], or by peroxide curing.

Accelerators derived from secondary amines (for example thiuram compounds, such as tetra methyl thiuram disulfide (TMTD)), when exposed to air, form nitrosamines in the presence of an atmospheric nitrosating agent [40], which generate carcinogenic nitrosamine (being potent carcinogens) [41, 42]. To prevent these problems, the academic community is studying new forms of additives to prevent harm to human health and the environment.

Alam *et al.* [43] analyzed the use of synthesized safe amine based accelerator Bis (N-benzyl piperazino) thiuram disulfide (BPTD) combined with MBTS in the replacement of the unsafe conventional TMTD, obtaining as a result an improvement of mechanical properties of the vulcanized NR. The same authors [40] analyzed the feasibility of exchanging conventional TMTD using safe amines. Various combinations of synthesized thiuram disulfides (TD) were studied in the NR vulcanization, namely Bis (N-phenyl piperazine) thiuram disulfide (PPTD) and Bis (N-ethyl piperazine) thiuram disulfide (EPTD) with mercapto benzothiazole disulfide (MBTS). The authors concluded that EPTD-MBTS (6:3) is a suitable choice of accelerator for replacing unsafe TMTD accelerators in the vulcanization of NR.

Regarding chemicals derived from secondary amines, Mariano *et al.* [42] investigated a new option to change them using tetrabutyl ammonium bis (4-methylphenyldithiocarbimatozincate(II)) (ZNIBU) in the NR vulcanization process. For commercial applications, this accelerator was found to be too slow and similar to other accelerators compared to

N-cyclohexyl-2-benzothiazole sulfenamide (CBS, MBTS, and TMTD). A possible solution is to use a recipe for the different types of accelerators that are available.

Abad *et al.* [44] investigated the antioxidant properties of some non-water soluble amino acids (cystine, tyrosine, asparagine phenyl alanine, and alanine), and keratin from chicken feathers on radiation vulcanization of NR latex. The results showed that some combinations of these amino acids improved antioxidant properties and that keratin acts as a natural antioxidant.

Zanchet *et al.* [30] adopted the use of a new material from sugar cane as an alternative activator in the replacement of zinc oxide and stearic acid in NR compounds. According to the authors, both the favorable results obtained and the considerable reduction of zinc content (74.4%) used in the compounds, with similar results compared to the standard sample, highlighted the possibility of changing conventional activators such as zinc oxide to a green option from sugar cane.

Concerning the environmental problems related to zinc, water pollution by the zinc contained in rubber formulations mainly occurs indirectly because the disposal of waste in the soil (mainly due to the natural wear of tires) is of great concern to the environment because of the contamination of surface water bodies or groundwater. Human health, aquatic fauna, and flora are polluted by inorganic pollutant residues. Organisms may also be sensitive to toxic actions, such as bioaccumulation, affecting the food chain and endangering the organisms located at the top of the chain [45, 46].

In a posterior study, Zanchet *et al.* [47] evaluated the thermal degradation behavior of a typical NR compound modified by a silanized activator derived from sugarcane (green additive). This activator was used to replace the conventional zinc oxide and stearic acid additives commonly used as vulcanization activators. The sample containing 3 phr of the green additive presented the highest kinetic parameter values combined with thermal weight loss parameters at different temperatures among the others analyzed. This sample had 81% less amount of zinc compared to the Control sample, a very promise result that pointed out to the viability of this green additive in new applications in the elastomer industry, resulting in eco-friendly materials.

7.9 Conclusions

Humans have always sought natural resources to meet their needs. Human evolution is directly linked to the evolution of materials. Alternative materials have emerged as substitutes for the depletion of natural resources, such as fossil fuels, as well as the exchange of conventional materials that are harmful to human health and the environment.

This study sought to show and discuss the importance of exchanging conventional additives widely used in the production of elastomeric compounds with other possibilities from renewable sources in the search for eco-friendly and sustainable consumer goods. These additives, in addition to protecting the environment and human health, still have the advantage of conserving non-renewable natural resources (such as oil, which is widely used as a raw material in the production of conventional additives) and reducing the cost of the final elastomeric product. The literature clearly demonstrates that there is still much to be studied, seeking to deeply understand the influence of such additives in the processing, vulcanization reaction, and final properties of the finished vulcanized compounds. But it can be concluded, without a doubt, that the materials from natural sources have advantages over synthetics because they are from abundant renewable sources.

References

1. de Sousa, F.D.B., Vulcanization of natural rubber: Past, present and future perspectives, in: *Natural Rubber: Properties, Behavior and Applications*, J.L. Hamilton (Ed.), pp. 47–88, Nova Science Publishers, New York, 2016.
2. Petchkaew, A., *Implications of non-carcinogenic pah-free extender oils in natural rubber based tire compounds*, Enschede, University of Twente, 2015.
3. Joseph, R., Alex, R., Madhusoodanan, K.N., Premalatha, C.K., Kuriakose, B., Use of epoxidized rubber seed oil as a coupling agent and a plasticizer in silica-filled natural rubber compounds. *J. Appl. Polym. Sci.*, 92, 3531–3536, 2004.
4. Gujel, A.A., Bandeira, M., Giovanela, M., Carli, L.N., Brandalise, R.N., Crespo, J.S., Development of bus body rubber profiles with additives from renewable sources: Part II - Chemical, physical-mechanical and aging characterization of elastomeric compositions. *Mater. Des.*, 53, 1119–1123, 2014.
5. Gujel, A.A., Bandeira, M., Veiga, V.D., Giovanela, M., Carli, L.N., Mauler, R.S., *et al.*, Development of bus body rubber profiles with additives from renewable sources: Part I - Additives characterization and processing and cure properties of elastomeric compositions. *Mater. Des.*, 53, 1112–1118, 2014.
6. Bezerra, A., Santos, A.C., Costa, H., Ramos, V.D., Efeito do óleo de linhaça e do óleo de amendoim sobre a vulcanização da borracha natural (NR). Parte I: Modelo generalizado. *Polimeros-Ciencia E Tecnol.*, 23, 395–401, 2013.

7. da Costa, H.M., Ramos, V.D., Furtado, C.R.G., Efeito do óleo de mamona em composições de borracha natural contendo sílica. *Polimeros-Ciencia E Tecnol.*, 14, 46–50, 2004.
8. Dasgupta, S., Agrawal, S.L., Bandyopadhyay, S., Mukhopadhyay, R., Malkani, R.K., Ameta, S.C., Eco-friendly processing oils: A new tool to achieve the improved mileage in tyre tread. *Polym. Test.*, 28, 251–263, 2009.
9. Dasgupta, S., Agrawal, S.L., Bandyopadhyay, S., Chakraborty, S., Mukhopadhyay, R., Malkani, R.K., *et al.*, Characterization of eco-friendly processing aids for rubber compound. *Polym. Test.*, 26, 489–500, 2007.
10. Zanchet, A., Garcia, P.S., Nunes, R.C.R., Crespo, J.S., Scuracchio, C.H., 10–19 Sustainable natural rubber compounds: Naphthenic oil exchange for another alternative from renewable source. *Int. Refereed J. Eng. Sci.*, 4, 10–19, 2016.
11. Dasgupta, S., Agrawal, S.L., Bandyopadhyay, R., Mukhopadhyay, S., MalkaniI, R.K., Ameta, S.C., Characterization of eco-friendly processing aids for rubber compounds: Part II. *Polym. Test.*, 27, 277–238, 2008.
12. Ismail, H., Salmiah, I., Tsukahara, Y., Palm oil fatty acid as an activator in carbon black filled natural rubber compounds: Effect of vulcanization system. *Polym. Int.*, 44, 523–529, 1997.
13. Raju, P., Nandanan, V., Sunil, K.N.K., A study on the use of coconut oil as plasticiser in natural rubber compounds. *J. Rubber Res.*, 10, 1–15, 2007.
14. Rahman, M. and Brazel, C.S., The plasticizer market: An assessment of traditional plasticizers and research trends to meet new challenges. *Prog. Polym. Sci.*, 29, 1223–1248, 2004.
15. Madaleno, E., Rosa, D.D.S., Zawadzki, S.F., Pedrozo, T.H., Ramos, L.P., Estudo do uso de plastificantes de fontes renovável em composições de PVC. *Polimeros-Ciencia E Tecnol.*, 19, 263–270, 2009.
16. Bocqué, M., Voirin, C., Lapinte, V., Caillol, S., Robin, J.J., Petro-based and bio-based plasticizers: Chemical structures to plasticizing properties. *J. Polym. Sci., Part A: Polym. Chem.*, 54, 11–33, 2015.
17. Vieira, M.G.A., Da Silva, M.A., dos Santos, L.O., Beppu, M.M., Natural-based plasticizers and biopolymer films: A review. *Eur. Polym. J.*, 47, 254–263, 2011.
18. Niska, K.O. and Sain, M., *Wood-Polymer Composites*, CRC Press, New York, 2008.
19. Santos, M., Lima, O., Nassar, E., Ciuffi, K., Calefi, P., Study of the storage conditions of the sugarcane bagasse through thermal analysis. *Quím. Nova*, 34, 507–511, 2011.
20. Li, F., Hanson, M., Larock, R., Soybean oil–divinylbenzene thermosetting polymers: synthesis, structure, properties and their relationships. *Polymer*, 42, 1567–1579, 2001.
21. Garcia, P.S., de Sousa, F.D.B., de Lima, J.A., Cruz, S.A., Scuracchio, C.H., Devulcanization of ground tire rubber: Physical and chemical changes after different microwave exposure times. *eXPRESS Polym. Lett.*, 9, 1015–1026, 2015.

22. Fernandez, S.S., Kunchandy, S., Ghosh, S., Linseed oil plasticizer based natural rubber/expandable graphite vulcanizates: Synthesis and characterizations. *J. Polym. Environ.*, 23, 526–533, 2015.
23. Jayewardhana, W.G.D., Perera, G.M., Edirisinghe, D.G., Karunanayake, L., Study on natural oils as alternative processing aids and activators in carbon black filled natural rubber. *J. Natl. Sci. Found. Sri Lanka*, 37, 187–193, 2009.
24. Rocha, E.C., Lovison, V.M.H., Pierozan, N.J., *Tecnologia de transformação dos elastômeros*, 2nd ed., SENAI-CETEPO, Centro Tecnológico de Polímeros, São Leopoldo, 2003.
25. Silva, I.B., Desenvolvimento de formulação de compostos elastoméricos, SENAI-CETEPO, Centro Tecnológico de Polímeros, São Leopoldo, 2005.
26. Leblanc, J.L., Rubber-filler interactions and rheological properties in filled compounds. *Prog. Polym. Sci.*, 27, 627–687, 2002.
27. da Costa, H.M., Visconte, L.L.Y., Nunes, R.C.R., Furtado, C.R.G., Cinética de vulcanização de composições de borracha natural com incorporação de cinza de casca de arroz. *Polimeros-Ciencia E Tecnol.*, 13, 102–106, 2003.
28. Hubbe, M.A., Rojas, O.J., Lucia, L.A., Sain, M., Cellulosic nanocomposites: A review. *BioResources*, 3, 929–980, 2008.
29. Poletto, M., Zattera, A.J., Santana, R.M.C., Thermal decomposition of wood: Kinetics and degradation mechanisms. *Bioresour. Technol.*, 126, 7–12, 2012.
30. Zanchet, A., de Sousa, F.D.B., Crespo, J.S., Scuracchio, C.H., Activator from sugar cane as a green alternative to conventional vulcanization additives. *J. Clean. Prod.*, 174, 437–446, 2018.
31. Carli, L.N., Roncato, C.R., Zanchet, A., Mauler, R.S., Giovanela, M., Brandalise, R.N., *et al.*, Characterization of natural rubber nanocomposites filled with organoclay as a substitute for silica obtained by the conventional two-roll mill method. *Appl. Clay Sci.*, 52, 56–61, 2011.
32. Asim, M., Abdan, K., Jawaid, M., Nasir, M., Dashtizadeh, Z., Ishak, M.R., *et al.*, A review on pineapple leaves fibre and its composites. *Int. J. Polym. Sci.*, 2015, 1–17, 2015.
33. Ornaghi, H.L., Poletto, M., Zattera, A.J., Amico, S.C., Correlation of the thermal stability and the decomposition kinetics of six different vegetal fibers. *Cellulose*, 21, 177–188, 2014.
34. Veigel, S., Muller, U., Keckes, J., Obersriebnig, M., Gindl-Altmutter, W., Cellulose nanofibrils as filler for adhesives: Effect on specific fracture energy of solid wood-adhesive bonds. *Cellulose*, 18, 1227–1237, 2011.
35. Castellano, M., Gandini, A., Fabbri, P., Belgacem, M.N., Modification of cellulose fibres with organosilanes: Under what conditions does coupling occur? *J. Colloid Interface Sci.*, 273, 505–511, 2004.
36. Trovatti, E., Carvalho, A.J.F., Gandini, A., A new approach to blending starch with natural rubber. *Polym. Int.*, 64, 605–610, 2015.
37. Chirinos, H., Yoshii, F., Makuuchi, K., Lugao, A., Radiation vulcanization of natural rubber latex using 250 keV electron beam machine. *Nucl. Instrum. Methods Phys. Res., Sect. B*, 208, 256–259, 2003.

38. Haque, M.E., Dafader, N.C., Akhtar, F., Ahmad, M.U., Radiation dose required for the vulcanization of natural rubber latex. *Radiat. Phys. Chem.*, 48, 505–510, 1996.
39. Hill, D.J.T., O'Donnell, J.H., Perera, M.C.S., Pomery, P.J., Smetsers, P., Mechanism of radiation vulcanization of natural rubber latex sensitized by monoacrylates. *J. Appl. Polym. Sci.*, 57, 1155–1171, 1995.
40. Alam, M.N., Mandal, S.K., Roy, K., Debnath, S.C., Synergism of novel thiuram disulfide and dibenzothiazyl disulfide in the vulcanization of natural rubber: curing, mechanical and aging resistance properties. *Int. J. Ind. Chem.*, 5, 8, 2014.
41. Spiegelhalder, B., Carcinogens in the workroom air in the rubber industry. *Scand. J. Work Environ. Health*, 9, Suppl 2, 15–26, 1983.
42. Mariano, R.M., Visconte, L.L.Y., Oliveira, M.R.L., Rubinger, M.M.M., Avaliação de bis(4-metilfenilditiocarbimato)zincato(II) de tetrabutilamônio como acelerador em composições de borracha natural. *Polimeros-Ciencia E Tecnol.*, 18, 297–301, 2008.
43. Alam, M.N., Mandal, S.K., Debnath, S.C., Bis(N-benzyl piperazino) thiuram disulfide and dibenzothiazyl disulfide as synergistic safe accelerators in the vulcanization of natural rubber. *J. Appl. Polym. Sci.*, 126, 1830–1836, 2012.
44. Abad, L.V., Relleve, L.S., Aranilla, C.T., Aliganga, A.K., San Diego, C.M., Dela Rosa, A.M., Natural antioxidants for radiation vulcanization of natural rubber latex. *Polym. Degrad. Stab.*, 76, 275–279, 2002.
45. Gualtieri, M., Andrioletti, M., Vismara, C., Milani, M., Camatini, M., Toxicity of tire debris leachates. *Environ. Int.*, 31, 723–730, 2005.
46. Nelson, S.M., Mueller, G., Hemphill, D.C., Identification of tire leachate toxicants and a risk assessment of water quality effects using tire reefs in canals. *Bull. Environ. Contam. Toxicol.*, 52, 574–581, 1994.
47. Zanchet, A., Demori, R., de Sousa, F.D.B., Ornaghi, H.L., Schiavo, L.S.A., Scuracchio, C.H., Sugar cane as an alternative green activator to conventional vulcanization additives in natural rubber compounds: Thermal degradation study. *J. Clean. Prod.*, 207, 248–260, 2019.

8

Natural Rubber Recycling: Advances, Limitations, and Applications

Fabiula Danielli Bastos de Sousa[1]* and Aline Zanchet[2]

[1]*Technology Development Center, Universidade Federal de Pelotas, Rua Gomes Carneiro, Pelotas, Brazil*
[2]*Universidade de Caxias do Sul, Rua Francisco Getúlio Vargas, Caxias do Sul, RS, Brazil*

Abstract
Owing to environmental problems and public health problems caused by the inadequate disposition of end-of-life vulcanized rubbers, especially natural rubber, which is widely used, their recycling has been encouraged by several sectors, seeking to minimize the above-mentioned problems, as well as to assist in sustainable development and to reduce the production cost of rubber goods. In this study, some rubber recycling methods are discussed, including their advantages and disadvantages. The final properties of some recycled rubbers devulcanized by different methods will be analyzed based on studies available in the literature, emphasizing the importance of deep knowledge of the parameters used in each process to obtain materials with low levels of degradation.

Keywords: Recycling, devulcanization, microwaves, eco-friendly, vulcanization, degradation

8.1 Introduction and General Aspects

Natural rubber (NR), an elastomer naturally produced through biosynthesis by the *Hevea brasiliensis* tree, is considered an engineering material [1]. NR are polymeric macromolecules formed by natural hydrocarbons that have mobility and movement when subjected to the action of a force [2]. It was discovered by native Americans, who have since then

**Corresponding author*: fabiuladesousa@gmail.com

Visakh. P.M. and Shiv Kumari Panda (eds.) Natural Rubber-Based Composites and Nanocomposites, (243–270) © 2025 Scrivener Publishing LLC

used NR as balls in organized games, as well as waterproofing in clothes and footwear, as flexible primitives containers and syringes [3]. Ancient civilizations from Central and South America used NR in rituals and as incense [4]. However, this rubber presented some problems: when it was solid, it was difficult to work, and when subjected to heat, the artifacts became soft and sticky. In cold weather, it becomes increasingly tough and rigid, and is almost completely inflexible during winter [3].

Such problems already verified by ancient civilizations when using NR are due to its predominantly plastic characteristics when in the raw state. In this state, it does not present useful properties, so the material cannot be used in technological applications. From vulcanization or curing, a complex process composed of many consecutive reactions with the joint performance of a wide range of additives previously incorporated into the elastomer (formulation), the elastomer changes from a predominantly plastic state to a predominantly elastic state owing to the cross-linking formation of its macromolecules.

Vulcanization is the name given to the process when the additive responsible for the cross-linking is sulfur (in this work, the vulcanization term is used, regardless of the additive used). It is a word derived from Roman mythology (*Vulcano*, God of fire and work with metals) [3]. The process was casually discovered by Charles Goodyear, who dropped a small amount of sulfur on heated rubber and perceived a clear variation in its behavior. This discovery is also due to another researcher, Thomas Hancock, at the same time, but in different places.

The result of vulcanization is that the elastomer becomes a three-dimensional network. The cross-linkings (intermolecular bonds) resulted from the process anchor the polymeric chains, free to move up to this point, being that, as a consequence, the vulcanized elastomer gains elasticity due to the "memory effect" generated by the cross-linkings and, thus, becoming predominantly elastic.

However, at the same time that vulcanization brings many benefits to the elastomer and provides its use in a wide range of applications, it also causes difficulties to recycle the vulcanized elastomer, since its polymeric chains "lose" mobility as a result of the vulcanization.

Vulcanized elastomers are materials that require long periods of time to degrade naturally because of their cross-linked structure, the presence of stabilizers and other additives in their formulation [5], making them infusible and difficult to reprocess. In addition, goods made of vulcanized elastomers, such as tires, are considered high value-added materials, since they possess on their formulation a range of chemical additives, which makes their disposal a waste of valuable materials.

In particular, end-of-life tires, when disposed in landfills, because of their large size and format, may be accumulators of rainwater, becoming a place able to proliferate vectors, such as the *Aedes aegypti* mosquito, transmitter of dengue, chikungunya, yellow fever, and zika, causing serious environmental and public health problems [6].

Thus, to minimize such problems and seek sustainable development, the recycling of vulcanized elastomeric materials has been encouraged, both through the many works discussed by the academy and by the creation of new laws in several countries addressing the subject. Recycling consists of a sustainable action, as it saves raw materials (often polymers derived from oil, which is a non-renewable natural resource [7]) and energy, does not harm the environment, and is still a source of income for many families who survive from the collection of recyclable materials [6]. Recycling is considered a category of green chemistry ("Green chemistry is the design of chemical products and processes that reduce or eliminate the generation of hazardous substances" [8]), i.e., the use of renewable or recycled material sources.

There are basically four different levels of vulcanized elastomer recycling: material recycling, monomer recycling, energy recycling, and fuel recycling. The names of the different levels are self-explanatory. Some recycling methods are briefly analyzed in the sequence.

8.2 Levels of Vulcanized Rubber Recycling

In certain aspects, energy recycling can be regarded as a waste. Approximately 128 J of energy is consumed to make a kilo of synthetic rubber tire. In contrast, the calorific power reclaimed by its burning is 27–33 J/kg, which is not much higher than that of low-cost coal, i.e., valuable rubber material is lost. Moreover, the burning of tires for energy production generates atmospheric pollution [9]. Toxic gases such as SO_2, H_2S, HCl, and HCN are generated, thus requiring additional systems for proper treatment [10].

In some countries, associations responsible for the collection and environmentally correct disposal of the end-of-life tires use them as replacement fuel in cement industries, among other forms of use [11, 12]. In Brazil, for example, in Reciclanip [13], a non-profitable association, a national reference in the collection, and environmentally correct disposal of end-of-life tires, approximately 70% of them are used as alternative fuel in cement ovens, by their calorific power, in replacement of petroleum coke [12].

The literature presents a number of different ways to perform material recycling with each method presenting advantages and disadvantages. Some of these aspects are discussed below.

The grinding of vulcanized rubbers, the direct use of the ground material, and its incorporation into other materials are the simplest ways to use the material. Among the applications, there is the possibility of application on the surfaces of sports courts, incorporation into asphalt and concrete, and the use of the powder in new compositions, among other forms of reuse. Among the alternatives mentioned, the only one used in considerable quantities is asphalt [14, 15]. The pavement exhibits increased durability and becomes more resistant to oxidation because the asphalt layer is thicker and the antioxidants and carbon black present in the rubber contribute to the improvement of the aging resistance. Pavement has better performance, is softer, and reduces noise [14–16].

On the whole, manuscripts about rubber recycling by the addition of vulcanized rubbers in the concrete usually analyze the workability of the mixture, i.e., the concrete's ability to be easily placed and homogeneously compacted in a specific place. Depending on the properties and size of the particles of the rubber added to the concrete, this ability could be improved or not, since vulcanized rubbers have high viscosity and do not flow. Many studies have focused on rubber reclaiming as a building and paving material [17–36].

Another promising application is the use of granulated vulcanized elastomers in equestrian soils [37]. The granules, incorporated into other materials and sand, provide horse cushioning during the race and impulse in jumps, due to the elasticity and flexibility of the material, so muscle and joint problems can be minimized in the animals. Unlike pure sand, the compound does not produce such an amount of dust during the exercise of the horses, in addition to avoiding excessive wear of horseshoes. The same can also be applied to the comfort of other large animals such as cattle.

Similarly, athletes of different sports have demonstrated the advantages of recycled rubber in multi-sport courts because of characteristics such as elasticity, shock absorption, drainage, and high strength [38]. However, some authors [39] claimed that the surfaces of pavements and playgrounds, where vulcanized rubbers are widely used, contain hazardous organic chemicals that can cause serious health problems.

There is also the possibility of producing polymeric blends from ground rubber. Polymeric blends are materials originating from a physical mixture of two or more polymers with different constitutional or configurational characteristics and have a low degree of chemical bonding between them. This is an economically viable and widely used route to improve the properties of existing polymeric materials. The properties of blends containing recycled polymers depend on the concentration of the recycled material, as well as the adhesion between the phases [40, 41].

Thus, it is necessary to create an important addendum. For elastomers, it is essential to know their vulcanization or cure behavior, which is accomplished through the analysis of vulcanization curves, resulting from the analysis of elastomeric formulations in specific rheometers, where important information is observed, vital to the processing of elastomeric formulations.

Figure 8.1 presents a typical vulcanization curve of the elastomers. It is divided into three major regions: (i) scorch time, induction, or reaction delay period (the period prior to the beginning of the reaction itself; most accelerator reactions occur [42]), (ii) cure or vulcanization (the period in which the network structure is formed [42]), and (iii) overcure (the period after the end of the reaction). During the third period, the network matures by overcuring reversion, equilibrium, or additional but slower cross-linking, depending on the nature of the compound [42]. The values obtained from the analysis are as follows:

- ML or R_{min}: the minimum torque. This parameter is related to the Mooney viscosity of a material;
- MH or R_{max}: maximum torque. This represents the final vulcanization state of rubber, which is proportional to the cross-linking density formed [43];
- ts_1: scorch time (time needed for the beginning of the reaction), also called scorch safety;
- t_{90}: time necessary to reach 90% of the total vulcanization (MH-ML), the optimum time of vulcanization [44–46]. This is the stage in which the physical properties yield optimal results [47].

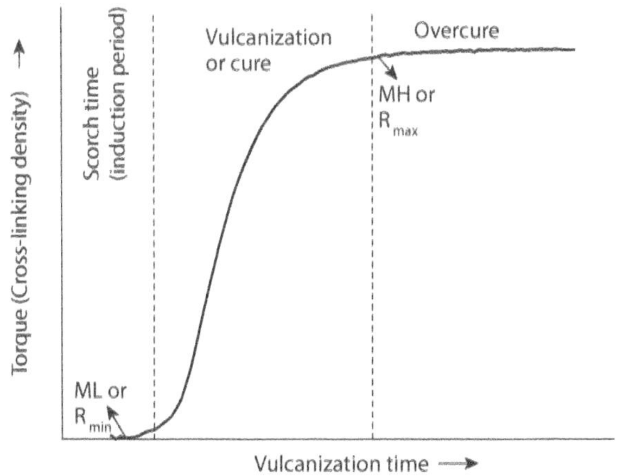

Figure 8.1 Rheometer chart showing the vulcanization characteristics of elastomers [6].

Overcure reversion may be ascribed to the maturation/degradation of the polysulfidic cross-linkings initially formed (which are more unstable) [48, 49]. Reversion may also be caused by the depolymerization of the polymeric chains. This phenomenon leads to a reduction in the crosslinking density and physical properties of the material [50]. This behavior is commonly observed in NR.

De Sousa and Zanchet [7] proposed the study of additional heating during compression molding to verify the vulcanization reversion of raw NR/recycled NR (containing different concentrations of the recycled phase), as well as to analyze the influence of vulcanization reversion as a result of the additional heating during compression molding on the mechanical properties of the samples. The recycled vulcanized NR was ground at room temperature using a knife mill. Because NR is strongly influenced by temperature, it is essential to determine the influence of additional heating on the mechanical properties of the final vulcanized materials.

Figure 8.2 depicts the correlation between reversion and the trend of the mechanical properties of all the blends, showing that the additional heating and, consequently, the vulcanization reversion influenced their final properties.

According to some authors [51], ground particles of vulcanized rubbers in the range of 28 to 35 mesh (0.425–0.600 mm) are ideal for incorporation

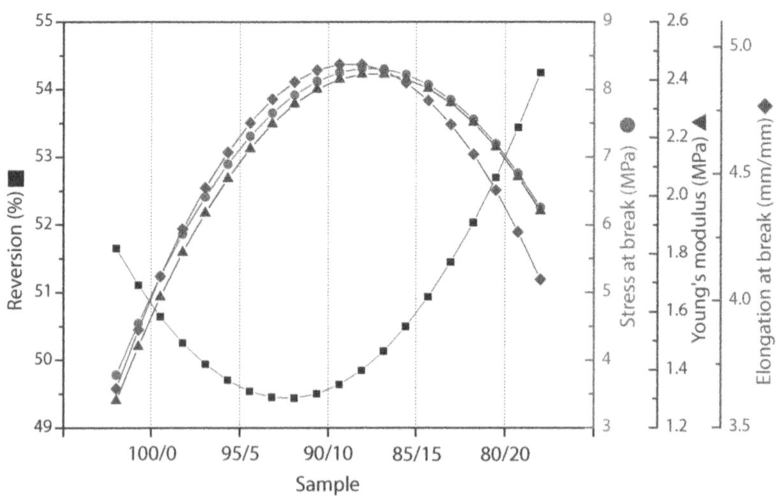

Figure 8.2 Influence of the phase concentration on the reversion and mechanical properties of the blends raw NR/recycled NR (% mass). Modified from [7].

into compounds. Milling at room temperature is a low-cost process that produces rough and irregular particles [52]. Recycled rubber powder contains voids in which the elastomeric matrix is trapped [53], which improves the mechanical properties of the blends or compounds containing recycled rubbers. This behavior was observed by Weber *et al.* [54] for ethylene-propylene-diene monomer rubber (EPDM) and styrene-butadiene rubber (SBR), as well as by de Sousa and Zanchet [7] for NR.

The results of this research [7] pointed to the potential application of ground NR to produce polymeric blends in the industry itself, owing to the satisfactory results obtained. On the other hand, according to the authors, "It is really important to know the consequences of additional heating in the final compounds, since the physical properties are degraded and energy is wasted, so the production process is not sustainable. It does not do any good to recycle a material, trying to contribute to the environment, if the process is not environmentally correct (unnecessary waste), resulting in a final product with no desirable properties."

Rattanasom *et al.* [55] studied the vulcanization characteristics and mechanical properties of polymeric blends raw NR/recycled NR (rubber powder). The results indicated that the scorch time and optimum vulcanization time decreased with an increase in the incorporated residue content. Regarding the mechanical properties, an increase in the modulus at 100% and 300% elongation was observed because of the increase in elastomeric stiffness resulting from the increase in the residue concentration. Consequently, the elongation at break decreased with increasing residue content. Owing to the weak interaction between the particles and elastomeric matrix, the tensile strength also decreased in compounds containing high residue contents.

Polymeric blends of raw NR and different concentrations of recycled rubber waste (rubber powder), product from the sanding process (polishing) of rubber balls and artificial eggs, were prepared and analyzed by Ismail *et al.* [56]. The presence of the residues accelerated the vulcanization reaction of the blends through a reduction on the scorch time. As observed in a previous study [55], this behavior is typical for recycled rubbers, occurring usually by the presence of unreacted additives from the initial vulcanization, or by sulfur diffusion from the rubber matrix phase to the recycled phase, which lowers the concentration of sulfur in the rubber matrix. In general, the results of the mechanical properties and morphology depicted the agglomeration of the recycled phase and its lack of adhesion in the matrix phase, both of which increased with an increase in the recycled phase in the blends. Therefore, these results highlight the importance of knowing the influence of the recycled phase in the polymeric blends to determine the optimum concentration to be used without general depreciation of their properties.

However, polymeric blends with at least one of the phases of a recycled elastomer, as previously mentioned, even though it is known that the surface of the ground vulcanized rubber is irregular, which facilitates its adhesion to the raw phase, it is hinted due to the cross-linkings present in the recycled phase, since they do not allow the movement of polymeric chains and harm the processability of the polymeric blend. Thus, the amount of recycled material that can be incorporated without degrading the mechanical properties of the blend is far below ideal.

A recycling method that can reduce this problem is devulcanization, which is described in the following sequence.

8.3 Devulcanization

Devulcanization is the process of cleaving total or partial sulfidic crosslinks formed during the initial vulcanization [57–60]. It is used to regain the fluidity of the rubber since the chains are free to move again. Thus, the recycled rubber can be remolded and revulcanized. The amount of recycled material used in polymeric blends, when devulcanized, can be higher without significant reduction in the mechanical properties of the final blend, and often the result is superior to neat raw rubber control.

As the devulcanized rubber can flow during processing, owing to the stresses applied in the various mixing equipment, the dispersed phase usually formed by the recycled rubber can break down into smaller particles, resulting in blends with more refined phase morphologies. Thus, the adhesion tends to increase because of the greater mobility of the chains of the recycled polymer. Furthermore, it increases the interaction between phases and the tension transfer between the phases during a mechanical request, resulting in superior mechanical properties to the same blend containing vulcanized recycled elastomer.

The literature presents a great range of studies addressing the different methods to devulcanize rubber, namely mechanical and mechanochemical, physical [60–86] (by microwaves [1, 59, 87–107], ultrasound [58, 108–141]), chemical [63, 142–154], and microbial [155–164]. Some available devulcanization methods are briefly introduced as follows.

In mechanical, mechano-chemical, or thermo-mechanical devulcanization, vulcanized rubber is subjected to shear and/or elongational stresses on suitable equipment such as mills and twin-screw extruders.

The characteristics of recycled rubber are strongly dependent on the processing equipment, its characteristics (e.g., local shear rate), and processing parameters (e.g., residence time) [165].

Ultrasound devulcanization is the breaking of the three-dimensional network of vulcanized rubber by the action of ultrasonic waves at certain levels in the presence of heat and pressure [166]. This process is based on the cavitation phenomenon caused by the acoustic wave propagation in a solid body. During the negative half-pressure cycle, the material is subjected to triaxial tensile stress and, during the positive half-cycle experience, to compression. One of the several types of equipment developed by Isayev [122, 125] comprises an ultrasound system connected to a single-screw extruder die. The material to be devulcanized is fed as pellets or grains in an extruder and forced to pass through the die where the ultrasound applicator is located. It is very fast and occurs in the order of a second or less [112].

Devulcanization by microwaves is considered an eco-friendly process because of its physical nature, i.e., it does not involve chemicals. The process takes advantage of the volumetric heating of the material by microwaves, promoting more uniform heating than that achieved by the more traditional methods of heating, which depend on conduction and/or convection [104, 105]. However, it is important to note that the capacity of rubber to be devulcanized by microwaves depends on its microwave absorption capacity. In rubbers such as NR, SBR, and EPDM, this capacity is low and can be improved by the incorporation of another material in the formulation with a higher microwave absorption capacity. Carbon black, a filler widely used in rubber goods such as tires, presents such ability for being a conductive filler [125, 167], that induces a phenomenon known as Maxwell–Wagner polarization [168]. In one of the possible apparatus used to devulcanize rubbers by microwaves, it is used a modified domestic microwave oven, adapted with a motorized stirring system with a speed control [1, 54, 90, 93–96, 98, 100, 104–106]. However, in this case, a disadvantage is that it is a batch process, in which small amounts of rubber are usually devulcanized per batch.

Chemical devulcanization is, as its name states, devulcanization through the use of chemicals. According to De *et al.* [146], a large number of chemical reclaiming agents, such as diphenyl disulfide, dibenzyl disulfide, diamyl disulfide, bis(alkoxy-aryl) disulfides, butyl mercaptans, thiophenols, xylene thiols, phenol sulfides and disulfides, iron oxide phenyl hydrazine-based catalysts, and copper chloride-tributyl amine catalysts, have been used for the treatment of scrap ground rubber crumbs or powders at elevated temperatures. However, this method is not widely used due to its lack of efficiency, since it produces only partial devulcanization and uses many dangerous chemicals to human health and to the environment, many of which are carcinogens. According to Isayev *et al.* [122], chemical recycling methods such as chemical devulcanization are usually very slow. In addition, "it creates a problem with the removal of the solvent and the additional waste that is generated in the forms of sludges" and thus, the total cost could be increased. In addition, chemical devulcanization methods may cause the release of poisonous gas [60].

Microbial devulcanization is the method in which the process is performed through microbial metabolism. It is usually known as desulfurization since the amount of sulfur in the devulcanized rubbers tends to be reduced in comparison to the vulcanized rubber. Desulfurization of ground rubber is performed by bacteria and fungi, which are capable of converting cross-linked sulfur to simple sulfur compounds or inorganic sulfates [161]. This method is considered eco-friendly because it does not use chemicals or energy. On the other hand, some disadvantages are that the process is slow, only superficial devulcanization/desulfurization occurs, some additives may inhibit the function of bacteria [165], and the culture responsible for the process may be reduced due to toxicity. Frequently, harmful additives are leached or extracted using solvents before the start of the process to avoid problems.

Overall, these available methods provide energy to destroy the sulfidic bonds originating from the initial vulcanization process. However, the parameters involved in the process must be carefully analyzed, since each bond requires a specific amount of energy, namely, C–S: 285 kJ/mol, S–S: 268 kJ/mol, S_X: 251 kJ/mol, and C–C: 346 kJ/mol [59, 71, 169].

To avoid degradation of the main chains of the rubber, it is necessary to provide only the required amount of energy to break the cross-links. The literature presents some studies in which the parameters involved in the devulcanization process are changed, and the final properties of the devulcanized rubbers are analyzed. Some of these are briefly discussed below.

As a consequence of the partial destruction of the three-dimensional network formed by cross-linking in vulcanized rubbers, the material becomes partially soluble (sol content >0%) and shows a lower cross-linking density than initially [165]. This is one of the typical characteristics of devulcanized rubber, which precisely presents a soluble phase (sol) and another insoluble phase (gel), thus presenting distinct properties to that of raw rubber.

Therefore, in order to analyze de quality of the employed process, some analytical techniques are widely used by researchers. The sol and gel content are measured by extraction techniques by using specific solvents for each rubber, and one of the extraction widely used is the Soxhlet one. By using the Soxhlet apparatus, the sol fraction can be separated (extracted) from the gel fraction. Another useful technique is the swelling by solvent. In this case, pressed samples are put in contact to the specific solvent during a period (according to standards), being that the swollen mass is measured and compared to the dry mass (before swelling). So, it is possible to obtain the swelling degree, as well as the cross-linking density by using Flory-Rehner equation [170].

De Sousa and Scuracchio [1] studied the influence of the amount of carbon black on devulcanization of NR by microwaves. Formulations of NR containing 0, 20, 40, 60, and 80 parts per hundred rubber (phr) of carbon black were analyzed. As a result, the degree of devulcanization increased as a function of the amount of carbon black present in the rubber and as a function of increase in the exposure time of NR to the microwaves. The final temperature reached by the sample is the determining factor for the degree of devulcanization, which is both dependent on the exposure time of the sample to the microwaves and the amount of carbon black present on it. Similar results were obtained by other authors by studying the influence of the amount of carbon black on the SBR [87–89].

Hong and Isayev [166] studied the influence of carbon black content on NR devulcanization by ultrasound and by chemicals and ultrasound. Contrary to what was observed by de Sousa and Scuracchio [1] in microwave devulcanization, the gel fraction and cross-linking density were not significantly affected by carbon black content. According to the authors, this result is due to the competition between cross-linking breakage and reformation during the ultrasound treatment.

Ghorai *et al.* [171] devulcanized NR *via* a mechano-chemical method using bis(3-triethoxysilyl propyl) tetrasulfide (TESPT) as a devulcanizing agent in a two-roll mill. The authors analyzed the influence of different milling times on the devulcanization degree of the samples. The sol content, glass transition temperature and heat capacity increased with milling time, indicating that the devulcanization degree increased with milling time.

De *et al.* [146] studied the mechano-chemical devulcanization of NR by using a vegetable product and diallyl disulfide under various reclaiming conditions. Similar to that observed by Ghorai *et al.* [171], the sol content increased as the milling time increased. The molecular weight of the sol fraction increased as the milling time increased, i.e., degradation increased with increasing milling time. According to the authors, this is because of the mechanical shearing of the vulcanized rubber network and the action of devulcanization agents present during the milling process. In addition, the Mooney viscosity of the devulcanized rubber decreased with increasing milling time. The authors presented a schema showing the steps for the formation of low-molecular-weight compounds in the sol fraction as a result of degradation during the devulcanization process.

Jiang *et al.* [162] analyzed the microbial devulcanization/desulfurization of NR by *Thiobacillus ferrooxidans*. The results of X-ray Photoelectron Spectroscopy (XPS) showed that the S peak of the desulfurized NR decreased, indicating desulfurization. The results also showed that the amount of C-S and S-S bonds decreased, and more oxidized sulfur was produced on the surface of the desulfurized NR. In addition, an increase in the SO_4^{2-} concentration was observed, indicating that some of the crosslinks on the surface of the ground rubber were fully oxidized.

In addition, it is necessary to understand in a deeper way the changes that occur in the chemical structure of the samples as a result of the devulcanization process. Fourier transform infrared spectroscopy (FTIR) is a technique widely used in the literature because it is a very useful technique with easy preparation of the samples. Many authors have adopted this technique for the analysis of several devulcanized rubber types [87, 90, 93, 98].

De Sousa *et al.* [98] analyzed the properties of the ground tire rubber (GTR) devulcanized by microwaves under different exposure times. In general, the results showed that the longer the exposure time of GTR to microwaves, the greater the devulcanization degree (demonstrated by reducing the gel fraction and the density of cross-linking of the samples). The FTIR results showed that during the devulcanization process, bonds were broken, and others were formed. It was observed that the formation of new S-S bonds and S-O bonds in the samples that were more exposed to the microwaves was a possible rearrangement of the sulfur-free radicals from devulcanization. The authors proposed a schema to show the chemical modifications suffered by the samples as consequence of the recycling process (Figure 8.3).

As described above, although the devulcanization process causing an increase in the material moldability, it is known that the higher the devulcanization degree, the higher the number of broken main chain links of the polymer, i.e., the higher the degradation of the main chains of the recycled material. The effect of this degradation in the polymer is a significant reduction in rigidity and other mechanical properties when the material is revulcanized. Depending on the final use, this may hinder its use. Therefore, the choice of the devulcanization process parameters must

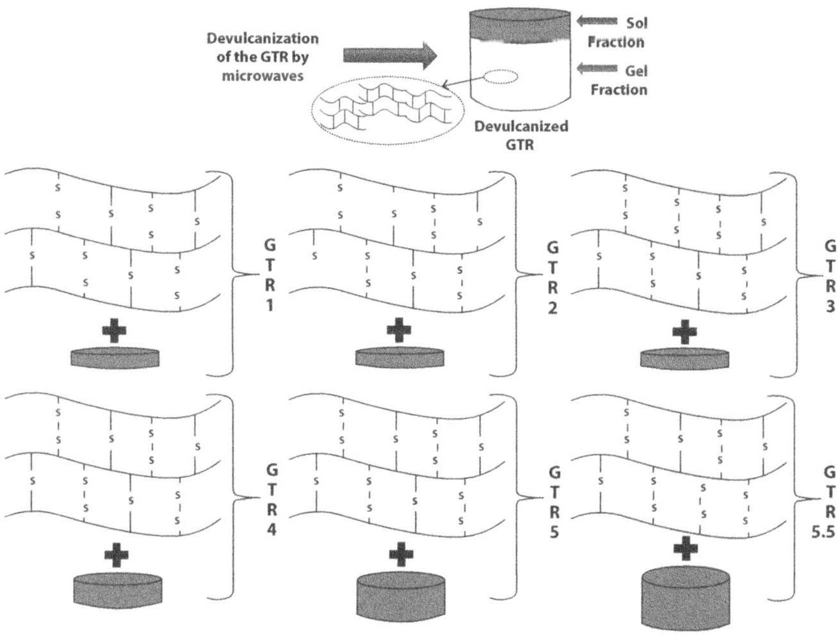

Figure 8.3 Scheme of structural changes (C–S and S–S bonds) and gel fraction as a function of the exposure time to the microwaves. Reprinted from [98] with permission from Elsevier.

consider the balance between the processability and mechanical properties of the final material [104].

Some authors [98] have analyzed various aspects involved in the devulcanization of rubber by microwaves, including the degradation that occurs during the process [98]. The study discusses a microwave devulcanization process for GTR. It is known that this rubber is formed by a natural rubber phase and another synthetic rubber, in this case, SBR. As previously mentioned, in the case of NR (a rubber with low microwave absorption capacity), devulcanization occurs owing to the presence of carbon black, a conductive filler that exhibits high microwave absorption capacity and induces the phenomenon known as Maxwell–Wagner polarization. Owing to its chemical structure and high affinity to carbon black, it is very susceptible to thermal degradation, which was depicted through thermo-oxidative degradation by the thermogravimetric analysis (TGA) results. According to them, it was observed that the mass loss percentage of each organic component of the GTR (processing oils and rubbers) tended to decrease as the exposure time of the GTR to the microwaves increased, especially for the NR phase, while the opposite trend was observed for the percentage of carbon black, showing that the degradation trend increased as the exposure time of the GTR to the microwaves increased.

In the same study [98], the authors proposed a study of the degradation of samples using Horikx's theory [172]. This theory investigates the break of the main chains of the sample (degradation), as well as the break of the cross-linkings (devulcanization). Figure 8.4 shows the results of Horikx's theory for GTR devulcanized by microwaves, where the nomenclature of the samples is of the type GTRX, in which 'X' represents the exposure time of the sample to the microwaves (in minutes). The results illustrated that the NR phases of all the samples were degraded at different levels, whereas in the SBR phase, only the cross-links were broken.

In conclusion, according to the authors [98], "It can be concluded that for the NR phase, in the samples more exposed to the microwaves, a random scission occurred, i.e., both breaking of the cross-links and the main chains took place. In other words, the higher the exposure time of the samples to microwaves and the higher the temperature achieved, the greater the degradation level of the NR, as corroborated by the FTIR and TGA results. However, the SBR phase was only subjected to the breaking of the cross-linkings, whatever the exposure time of the sample to the microwaves."

In another study [93], the authors demonstrated that the long exposure of GTR to microwaves changed its structure, including changes in morphology at the nanoscale. This change in morphology occurred because of the degradation of the NR phase, as described previously by de

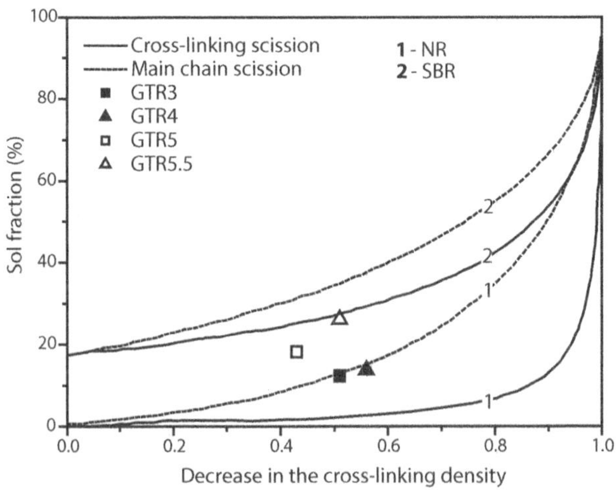

Figure 8.4 Sol fractions of GTR3, GTR4, GTR5, and GTR5.5 *versus* relative decrease in the cross-linking density. Reprinted from [98] with permission from Elsevier.

Sousa *et al.* [98]. In Figure 8.5, the darkest domains are the softer materials (NR phase), the clearest domains are the hardest materials (carbon black), and the middle-color domains are the SBR phase. The images (Figure 8.5) show the alteration of the morphologies as a consequence of devulcanization by microwaves, in which the darkest phase (NR) is consumed and the clearest phase (carbon black) increases as the exposure time of the GTR to the microwaves increases.

As described previously, by devulcanization, recycled rubber can be remolded and revulcanized. However, the studies discussed above clearly showed that different degradation levels occurred during the process, which may reduce the quality of the final material and its applicability. A commonly used way to verify the degradation occurred along the different devulcanization processes is the analysis of the revulcanization process, through the analysis of the vulcanization curve previously presented (Figure 8.1), as well as the study of some physical properties of the revulcanized samples, being that the study of mechanical properties is very usual. As the mechanical properties are the result of factors such as morphology, the morphology of the samples is also frequently analyzed using microscopes.

In the case of the revulcanization study using rheometry, the viscosity of the devulcanized rubber is generally lower than that of the raw rubber. However, the lower viscosity can be indicative of the degree of devulcanization and/or degradation of the rubber. Thus, the mechanical properties are generally analyzed together with the revulcanization study, and other analytical techniques can be employed.

De Sousa *et al.* [100] analyzed the influence of devulcanization by microwaves/degradation, as well as the additional heating during the compression molding in the blends composed of raw NR/NR devulcanized by microwaves. In conclusion, the high levels of reversion observed were the

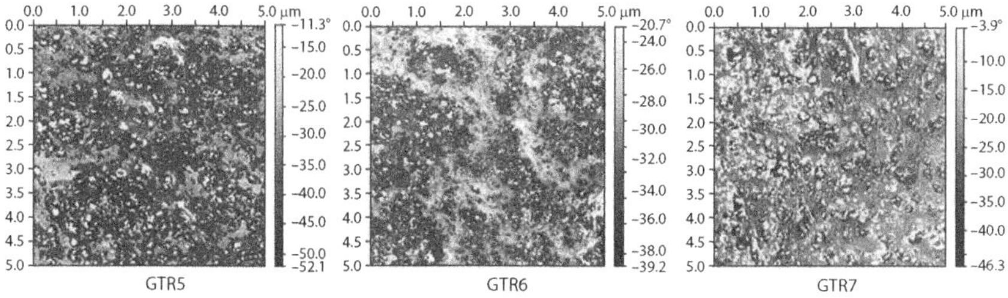

Figure 8.5 AFM tapping mode phase images of the samples. Reprinted from [93] with permission from Express Polymer Letters.

result of devulcanization/degradation of the recycled phase and additional heating.

De *et al.* [146] devulcanized NR *via* mechano-chemical method (as previously discussed) and analyzed the mechanical properties of the revulcanized samples. In general, it is observed that the tensile strength, elongation at break, and Shore A hardness were lower in the case of revulcanized samples when compared to those of the neat fresh vulcanized sample. According to them, in the case of the revulcanized samples, the swelling value was much higher, which means that the cross-linking density was very low. "Since tensile strength is a function of cross-linking density, tensile strength of revulcanized NR is lower due to its low cross-linking density (high swelling value)."

After analyzing the microbial devulcanization/desulfurization of NR by *T. ferrooxidans*, Jiang *et al.* [162] investigated the influence of its use as a filler in an SBR matrix. At 40 phr, SBR filled with desulfurized NR showed improvements in tensile strength and elongation at break. Scanning electron microscopy and dynamic-mechanical results indicated good interface adhesion between the desulfurized NR and SBR matrices.

Massey *et al.* [113] devulcanized NR by ultrasound and analyzed the devulcanized samples by Nuclear Magnetic Resonance (NMR). Based on the results, an important observation was depicted: "It appears that devulcanization degrades the network by detaching molecular fragments of at least intermediate size but leaves substantial amounts of dangling fragments."

De Sousa *et al.* [173] investigated the properties of revulcanized GTRs previously devulcanized by microwaves. In general, the results pointed out to modifications as result of the revulcanization process, such as in the chemical structure, composition, thermal stability, and morphology. Regarding degradation, the FTIR results showed a change in the chemical structure of the samples during the revulcanization process as a result of the partial oxidation of the main chains. The TGA results showed that the NR phase was even more degraded, as well as a significant decrease in the thermal stability of the SBR phase during revulcanization. A new peak, compared to the devulcanized samples, was observed, probably, due to the degradation processes.

8.4 Conclusions

This study was concerned with NR recycling, discussing the advances, limitations, and applications of some available techniques. Emphasis was given to the devulcanization process, and the importance of deep knowledge of the influence of the parameters used during processing on the final properties of the devulcanized rubber, with the aim of obtaining materials with low levels of degradation.

References

1. de Sousa, F.D.B. and Scuracchio, C.H., The role of carbon black on devulcanization of natural rubber by microwaves. *Mater. Research-Ibero-American J. Mater.*, 18, 791–797, 2015.
2. Rocha, E.C., Lovison, V.M.H., Pierozan, N.J., *Tecnologia de transformação dos elastômeros*, 2nd ed., SENAI-CETEPO, Centro Tecnológico de Polímeros, São Leopoldo, 2003.
3. da Costa, H.M., Visconte, L.L.Y., Nunes, R.C.R., Furtado, C.R.G., Aspectos históricos da vulcanização. *Polimeros-Ciencia E Tecnol.*, 13, 125–129, 2003.
4. Sethuraj, M.R. and Mathew, N.M., *Natural rubber: Biology, cultivation and technology*, Elsevier, Amsterdam, 1992.
5. Roche, N., Ichchou, M.N., Salvia, M., Chettah, A., Dynamic damping properties of thermoplastic elastomers based on EVA and recycled ground tire rubber. *J. Elastomers Plast.*, 43, 317–340, 2011.
6. de Sousa, F.D.B., Vulcanization of natural rubber: Past, present and future perspectives, in: *Natural Rubber: Properties, Behavior and Applications*, J.L. Hamilton (Ed.), pp. 47–88, Nova Science Publishers, New York, 2016.
7. de Sousa, F.D.B. and Zanchet, A., In the search for sustainable processing in compounds containing recycled natural rubber: The role of the reversion process. *Recycling*, 3, 47, 2018.
8. US EPA, www.epa.gov/greenchemistry (accessed March 2, 2018).
9. Isayev, A.I., Recycling of natural and synthetic isoprene rubbers, in: *Chemistry, Manufacture and Applications of Natural Rubber*, S. Kohjiya and Y. Ikeda (Eds.), pp. 395–435, Elsevier, United Kingdom, 2014.
10. Sharma, V.K., Fortuna, F., Mincarini, M., Berillo, M., Cornacchia, G., Disposal of waste tyres for energy recovery and safe environment. *Appl. Energy*, 65, 381–394, 2000.
11. Vimeo, https://vimeo.com/194347842 (accessed May 27, 2017).
12. Reciclanip, http://www.reciclanip.org.br/v3/perguntas-frequentes (accessed May 27, 2017).
13. Reciclanip, http://www.reciclanip.org.br/v3/ (accessed May 27, 2017).
14. Myhre, M. and Mackillop, D.A., Rubber recycling. *Rubber Chem. Technol.*, 75, 429–474, 2002.
15. Adhikari, B., De, D., Maiti, S., Reclamation and recycling of waste rubber. *Prog. Polym. Sci.*, 25, 909–948, 2000.
16. Fukumori, K., Matsushita, M., Okamoto, H., Sato, N., Suzuki, Y., Takeuchi, K., Recycling technology of tire rubber. *JSAE Rev.*, 23, 259–264, 2002.

17. Celauro, B., Celauro, C., Lo Presti, D., Bevilacqua, A., Definition of a laboratory optimization protocol for road bitumen improved with recycled tire rubber. *Constr. Build. Mater.*, 37, 562–572, 2012.
18. Navarro, F.J., Partal, P., Martinez-Boza, F.J., Gallegos, C., Novel recycled polyethylene/ground tire rubber/bitumen blends for use in roofing applications: Thermo-mechanical properties. *Polym. Test.*, 29, 588–595, 2010.
19. Chiu, C.T., Hsu, T.H., Yang, W.F., Life cycle assessment on using recycled materials for rehabilitating asphalt pavements. *Resour. Conserv. Recycl.*, 52, 545–556, 2008.
20. Chiu, C.T., Use of ground tire rubber in asphalt pavements: Field trial and evaluation in Taiwan. *Resour. Conserv. Recycl.*, 52, 522–532, 2008.
21. Hernandez-Olivares, F., Barluenga, G., Bollati, M., Witoszek, B., Static and dynamic behaviour of recycled tyre rubber-filled concrete. *Cem. Concr. Res.*, 32, 1587–1596, 2002.
22. Oliveira, J.R.M., Silva, H., Abreu, L.P.F., Fernandes, S.R.M., Use of a warm mix asphalt additive to reduce the production temperatures and to improve the performance of asphalt rubber mixtures. *J. Clean. Prod.*, 41, 15–22, 2013.
23. Centonze, G., Leone, M., Aiello, M.A., Steel fibers from waste tires as reinforcement in concrete: A mechanical characterization. *Constr. Build. Mater.*, 36, 46–57, 2012.
24. Azevedo, F., Pacheco-Torgal, F., Jesus, C., de Aguiar, J.L.B., Camoes, A.F., Properties and durability of HPC with tyre rubber wastes. *Constr. Build. Mater.*, 34, 186–191, 2012.
25. Pacheco-Torgal, F., Ding, Y.N., Jalali, S., Properties and durability of concrete containing polymeric wastes (tyre rubber and polyethylene terephthalate bottles): An overview. *Constr. Build. Mater.*, 30, 714–724, 2012.
26. Bravo, M. and de Brito, J., Concrete made with used tyre aggregate: durability-related performance. *J. Clean. Prod.*, 25, 42–50, 2012.
27. Richardson, A., Coventry, K., Dave, U., Pienaar, J., Freeze/thaw performance of concrete using granulated rubber crumb. *J. Green Build.*, 6, 83–92, 2011.
28. Segre, N., Monteiro, P.J.M., Sposito, G., Surface characterization of recycled tire rubber to be used in cement paste matrix. *J. Colloid Interface Sci.*, 248, 521–523, 2002.
29. Navarro, F.J., Partal, P., Martinez-Boza, F., Valencia, C., Gallegos, C., Rheological characteristics of ground tire rubber-modified bitumens. *Chem. Eng. J.*, 89, 53–61, 2002.
30. Peralta, J., Silva, H., Hilliou, L., Machado, A.V., Pais, J., Williams, R.C., Mutual changes in bitumen and rubber related to the production of asphalt rubber binders. *Constr. Build. Mater.*, 36, 557–565, 2012.
31. Fattuhi, N.I. and Clark, L.A., Cement-based materials containing shredded scrap truck tyre rubber. *Constr. Build. Mater.*, 10, 229–236, 1996.
32. Topcu, I.B. and Bilir, T., Analysis of rubberized concrete as a three-phase composite naterial. *J. Compos. Mater.*, 43, 1251–1263, 2009.

33. Topcu, I.B. and Avcular, N., Analysis of rubberized concrete as a composite material. *Cem. Concr. Res.*, 27, 1135–1139, 1997.
34. Pierce, C.E. and Blackwell, M.C., Potential of scrap tire rubber as lightweight aggregate in flowable fill. *Waste Manage. (Oxford)*, 23, 197–208, 2003.
35. Benazzouk, A., Douzane, O., Langlet, T., Mezreb, K., Roucoult, J.M., Queneudec, M., Physico-mechanical properties and water absorption of cement composite containing shredded rubber wastes. *Cem. Concr. Compos.*, 29, 732–740, 2007.
36. Kocevski, S., Yagneswaran, S., Xiao, F.P., Punith, V.S., Smith, D.W., Amirkhanian, S., Surface modified ground rubber tire by grafting acrylic acid for paving applications. *Constr. Build. Mater.*, 34, 83–90, 2012.
37. Aliapur, https://www.aliapur.fr/fr/les-applications/applications/sol-equestre (accessed May 27, 2017).
38. Ecopneus, http://www.ecopneus.it/en/spazio-sensoriale/sensory-space-the-project.html (accessed May 27, 2017).
39. Llompart, M., Sanchez-Prado, L., Lamas, J.P., Garcia-Jares, C., Roca, E., Dagnac, T., Hazardous organic chemicals in rubber recycled tire playgrounds and pavers. *Chemosphere*, 90, 423–431, 2013.
40. Anandhan, S., De, P.P., Bhowmick, A.K., De, S.K., Bandyopadhyay, S., Thermoplastic elastomeric blend of nitrile rubber and poly(styrene-co-acrylonitrile). II. Replacement of nitrile rubber by its vulcanizate powder. *J. Appl. Polym. Sci.*, 90, 2348–2357, 2003.
41. Anandhan, S. and Bhowmick, A.K., Thermoplastic vulcanizates from post consumer computer plastics/nitrile rubber blends by dynamic vulcanization. *J. Mater. Cycles Waste Manage.*, 15, 300–309, 2013.
42. Ding, R., Leonov, A.I., Coran, A.Y., A study of the vulcanization kinetics of an accelerated-sulfur SBR compound. *Rubber Chem. Technol.*, 69, 81–91, 1996.
43. Santos, G.C., Carmo, D.M., Rezende, C.G.F., Zattera, A.J., Oliveira, M.G., Oliveira, P.J., Use of EPDMSDD as compatibilizer agent for EPDM/EPDMR blends: Rheologic, mechanical, and morphologic properties. *J. Appl. Polym. Sci.*, 122, 948–955, 2011.
44. Mariano, R.M., Picciani, P.S., Nunes, R.C.R., Visconte, L.L.Y., Preparation, structure, and properties of montmorillonite/cellulose II/natural rubber nanocomposites. *J. Appl. Polym. Sci.*, 120, 458–465, 2011.
45. Karaağaç, B., İnal, M., Deniz, V., Artificial neural network approach for predicting optimum cure time of rubber compounds. *Mater. Des.*, 30, 1685–1690, 2009.
46. Dick, J.S. and Pawlowski, H., Applications for the curemeter maximum cure rate in rubber compound development process control and cure kinetic studies. *Polym. Test.*, 15, 207–243, 1996.
47. Ignatz-Hoover, F. and To, B.H., Vulcanization, in: *Rubber Compounding: Chemistry and Applications*, B. Rodgers (Ed.), Marcel Dekker, New York, 2004.

48. Bandyopadhyay, P.K. and Banerjee, S., Studies on sulfur vulcanization of natural rubber accelerated combinedly with 2-mercaptobenzothiazole and diphenylguanidine both in presence and absence of dicumyl peroxide. *Die Angew. Makromol. Chem.*, 64, 59–74, 1977.
49. López-Manchado, M.A., Arroyo, M., Herrero, B., Biagiotti, J., Vulcanization kinetics of natural rubber-organoclay nanocomposites. *J. Appl. Polym. Sci.*, 89, 1–15, 2003.
50. Kok, C.M., The effects of compounding variables on the reversion process in the sulphur vulcanization of natural rubber. *Eur. Polym. J.*, 23, 611–615, 1987.
51. Gomide, R., *Operações unitárias: operações com sistemas sólidos granulares*, Câmara Brasileira do Livro, São Paulo, 1983.
52. Zanchet, A., Dal'Acqua, N., Weber, T., Crespo, J.S., Brandalise, R.N., Nunes, R.C.R., Propriedades reométricas e mecânicas e morfologia de compósitos desenvolvidos com resíduos elastoméricos vulcanizados. *Polimeros-Ciencia E Tecnol.*, 17, 23–27, 2007.
53. Jacob, C., De, P.P., Bhowmick, A.K., De, S.K., Recycling of EPDM waste. I. Effect of ground EPDM vulcanizate on properties of EPDM rubber. *J. Appl. Polym. Sci.*, 82, 3293–3303, 2001.
54. Weber, T., Zanchet, A., Brandalise, R.N., Crespo, J.S., Nunes, R.C.R., Grinding and characterization of scrap rubbers powders. *J. Elastomers Plast.*, 40, 147–159, 2008.
55. Rattanasom, N., Poonsuk, A., Makmoon, T., Effect of curing system on the mechanical properties and heat aging resistance of natural rubber/tire tread reclaimed rubber blends. *Polym. Test.*, 24, 728–732, 2005.
56. Ismail, H., Nordin, R., Noor, A.M., Cure characteristics, tensile properties and swelling behaviour of recycled rubber powder-filled natural rubber compounds. *Polym. Test.*, 21, 565–569, 2002.
57. Sienkiewicz, M., Kucinska-Lipka, J., Janik, H., Balas, A., Progress in used tyres management in the European Union: A review. *Waste Manage. (Oxford)*, 32, 1742–1751, 2012.
58. Oh, J.S. and Isayev, A.I., Continuous ultrasonic devulcanization of unfilled butadiene rubber. *J. Appl. Polym. Sci.*, 93, 1166–1174, 2004.
59. Kleps, T., Piaskiewicz, M., Parasiewicz, W., The use of thermogravimetry in the study of rubber devulcanization. *J. Therm. Anal. Calorim.*, 60, 271–277, 2000.
60. Warner, W.C., Methods of devulcanization. *Rubber Chem. Technol.*, 67, 559–566, 1994.
61. Zhang, X.X., Lu, C.H., Liang, M., Properties of natural rubber vulcanizates containing mechanochemically devulcanized ground tire rubber. *J. Polym. Res.*, 16, 411–419, 2009.
62. Li, S.Y., Lamminmaki, J., Hanhi, K., Effect of ground rubber powder and devulcanizates on the properties of natural rubber compounds. *J. Appl. Polym. Sci.*, 97, 208–217, 2005.

63. Rios, R., Gontijo, M., Ferraz, V.P., Lago, R.M., Araujo, M.H., Devulcanization of styrene butadiene (SBR) waste tire by controlled oxidation. *J. Braz. Chem. Soc.*, 17, 603–608, 2006.
64. Maridass, B. and Gupta, B.R., Performance optimization of a counter rotating twin screw extruder for recycling natural rubber vulcanizates using response surface methodology. *Polym. Test.*, 23, 377–385, 2004.
65. Maridass, B. and Gupta, B.R., Process optimization of devulcanization of waste rubber powder from syringe stoppers by twin screw extruder using response surface methodology. *Polym. Compos.*, 29, 1350–1357, 2008.
66. Yazdani, H., Karrabi, M., Ghasmi, I., Azizi, H., Bakhshandeh, G.R., Devulcanization of waste tires using a twin-screw extruder: The effects of processing conditions. *J. Vinyl Addit. Technol.*, 17, 64–69, 2011.
67. Sutanto, P., Picchioni, F., Janssen, L., Dijkhuis, K.A.J., Dierkes, W.K., Noordermeer, J.W.M., EPDM rubber reclaim from devulcanized EPDM. *J. Appl. Polym. Sci.*, 102, 5948–5957, 2006.
68. Sutanto, P., Laksmana, F.L., Picchioni, E., Janssen, L., Modeling on the kinetics of an EPDM devulcanization in an internal batch mixer using an amine as the devulcanizing agent. *Chem. Eng. Sci.*, 61, 6442–6453, 2006.
69. Sutanto, P., Picchioni, F., Janssen, L.P.B.M., Modelling a continuous devulcanization in an extruder. *Chem. Eng. Sci.*, 61, 7077–7086, 2006.
70. Sutanto, P., Picchioni, F., Janssen, L., The use of experimental design to study the responses of continuous devulcanization processes. *J. Appl. Polym. Sci.*, 102, 5028–5038, 2006.
71. Rooj, S., Basak, G.C., Maji, P.K., Bhowmick, A.K., New route for devulcanization of natural rubber and the properties of devulcanized rubber. *J. Polym. Environ.*, 19, 382–390, 2011.
72. Bilgili, E., Dybek, A., Arastoopour, H., Bernstein, B., A new recycling technology: Compression molding of pulverized rubber waste in the absence of virgin rubber. *J. Elastomers Plast.*, 35, 235–256, 2003.
73. Maridass, B. and Gupta, B.R., Effect of carbon black on devulcanized ground rubber tire-natural rubber vulcanizates: Cure characteristics and mechanical properties. *J. Elastomers Plast.*, 38, 211–229, 2006.
74. Dijkhuis, K.A.J., Noordermeer, J.W.M., Dierkes, W.K., The relationship between crosslink system, network structure and material properties of carbon black reinforced EPDM. *Eur. Polym. J.*, 45, 3302–3312, 2009.
75. Yehia, A.A., Ismail, M.N., Hefny, Y.A., Abdel-Bary, E.M., Mull, M.A., Mechano-chemical reclamation of waste rubber powder and its effect on the performance of NR and SBR vulcanizates. *J. Elastomers Plast.*, 36, 109–123, 2004.
76. Yazdani, H., Ghasemi, I., Karrabi, M., Azizi, H., Bakhshandeh, G.R., Continuous devulcanization of waste tires by using a co-rotating twin screw extruder: Effects of screw configuration, temperature profile, and devulcanization agent concentration. *J. Vinyl Addit. Technol.*, 19, 65–72, 2013.

77. Mandal, S.K., Alam, N., Debnath, S.C., Reclaiming of ground rubber tire by safe multifunctional rubber additives: I. Tetra benzylthiuram disulfide. *Rubber Chem. Technol.*, 85, 629–644, 2012.
78. Tripathy, A.R., Morin, J.E., Williams, D.E., Eyles, S.J., Farris, R.J., A novel approach to improving the mechanical properties in recycled vulcanized natural rubber and its mechanism. *Macromolecules*, 35, 4616–4627, 2002.
79. Verbruggen, M.A.L., van der Does, L., Noordermeer, J.W.M., van Duin, M., Manuel, H.J., Mechanisms involved in the recycling of NR and EPDM. *Rubber Chem. Technol.*, 72, 731–740, 1999.
80. Sadaka, F., Campistron, I., Laguerre, A., Pilard, J.F., Controlled chemical degradation of natural rubber using periodic acid: Application for recycling waste tyre rubber. *Polym. Degrad. Stab.*, 97, 816–828, 2012.
81. Kumar, C.R., Fuhrmann, I., Karger-Kocsis, J., LDPE-based thermoplastic elastomers containing ground tire rubber with and without dynamic curing. *Polym. Degrad. Stab.*, 76, 137–144, 2002.
82. Hassan, M.M., Badway, N.A., Elnaggar, M.Y., Hegazy, E.A., Effects of peroxide and gamma radiation on properties of devulcanized rubber/polypropylene/ethylene propylene diene monomer formulation. *J. Appl. Polym. Sci.*, 131, 40611, 2014.
83. Lee, S.H., Zhang, Z.X., Xu, D., Chung, D., Oh, G.J., Kim, J.K., Dynamic reaction involving surface modified waste ground rubber tire powder/polypropylene. *Polym. Eng. Sci.*, 49, 168–176, 2009.
84. Zhang, X.X., Lu, C.H., Liang, M., Preparation of thermoplastic vulcanizates based on waste crosslinked polyethylene and ground tire rubber through dynamic vulcanization. *J. Appl. Polym. Sci.*, 122, 2110–2120, 2011.
85. Zhang, X.X., Lu, Z.X., Tian, D., Li, H., Lu, C.H., Mechanochemical devulcanization of ground tire rubber and its application in acoustic absorbent polyurethane foamed composites. *J. Appl. Polym. Sci.*, 127, 4006–4014, 2013.
86. Zhang, S.L., Zhang, Z.X., Kim, J.K., Study on thermoplastic elastomers (TPEs) of waste polypropylene/waste ground rubber tire powder. *J. Macromol. Sci. Part B Phys.*, 50, 762–768, 2011.
87. Hirayama, D. and Saron, C., Chemical modifications in styrene-butadiene rubber after microwave devulcanization. *Ind. Eng. Chem. Res.*, 51, 3975–3980, 2012.
88. Hirayama, D., Scuracchio, C.H., Saron, C., Microwave devulcanization of SBR containing carbon black. *J. Res. Updates Polym. Sci.*, 5, 52–59, 2016.
89. Paulo, G.D., Hirayama, D., Saron, C., Microwave devulcanization of waste rubber with inorganic salts and nitric acid. *Adv. Mater. Res.*, 418–420, 1072–1075, 2012.
90. Zanchet, A., Carli, L.N., Giovanela, M., Brandalise, R.N., Crespo, J.S., Use of styrene butadiene rubber industrial waste devulcanized by microwave in rubber composites for automotive application. *Mater. Des.*, 39, 437–443, 2012.

91. Landini, L., de Araujo, S.G., Lugao, A.B., Wiebeck, H., Preliminary analysis to BIIR recovery using the microwave process. *Eur. Polym. J.*, 43, 2725–2731, 2007.
92. Sombatsompop, N. and Kumnuantip, C., Comparison of physical and mechanical properties of NR/carbon black/reclaimed rubber blends vulcanized by conventional thermal and microwave irradiation methods. *J. Appl. Polym. Sci.*, 100, 5039–5048, 2006.
93. Garcia, P.S., de Sousa, F.D.B., de Lima, J.A., Cruz, S.A., Scuracchio, C.H., Devulcanization of ground tire rubber: Physical and chemical changes after different microwave exposure times. *eXPRESS Polym. Lett.*, 9, 1015–1026, 2015.
94. de Sousa, F.D.B., Gouveia, J.R., de Camargo Filho, P.M.F., Vidotti, S.E., Scuracchio, C.H., Amurin, L.G., *et al.*, Blends of ground tire rubber devulcanized by microwaves/HDPE - Part A: Influence of devulcanization process. *Polimeros-Ciencia E Tecnol.*, 25, 256–264, 2015.
95. de Sousa, F.D.B., Gouveia, J.R., de Camargo Filho, P.M.F., Vidotti, S.E., Scuracchio, C.H., Amurin, L.G., *et al.*, Blends of ground tire rubber devulcanized by microwaves/HDPE-Part B: Influence of clay addition. *Polimeros-Ciencia E Tecnol.*, 25, 382–391, 2015.
96. de Sousa, F.D.B., Scuracchio, C.H., Hu, G.H., Hoppe, S., Effects of processing parameters on the properties of microwave-devulcanized ground tire rubber/polyethylene dynamically revulcanized blends. *J. Appl. Polym. Sci.*, 133, 43503, 2016.
97. Bani, A., Polacco, G., Gallone, G., Microwave-induced devulcanization for poly(ethylene-propylene-diene) recycling. *J. Appl. Polym. Sci.*, 120, 2904–2911, 2011.
98. de Sousa, F.D.B., Scuracchio, C.H., Hu, G.H., Hoppe, S., Devulcanization of waste tire rubber by microwaves. *Polym. Degrad. Stab.*, 138, 169–181, 2017.
99. de Sousa, F.D.B., Devulcanization of elastomers and applications, in: *Elastomers*, N. Cankaya (Ed.), InTech, Rijeka, 2017.
100. de Sousa, F.D.B., Zanchet, A., Scuracchio, C.H., Influence of reversion in compounds containing recycled natural rubber: In search of sustainable processing. *J. Appl. Polym. Sci.*, 134, 45325, 2017.
101. Pistor, V. and Zattera, A.J., Degradation kinetics of ethylene propylene diene terpolymer residues devulcanized by microwaves. *J. Elastomers Plast.*, 46, 1–15, 2012.
102. Pistor, V., Scuracchio, C.H., Oliveira, P.J., Fiorio, R., Zatteral, A.J., Devulcanization of ethylene-propylene-diene polymer residues by microwave-Influence of the presence of paraffinic oil. *Polym. Eng. Sci.*, 51, 697–703, 2011.
103. Pistor, V., Ornaghi, F.G., Fiorio, R., Zattera, A.J., Oliveira, P.J., Scuracchio, C.H., Devulcanization of ethylene-propylene-diene polymer residues (EPDM-r) by microwaves. *Polimeros-Ciencia E Tecnol.*, 20, 165–169, 2010.

104. Scuracchio, C.H., Waki, D.A., Bretas, R.E.S., Caracterização térmica e reológica de borracha de pneu desvulcanizada por microondas. *Polimeros-Ciencia E Tecnol.*, 16, 46–52, 2006.
105. Scuracchio, C.H., Waki, D.A., da Silva, M.L.C.P., Thermal analysis of ground tire rubber devulcanized by microwaves. *J. Therm. Anal. Calorim.*, 87, 893–897, 2007.
106. Zanchet, A., Carli, L.N., Giovanela, M., Crespo, J.S., Scuracchio, C.H., Nunes, R.C.R., Characterization of microwave-devulcanized composites of ground SBR scraps. *J. Elastomers Plast.*, 41, 497–507, 2009.
107. Hirayama, D. and Saron, C., Moagem e desvulcanização por micro-ondas da borracha estireno-butadieno (SBR). *Rev. Eletrônica Mater. Processos*, 7, 1, 34–38, 2012.
108. Scuracchio, C.H., Bretas, R.E.S., Isayev, A.I., Blends of PS with SBR devulcanized by ultrasound: Rheology and morphology. *J. Elastomers Plast.*, 36, 45–75, 2004.
109. Yun, J., Isayev, A.I., Kim, S.H., Tapale, M., Comparative analysis of ultrasonically devulcanized unfilled SBR, NR, and EPDM rubbers. *J. Appl. Polym. Sci.*, 88, 434–441, 2003.
110. Zhu, S.H., Penlidis, A., Tzoganakis, C., Ginzel, E., Ultrasonic properties and morphology of devulcanized rubber blends. *J. Appl. Polym. Sci.*, 124, 2062–2070, 2012.
111. Isayev, A.I., Yushanov, S.P., Chen, J., Ultrasonic devulcanization of rubber vulcanizates .2. Simulation and experiment. *J. Appl. Polym. Sci.*, 59, 815–824, 1996.
112. Isayev, A.I., Yushanov, S.P., Chen, J., Ultrasonic devulcanization of rubber vulcanizates .1. Process model. *J. Appl. Polym. Sci.*, 59, 803–813, 1996.
113. Massey, J.L., Parr, J.C., Wagler, T.A., von Meerwall, E., Hong, C.K., Isayev, A.I., Ultrasound devulcanization of unfilled natural rubber networks, studied *via* component molecular mobility. *Polym. Int.*, 56, 860–869, 2007.
114. Oh, J.S., Ghose, S., Isayev, A.I., Effects of ultrasonic treatment on unfilled butadiene rubber. *J. Polym. Sci., Part B: Polym. Phys.*, 41, 2959–2968, 2003.
115. Shim, S.E. and Isayev, A.I., Effects of the presence of water on ultrasonic devulcanization of polydimethylsiloxane. *J. Appl. Polym. Sci.*, 88, 2630–2638, 2003.
116. Shim, S.E., Ghose, S., Isayev, A.I., Formation of bubbles during ultrasonic treatment of cured poly(dimethyl siloxane). *Polymer*, 43, 5535–5543, 2002.
117. Shim, S.E., Parr, J.C., von Meerwall, E., Isayev, A.I., NMR relaxation and pulsed gradient NMR diffusion measurements of ultrasonically devulcanized poly (dimethylsiloxane). *J. Phys. Chem. B*, 106, 12072–12078, 2002.
118. Oh, J.S. and Isayev, A.I., Ultrasonically treated polypropylene/ground tire rubber blends. *Rubber Chem. Technol.*, 75, 617–625, 2002.
119. Ghose, S. and Isayev, A.I., Ultrasonic devulcanization of carbon black filled polyurethane rubber. *J. Elastomers Plast.*, 36, 213–239, 2004.

120. Luo, T. and Isayev, A.I., Rubber/plastic blends based on devulcanized ground tire rubber. *J. Elastomers Plast.*, 30, 133–160, 1998.
121. Diao, B., Isayev, A.I., Levin, V.Y., Basic study of continuous ultrasonic devulcanization of unfilled silicone. *Rubber Chem. Technol.*, 72, 152–164, 1999.
122. Isayev, A.I., Chen, J., Tukachinsky, A., Novel ultrasonic technology for devulcanizatio of waste rubbers. *Rubber Chem. Technol.*, 68, 267–280, 1995.
123. Isayev, A.I., Yushanov, S.P., Kim, S.H., Levin, V.Y., Ultrasonic devulcanization of waste rubbers: Experimentation and modeling. *Rheol. Acta*, 35, 616–630, 1996.
124. Levin, V.Y., Kim, S.H., Isayev, A.I., Massey, J., VonMeerwall, E., Ultrasound devulcanization of sulfur vulcanized SBR: Crosslink density and molecular mobility. *Rubber Chem. Technol.*, 69, 104–114, 1996.
125. Sun, X., Isayev, A.I., Joshi, T.R., von Meerwall, E., Molecular mobility of unfilled and carbon-black-filled isoprene rubber: Proton NMR transverse relaxation and diffusion. *Rubber Chem. Technol.*, 80, 854–872, 2007.
126. Sun, X. and Isayev, A.I., Continuous ultrasonic devulcanization: Comparison of carbon black filled synthetic isoprene and natural rubbers. *Rubber Chem. Technol.*, 81, 19–46, 2008.
127. Yun, J., Oh, J.S., Isayev, A.I., Ultrasonic devulcanization reactors for recycling of GRT: Comparative study. *Rubber Chem. Technol.*, 74, 317–330, 2001.
128. Yushanov, S.P., Isayev, A.I., Levin, V.Y., Percolation simulation of the network degradation during ultrasonic devulcanization. *J. Polym. Sci., Part B: Polym. Phys.*, 34, 2409–2418, 1996.
129. Yashin, V.V. and Isayev, A.I., A model for rubber degradation under ultrasonic treatment: Part II. Rupture of rubber network and comparison with experiments. *Rubber Chem. Technol.*, 73, 325–339, 2000.
130. Yashin, V.V. and Isayev, A.I., The effect of polydispersity on structure of ultrasonically treated rubbers. *Polymer*, 45, 6083–6094, 2004.
131. Yashin, V.V. and Isayev, A.I., A model for rubber degradation under ultrasonic treatment: Part I. Acoustic cavitation in viscoelastic solid. *Rubber Chem. Technol.*, 72, 741–757, 1999.
132. Yun, J.S. and Isayev, A.I., The effect of high power ultrasound on EPDM gum in extrusion process. *J. Appl. Polym. Sci.*, 105, 3698–3707, 2007.
133. Ghose, S., Isayev, A.I., von Meerwall, E., Effect of ultrasound on thermoset polyurethane: NMR relaxation and diffusion measurements. *Polymer*, 45, 3709–3720, 2004.
134. Feng, W.L., Isayev, A.I., von Meerwall, E., Molecular mobility in ultrasonically treated butyl gum and devulcanized butyl rubber. *Polymer*, 45, 8459–8467, 2004.
135. Tukachinsky, A., Schworm, D., Isayev, A.I., Devulcanization of waste tire rubber by powerful ultrasound. *Rubber Chem. Technol.*, 69, 92–103, 1996.
136. Feng, W.L. and Isayev, A.I., Blends of ultrasonically devulcanized tire-curing bladder and butyl rubber. *J. Mater. Sci.*, 40, 2883–2889, 2005.

137. Feng, W.L. and Isayev, A.I., High-power ultrasonic treatment of butyl rubber gum: Structure and properties. *J. Polym. Sci., Part B: Polym. Phys.*, 43, 334–344, 2005.
138. Isayev, A.I. and Sujan, B., Nonisothermal vulcanization of devulcanized GRT with reversion type behavior. *J. Elastomers Plast.*, 38, 291–318, 2006.
139. Kim, J.K., Lee, S.H., Hwang, S.H., Study on the thermoplastic vulcanizate using ultrasonically treated rubber powder. *J. Appl. Polym. Sci.*, 90, 2503–2507, 2003.
140. Sun, X. and Isayev, A.I., Ultrasound devulcanization: comparison of synthetic isoprene and natural rubbers. *J. Mater. Sci.*, 42, 7520–7529, 2007.
141. Hong, C.K. and Isayev, A.I., Plastic/rubber blends of ultrasonically devulcanized GRT with HDPE. *J. Elastomers Plast.*, 33, 47–71, 2010.
142. Alexandre-Franco, M., Fernandez-Gonzalez, C., Alfaro-Dominguez, M., Latasa, J.M.P., Gomez-Serrano, V., Devulcanization and demineralization of used tire rubber by thermal chemical methods: A study by X-ray diffraction. *Energy Fuels*, 24, 3401–3409, 2010.
143. Lai, S.K., Batra, A., Cohen, C., Characterization of polydimethylsiloxane elastomer degradation *via* cross-linker hydrolysis. *Polymer*, 46, 4204–4211, 2005.
144. Masaki, K., Ohkawara, S.I., Hirano, T., Seno, M., Sato, T., Devulcanization of nitrile butadiene rubber in nitrobenzene. *J. Appl. Polym. Sci.*, 91, 3342–3353, 2004.
145. De, D., Das, A., De, D., Dey, B., Chandra, S., Chandra, B., Reclaiming of ground rubber tire (GRT) by a novel reclaiming agent. *Eur. Polym. J.*, 42, 917–927, 2006.
146. De, D., Maiti, S., Adhikari, B., Reclaiming of rubber by a renewable resource material (RRM). II. Comparative evaluation of reclaiming process of NR vulcanizate by RRM and diallyl disulfide. *J. Appl. Polym. Sci.*, 73, 2951–2958, 1999.
147. Shanmugharaj, A.M., Kim, J.K., Ryu, S.H., UV surface modification of waste tire powder: Characterization and its influence on the properties of polypropylene/waste powder composites. *Polym. Test.*, 24, 739–745, 2005.
148. Sonnier, R., Leroy, E., Clerc, L., Bergeret, A., Lopez-Cuesta, J.M., Polyethylene/ground tyre rubber blends: Influence of particle morphology and oxidation on mechanical properties. *Polym. Test.*, 26, 274–281, 2007.
149. Ishiaku, U.S., Chong, C.S., Ismail, H., Determination of optimum De-Link R concentration in a recycled rubber compound. *Polym. Test.*, 18, 621–633, 1999.
150. Jana, G.K., Mahaling, R.N., Das, C.K., A novel devulcanization technology for vulcanized natural rubber. *J. Appl. Polym. Sci.*, 99, 2831–2840, 2006.
151. Kojima, M., Tosaka, M., Ikeda, Y., Chemical recycling of sulfur-cured natural rubber using supercritical carbon dioxide. *Green Chem.*, 6, 84–89, 2004.

152. Tripathy, A.R., Williams, D.E., Farris, R.J., Rubber plasticizers from degraded/devulcanized scrap rubber: A method of recycling waste rubber. *Polym. Eng. Sci.*, 44, 1338–1350, 2004.
153. Vega, B., Montero, L., Lincoln, S., Agullo, N., Borros, S., Control of vulcanizing/devulcanizing behavior of diphenyl disulfide with microwaves as the heating source. *J. Appl. Polym. Sci.*, 108, 1969–1975, 2008.
154. Dubkov, K.A., Semikolenov, S.V., Ivanov, D.P., Babushkin, D.E., Panov, G.I., Parmon, V.N., Reclamation of waste tyre rubber with nitrous oxide. *Polym. Degrad. Stab.*, 97, 1123–1130, 2012.
155. Li, Y.H., Zhao, S.H., Zhang, L.Q., Wang, Y.Q., Yu, W.Y., The effect of different Fe2+ concentrations in culture media on the recycling of ground tyre rubber by acidithiobacillus ferrooxidans YT-1. *Ann. Microbiol.*, 63, 315–321, 2013.
156. Li, Y.H., Zhao, S.H., Wang, Y.Q., Microbial desulfurization of ground tire rubber by sp.: A novel technology for crumb rubber composites. *J. Polym. Environ.*, 20, 372–380, 2012.
157. Li, Y.H., Zhao, S.H., Wang, Y.Q., Microbial desulfurization of ground tire rubber by Thiobacillus ferrooxidans. *Polym. Degrad. Stab.*, 96, 1662–1668, 2011.
158. Bredberg, K., Andersson, B.E., Landfors, E., Holst, O., Microbial detoxification of waste rubber material by wood-rotting fungi. *Bioresour. Technol.*, 83, 221–224, 2002.
159. Christiansson, M., Stenberg, B., Wallenberg, L.R., Holst, O., Reduction of surface sulphur upon microbial devulcanization of rubber materials. *Biotechnol. Lett.*, 20, 637–642, 1998.
160. Holst, O., Stenberg, B., Christiansson, M., Biotechnological possibilities for waste tyre-rubber treatment. *Biodegradation*, 9, 301–310, 1998.
161. Jiang, G.M., Zhao, S.H., Li, W.J., Luo, J.Y., Wang, Y.Q., Zhou, Q.S., *et al.*, Microbial desulfurization of SBR ground rubber by Sphingomonas sp. and its utilization as filler in NR compounds. *Polym. Adv. Technol.*, 22, 2344–2351, 2011.
162. Jiang, G.M., Zhao, S.H., Luo, J.Y., Wang, Y.Q., Yu, W.Y., Zhang, C.R., Microbial desulfurization for NR ground rubber by Thiobacillus ferrooxidans. *J. Appl. Polym. Sci.*, 116, 2768–2774, 2010.
163. Sato, S., Honda, Y., Kuwahara, M., Kishimoto, H., Yagi, N., Muraoka, K., *et al.*, Microbial scission of sulfide linkages in vulcanized natural rubber by a white rot basidiornycete, Ceriporiopsis subvermispora. *Biomacromolecules*, 5, 511–515, 2004.
164. Tsuchii, A. and Tokiwa, Y., Microbial degradation of tyre rubber particles. *Biotechnol. Lett*, 23, 963–969, 2001.
165. Karger-Kocsis, J., Meszaros, L., Barany, T., Ground tyre rubber (GTR) in thermoplastics, thermosets, and rubbers. *J. Mater. Sci.*, 48, 1–38, 2013.
166. Hong, C.K. and Isayev, A.I., Continuous ultrasonic devulcanization of carbon black-filled NR vulcanizates. *J. Appl. Polym. Sci.*, 79, 2340–2348, 2001.

167. Amari, T., Themelis, N.J., Wernick, I.K., Resource recovery from used rubber tires. *Resour. Policy*, 25, 179–188, 1999.
168. Menendez, J.A., Arenillas, A., Fidalgo, B., Fernandez, Y., Zubizarreta, L., Calvo, E.G., *et al.*, Microwave heating processes involving carbon materials. *Fuel Process. Technol.*, 91, 1–8, 2010.
169. Loganathan, K.S., *Rubber engineering*, McGraw-Hill, New York, 2000.
170. Flory, P.J. and Rehner Jr., J., Statistical mechanics of cross-linked polymer networks. II. Swelling. *J. Chem. Phys.*, 11, 521–526, 1943.
171. Ghorai, S., Bhunia, S., Roy, M., De, D., Mechanochemical devulcanization of natural rubber vulcanizate by dual function disulfide chemicals. *Polym. Degrad. Stab.*, 129, 34–46, 2016.
172. Horikx, M.M., Chain scissions in a polymer network. *J. Polym. Sci., Part A: Polym. Chem.*, 19, 445–454, 1956.
173. de Sousa, F.D.B., Zanchet, A., Scuracchio, C.H., From devulcanization to revulcanization: Challenges in getting recycled tire rubber for technical applications. *ACS Sustain. Chem. Eng.*, 7, 8755–8765, 2019.

9

Green Development of Natural Rubber-Based Biocomposites

Azura A. Rashid* **and Norfatirah Muhamad Sarih**

School of Materials and Mineral Resources Engineering, Universiti Sains Malaysia, Engineering Campus, Seberang Perai Selatan, Pulau Pinang, Malaysia

Abstract

The present review relied on assessing the impact of natural fibers, especially cellulose and starch, as additives on the biodegradability of the resultant NR composites. Despite the advantageous properties of biodegradable materials, certain limitations significantly impede their broader potential applications. For example, cellulose, glucose, and chitin are incompatible with NR; thus, surface modification of the fillers or NR matrix is required to circumvent this drawback.

***Keywords*:** Biodegradation, polymers, natural rubber, composites, starch, cellulose, nanocellulose, starch nanocrystal

9.1 Introduction

Natural rubber (NR) is widely used in many applications, including tires, gloves, medical devices, and rubber bands, owing to its superior mechanical properties and elasticity. Global demand for NR continues to increase. Consequently, managing excessive rubber waste presents significant risks to ecosystems, primarily because of its uncontrolled and direct disposal into the environment. Over the years, various strategies have been proposed or implemented to address rubber waste disposal, including burning, dumping, stacking, and landfilling. Unfortunately, these methods are not environmentally friendly and are often performed illegally, leading to additional pollution or other adverse effects. For example, rubber waste

**Corresponding author*: srazura@usm.my

Visakh. P.M. and Shiv Kumari Panda (eds.) Natural Rubber-Based Composites and Nanocomposites, (271–304) © 2025 Scrivener Publishing LLC

disposal in agricultural soils can lead to soil infertility, causing severe economic impacts on plantations. Therefore, biodegradable rubber presents a green solution for mitigating rubber waste management issues, although economic concerns limit its broader adoption.

This chapter presents an introduction to the biodegradation of polymers, especially NR. Next, we review the biodegradation analysis studies of NR composites and nanocomposites that focus on NR composites with natural fillers. Natural rubber (NR) is commonly employed as a polymer matrix in composites, owing to its significance and wide range of applications. However, their extensive use results in a significant amount of waste. Special consideration is given to NR composites made with natural fillers, such as cellulose and starch, as they offer a more sustainable approach. Natural fillers are particularly significant in the realm of environmentally friendly composite applications because of their widespread availability, low cost, and straightforward manufacturing processes [1]. The advancements achieved thus far in producing and studying the properties of NR composites with natural fillers show promising potential for the development of biodegradable materials. These materials exhibit operational characteristics that can enable their use as green and sustainable options in the near future.

9.2 Biodegradation

Although many organisms carry out biodegradation in general, microbiological biodegradation is the most significant process from an environmental perspective. Microorganisms can metabolize a wide range of synthetic chemicals, rendering them necessary for the removal of polluted surroundings [2]. Several factors influence the rate and extent of a contaminant's biodegradation in soil, including co-metabolites and contaminant bioavailability [3, 4]. Moreover, it also affects various environmental factors, such as temperature, pH level, nutrient availability, soil humidity, the types and concentrations of co-contaminants, predation, and the availability or absence of plants. Moreover, the impacts of climate change on biodegradation processes are difficult to predict, yet rising temperatures are expected to influence various variables, including water, nutrients, and oxygen availability, which can alter the speed, extent, and route of pollutant biodegradability [5].

In addition, degradation occurs through abiotic or biotic degradation mechanisms. Abiotic degradation of polymers does not involve microorganisms, but the degradation needs to reduce the polymeric substance to smaller fragments so that it can biodegrade. This can be caused by sunlight, heat, water, or other factors. A common measure of abiotic degradation is half-life, which is the time required for half of a substance to degrade. The main abiotic processes are oxidation and hydrolysis. One example of abiotic degradation of NR is the oxidation of rubber by oxygen in air. Oxidation is a chemical reaction involving the loss of electrons from a substance. When NR is exposed to oxygen in an atmosphere that can attack the covalent bonds of NR, it can lose electrons and form carbonyl groups, which are carbon atoms double-bonded to oxygen atoms. Carbonyl groups can make rubber brittle and weak and reduce its elasticity and strength [6].

Another example of the abiotic degradation of NR is the hydrolysis of rubber by water. Hydrolysis in rubber refers to the chemical breakdown of the polymer and the resulting physical breakdown or crumbling of rubber. This process was accelerated by warmth and high humidity. NR can undergo hydrolysis in water under certain conditions. One method for degrading NR by hydrolysis is hydrous pyrolysis [7]. This process involves heating NR in water at high temperature and pressure. Hydrous pyrolysis, using Bio oil and biogas, can produce liquid fuels from NR. The biotic degradation of polymers is the process of breaking down a material by living organisms, such as bacteria, fungi, and algae.

The biotic degradation of rubber can be affected by various factors, such as the type and composition of rubber, environmental conditions, and microbial diversity. Some microorganisms can use NR as a carbon source and produce enzymes that cleave polyisoprene chains [8]. Biotic degradation can be studied in different ways, such as by using soil microcosms, composting systems, or bioreactors. Biotic degradation can also be combined with other methods, such as hydrolysis or ozonolysis, to enhance the degradation rate and produce valuable products from natural rubber.

9.2.1 Biodegradation Polymer Overview

Polymeric materials exhibit distinct characteristics in terms of their chemical composition, physical structure, mechanical properties, and practical applications. The ability of carbon atoms to form diverse bonds with other carbon atoms and non-carbon atoms and the presence of substituent groups allow for a wide range of chemical structures and stereochemical variations [9]. Slight deviations in chemical composition can lead to significant disparities in terms of biodegradability. Biodegradation is influenced by various factors, such as polymer properties, the type of organism involved, and the pretreatment method. Biodegradable polymers are a special class of polymers that break down after their intended use through the action of microorganisms, enzymes, or other environmental factors. This results in natural byproducts, including gases, biomass, water, and inorganic salts. Biodegradable polymers can be sourced from natural materials such as starch, cellulose, or proteins, or synthesized from monomers that are either natural or synthetic. They have many advantages such as reducing waste accumulation, saving energy and resources, and preventing environmental pollution.

Polymer degradability is affected by several factors, including polymer chemical properties, branching and flexibility of macrochains, molecular weight, and functional groups and substituents that are included in its structure, as well as any additives incorporated into the polymer. The molecular weight significantly affects polymer properties by decreasing biodegradability with increasing molecular weight. Polymers with greater surface areas generally experience faster degradation. Nonpolymeric additives such as colorants, fillers, waste, and catalyst residues can affect polymer degradability. Bio-based additives can accelerate polymer degradation by increasing susceptibility to microorganisms. In addition, biosurfactants, which are microbial surface-active compounds, enhance polymer biodegradability because of their low toxicity and excellent biodegradability. In addition, the existence of specific functional groups in biosurfactants facilitates the detection of biological activity, regardless of the highly challenging environments characterized by extreme temperatures, pH levels, and salinity.

Polymer biodegradation occurs in four stages. The first stage is biodeterioration, in which the polymer's physical and chemical properties change. The second stage is bio-fragmentation, in which the polymer breaks down into its lowest form *via* enzymatic cleavage, followed by assimilation, in which molecules are taken up by microorganisms in the third step. The last step is mineralization, meaning that after degradation, oxidized metabolites such as CO_2, CH_4, and H_2O are produced.

9.2.2 Natural Rubber (NR) Biodegradation

NR biodegradation can be affected by various factors, particularly polymer characteristics (such as substituents, presence of functional groups, molecular weight, additives, or crystallinity), the specific microorganisms engaged, and any pre-treatments applied to the polymer [9]. NR degrades slowly in nature because of certain microorganisms, and natural environments contain a few bacteria and fungi that can decompose NR as a source of carbon and energy. The majority of studies on rubber degradation by microorganisms have been carried out under laboratory conditions using liquid cultures [10–14]. These cultures frequently comprise a small mineral medium with rubber serving as the sole carbon source. The rate at which rubber is broken down by microorganisms is affected by both the composition of the rubber and the specific types of bacteria involved as well as the way they interact with the surrounding environment [15]. Moreover, NR biodegrades at a slower rate than other natural polymers, because of the rubber's branching chemical structure [8, 16]. Conducting laboratory-scale studies using soil microcosms that utilize natural soil microflora can be beneficial for several reasons. This approach allows for the assessment of (i) the microbial potential to digest rubber, (ii) the likely activity of the present soil biomass, (iii) longer incubation periods, and (iv) the most effective potential bioremediation approach [8].

Rubber degradation occurs gradually, which necessitates extended incubation durations, often ranging from several weeks to several months. This extended timeframe allows for the accumulation of biomass and significant alterations in the mass and structure of the rubber material [9]. The soil burial test, for example, has shown that NR is significantly biodegradable when exposed to natural soil microflora. The test found that the mass decreased by 16.2% in 45 days, and by the end of the 90-day incubation period, it decreased by an additional 38.3%. To accomplish this, thin probes were buried in nutrient-rich soil and exposed to the soil microorganisms [17]. Experiments conducted in microcosms at the laboratory scale may not always produce consistent results under real-world conditions because of the impact of various biological, chemical, and physical factors. However, these experiments provide an opportunity to consider and simulate these factors under controlled conditions, allowing the assessment of rubber biodegradability [18]. The exposure of rubber to external factors, such as heat, light, moisture, and oxygen, alters its unsaturated backbone. Specifically, oxygen can readily react with rubber, leading to deterioration of the polymer and resulting in either hardening or softening of the rubber [19]. In addition, environmental conditions have a direct impact on the types of microorganisms and the pathways they use to degrade materials. For example, when oxygen is available, aerobic microorganisms dominate the degradation of polymeric materials. They grow and produce new microbial masses with carbon dioxide and water as the final products.

9.2.2.1 *Microorganisms Involved in the Degradation of NR*

In the last century, extensive studies have been conducted to identify and describe effective microorganisms that break down rubber and to understand the molecular mechanisms underlying NR degradation. Rubber-degrading bacteria can be classified into two different types: those that can generate transparent halo zones by the secretion of extracellular polyisoprene-cleaving enzymes, and those that lack the capacity to make such translucent halo zones [20]. Several bacterial species that can break down NR have been identified and are listed in Table 9.1. The bacteria under consideration consisted predominantly of actinobacteria containing mycolic acid, corresponding to the genera Nocardia, Gordonia, and Mycobacterium.

Table 9.1 NR degradation by bacteria.

Bacterium	**Rubber oxygenases involved**	**Culture time**	**Weight loss (%)**	**References**
Steroidobacter cummioxidans 35Y	RoxA, RoxB	–	–	[26–29]
Rhizobacter gummiphilus NS21	RoxA, RoxB	–	–	[30]
Pseudomonas aeruginosa AL98	Unknown	6 weeks	26%	[31]
Pseudomonas citronellolis	Lcp	10 weeks	13%	[32]
Pseudomonas sp.	Unknown	2 months	10.8%	[25]
Acinetobacter calcoaceticus	Lcp	10 weeks	12%	[32]
Actinoplanes sp. *strain OR16*	Lcp	30 days	–	[33]
Streptomyces sp. strain K30	Lcp	–	–	[34]
Streptomyces sp. strain CFMR 7	Lcp	3–9 months	–	[35]
Streptomyces griseus 1D	Lcp	10 weeks	18%	[32]

(Continued)

Table 9.1 NR degradation by bacteria. (*Continued*)

Bacterium	**Rubber oxygenases involved**	**Culture time**	**Weight loss (%)**	**References**
Streptomyces coelicolor 1A	Lcp	10 weeks	18%	[32]
Streptomyces sp. *strain LCIC4*	Lcp	30 days	–	[33]
Rhodococcus rhodochrous RPK1	Lcp	–	–	[36]
Gordonia polyisoprenivorans VH2	Lcp	4 weeks	>50%	[37]
Nocardia nova SH22a	Lcp	–	–	[38]
Nocardia sp. *BSTN01*	Lcp	6 weeks	55.3%	[39]
Nocardia farcinica E1	Lcp	50 days	15%	[40]
Nocardia farcinica NVL3	Lcp	30 days	–	[41]
Xanthomonas sp.	Lcp	10 weeks	12%	[32]

Two different enzymes have currently been identified in bacteria that are Gram-positive and Gram-negative, as crucial in the earliest degradation of poly(cis-1,4-isoprene) namely, rubber oxygenase (RoxA/RoxB) and latex clearing protein (Lcp) [21]. The rubber oxygenase RoxA is responsible for the breakdown of cis-1,4-polyisoprene in Gram-negative NR degrading bacteria, including certain mycobacteria. RoxA, encoded by the roxA gene, is an exo-type and highly descriptive rubber oxygenase that plays a critical role in Lcp. RoxA binds oxygen molecules and subsequently cleaves the carbon–carbon double bond of cis-1,4-polyisoprene, producing isoprene-containing aldehyde and ketone groups at the end of the molecule [22]. In contrast, RoxB decays NR through endo-cleavage of cis-1,4-polyisoprene. Lcp, expressed by the lcp gene in rubber-utilizing actinomycetes, is a type B cytochrome capable of cleaving cis-1,4-polyisoprene into a product of low-molecular-weight product containing isoprene with aldehyde and ketone termini [20]. Subsequently, the resultant compound undergoes additional degradation to its constituent acid through the action of the oxidoreductase OxiAB, enters the β-oxidation pathway, and finally undergoes oxidative catabolism.

Rubber degradation by fungi was initially examined by De Vries using several strains of Penicillium and Aspergillus in a liquid medium consisting of 10% (w/v) NaCl. The findings showed that throughout an incubation period of 19 months to 5 years, the rubber material biomass increased by 6% and its weight decreased by 15.5% to 30.9% [23]. However, in 2013, an investigation into the isolation of two fungal species, Penicillium and *Aspergillus niger*, degraded the NR sample by 20% and 25.9% of its weight loss, respectively, over a two-month period [24]. From a liquid culture containing NR, Bosco *et al.* isolated a strain of *Rhodotorula mucilaginosa* and discovered *Alternaria alternata*, a filamentous fungus that used polyisoprene as the sole carbon source to break down the rubber [25]. Table 9.2 presents some studies on NR degradation by fungi.

In a recent investigation of NR biodegradation, a naturally selected microbial consortium from a tire disposal site was executed [8]. Using natural cycles of light and dark, soil microcosms were created and cultured at room temperature for 236 days. Laboratory-scale experimental studies were conducted to assess biodegradation, and tests were performed in microcosms comprising soil obtained from a tire dump and NR sheets. The biodegradation experiment evaluated the degree of colonization by

Table 9.2 NR degradation by fungi.

Taxa	Culture time	Weight loss (%)	References
Alternaria alternata	160 days	–	[25]
Aspergillus niger	60 days	28%	[24]
Ceriporiopsis subvermispora Quercus sp.	200 days	2.70%	[42]
Cladosporium cladosporioides	6 months	–	[43]
Fusarium solani	10 days	–	[43]
Paecilomyces lilacinus	20 days	–	[43]
Penicillium chrysogenum	60 days	40%	[24]
Phoma eupyrena	20 days	–	[43]
Rhodotorula mucilaginosa	160 days	–	[25]

microbes, particularly fungi, using SEM (Figure 9.1). NR sheets were sampled periodically from the microcosms and analyzed either untreated or after washing to remove all contaminated biomass. The mycelia appeared in limited areas, mostly superficial patches on the rubber after 42 days of incubation. By 236 days, the mycelium layer had thickened, covering the entire surface, and became strongly anchored to it. The washing process during the two incubation periods also demonstrated the degree of fungal adhesion to the rubber surfaces. The rubber appeared clean after 42 days, with only the footprints of hyphae vanishing, whereas after 236 days, the mycelium remained firmly attached to the surface. Additionally, after 236 days, a dry weight loss of 15.6% in the NR was recorded in the BD microcosms, which was nearly four times higher than the 3.7% loss observed in the BC control. The findings demonstrated that the naturally chosen microbial consortia were capable of biodegrading NR as the sole C source. Moreover, soil mixing proved beneficial by enhancing the biodegradation process, which was predominantly driven by aerobic biomass, particularly filamentous fungi. Nonetheless, NR biodegradation *via* a microbial consortium remains a slow process, often extending over several months or years.

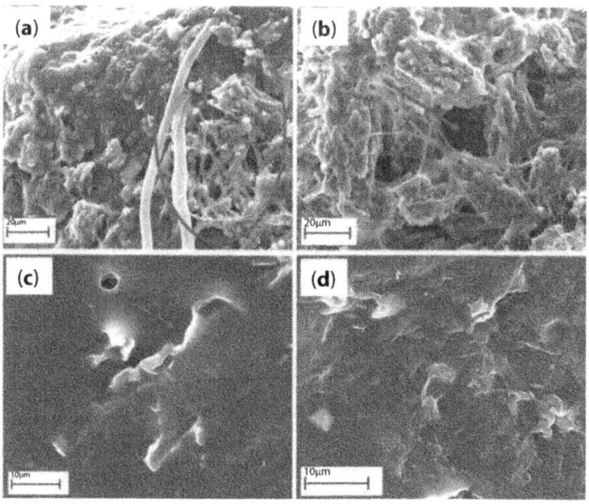

Figure 9.1 SEM images of the mycelium that has grown on the NR surface (SE mode). (a, b) Non-washed NR after 42 and 236 days of incubation respectively, whereas (c, d) are the washed samples. The figure is reprinted from [8]. Copyright (2021), which is distributed under the terms of the Creative Commons CC BY license.

Although NR is a biopolymer, the vulcanization process involving NR latex (NRL) film or rubber production renders it resistant to microbial degradation. Consequently, the waste generated from NR requires an extended period for biodegradation. A recent study by Rosendo *et al.* assessed the impact of crosslink bonding on the degradation of NR latex films containing two accelerators, zinc diethyldithiocarbamate (ZDEC) and tetramethyl thiuram disulfide (TMTD), in different sulfur-based curing systems [44]. After 180 days of soil burial, the NR latex films' mechanical, chemical, and physical properties were assessed to verify their biodegradability. This study examines the NR latex films biodegradation in soil and the relationship between poly-, mono-, and di-crosslinking concentrations. As shown in Figure 9.2, the visible alterations are more noticeable after 90 days of soil burial, regardless of the accelerator content. Considering only the past 90 days, the 0.6/2 sample appears to have been the most degraded at this stage, as it is darker, rougher, and more distorted than the others. All samples showed a decrease in tensile strength with time; however, only the 0.6/0 (ZDEC/TMTD ratio) NR latex film experienced a loss exceeding 50% after 30 days. Moreover, decreased levels of mono- and disulfide bonds led to a more significant decline in tensile strength. The findings demonstrated that films with decreased crosslink density and increased polysulfide bond concentration displayed notable declines in mechanical characteristics, suggesting that these parameters play a role in the biodegradability of the films.

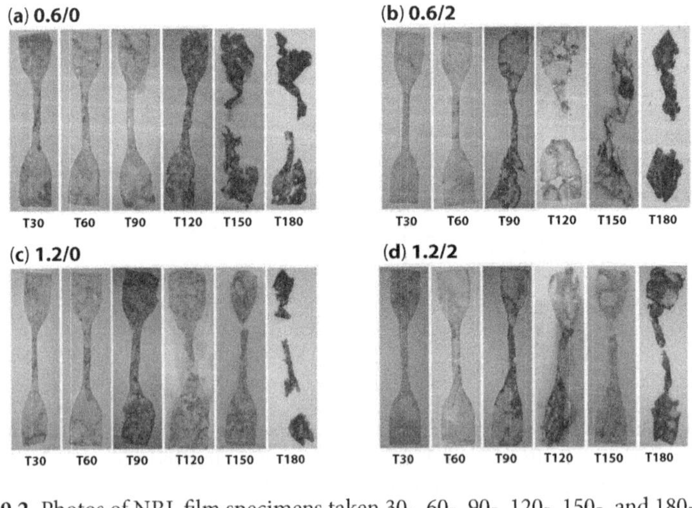

Figure 9.2 Photos of NRL film specimens taken 30-, 60-, 90-, 120-, 150-, and 180-days following soil burial. According to ZDEC/TMTD ratio: (a) 0.6/0; (b) 0.6/2; (c) 1.2/0; and (d) 1.2/2 respectively. The figure is reprinted from [44]. Copyright (2023), with permission from Elsevier.

9.2.3 Analysis of NR Degradation

The degradation of NR is commonly assessed through the utilization of Schiff's reagent, scanning electron microscopy (SEM), Fourier Transform Infrared spectroscopy (FTIR), CO_2 formation analysis, and the effect of pretreatment [9]. The application of Schiff's reagent resulted in the detection of aldehyde groups, as evidenced by the appearance of purple color. This implies the presence of aldehyde-containing degradation products. When *G. polyisoprenivorans* was incubated on NRL gloves, isoprene oligomers, aldehyde, and ketone groups were produced. Visual observations, SEM analysis, Atomic Force Microscopy (AFM) analysis, and modifications in mechanical properties and molar mass are among the techniques used to track changes in rubber specimens after microbial treatment. For example, a comprehensive visual examination of the growth patterns of the Pseudomonas strain (PS) on NRL gloves was conducted using SEM [45]. The proliferation of this variant was accompanied by the emergence of fissures outside several latex particles (Figure 9.3B). After one month, it was observed that PS cells attached to the latex particles and became embedded in the latex structure due to degradation around the cells. This resulted in the formation of cell-shaped cavities (Figures 9.3C and 9.4D).

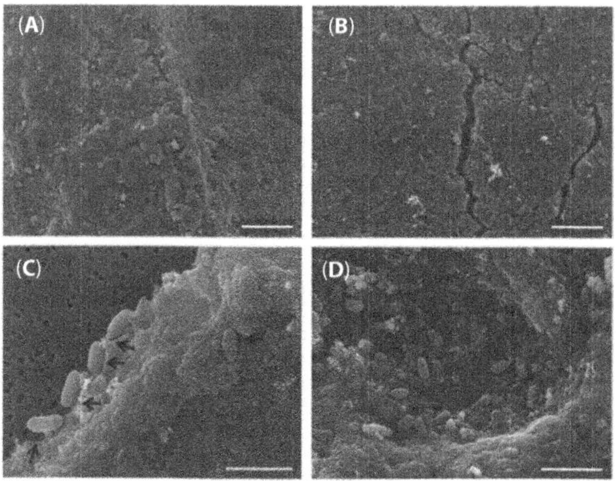

Figure 9.3 SEM photographs of latex deteriorate carried out by Pseudomonas strain (PS). The images display particles from (A) a latex media that was not infected, as well as particles from a 15-day culture (B) and a one-month culture (C and D) of strain PS in the same medium. The arrows denote the cell-shaped niches that are created on the surface of the latex. The scale bars for A and B are 10 mm, while the scale bars for C and D are 2 mm. The figure is reprinted from [45]. Copyright (2014) with permission from PLOS ONE.

The chemical changes that occurred on the rubber surface as a direct result of the biodegradation process were identified using FTIR. For instance, FTIR analysis of NRL gloves after treatment with *Bacillus* sp. AF-666 has also been reported [46]. The rubber glove parts that were subjected to bacterial culture exhibited a reduction in peaks approximately 1,200^{-1},400 cm^{-1}, as compared to the control. This suggests the degradation of significant functional groups such as C double bond C, carbonyl, methyl, and ester bonds. At the peak of 3,034 cm^{-1}, observed in the control, it was no longer detected after microbial treatment in the test sample. This peak was also associated with CH_2 stretching. The peak at 835 cm^{-1}, which corresponds to the cis-1,4 double bond, decreased to 834 cm^{-1}. The presence of aldehydes in the test pieces can be inferred from the observed changes in the range of 1,160 cm^{-1} compared to the control. When compared to the uninoculated controls, the spectra revealed the following variations: (1) ketone and aldehyde group formation, (2) the quantity of cis-1,4 double bonds decreased, and (3) two different bonding conditions were formed [47]. The investigation revealed that the degradation process exhibited oxidative characteristics resembling those of chemical aging. The presence of C–O and C double bond O groups in the spectra of the biodeteriorated rubber samples suggests that the biodegradation process was initiated by monooxygenase.

The next analysis was CO_2 evolution analysis using the Sturm test, a conventional method for assessing biodegradation in laboratory tests [48], providing direct evidence for the bioconversion of the polymer's carbon backbone into a metabolic product. Anaerobic microorganisms generate a mixture of CO_2 and methane as extracellular byproducts of metabolic reactions. The amount of CO_2 released was measured during the cultivation of cells in mineral salt latex medium, where latex served as the sole source of carbon. The CO_2 was captured in a 1 N NaOH solution and quantified by titrating the remaining NaOH with 0.1 N HCl. The Sturm test was used to evaluate CO_2 evolution during rubber biodegradation by *Bacillus* sp. AF-666. The test entailed measuring the change in viable cell count and CO_2 production in both the test and control bottles during a 30-day period at room temperature (30°C). An increase in CFU/ml in the test confirmed the microorganisms' ability to utilize rubber and their growth. Although the Sturm test is inexpensive and straightforward, it is labor-intensive because of the need for regular monitoring to ensure a consistent and regulated flow of CO_2-free air to the test and control vessels.

9.3 Biodegradation of Natural Rubber and Their Composites

Biodegradable NR composites, which include bio-based fillers or natural polymer fillers, exhibit beneficial properties, such as renewability, non-toxicity, and biodegradability. The demand for biodegradable rubber production has been increasing and these materials will coexist with conventional rubber for many years. Moreover, many researchers have verified the substantial potential of natural fillers as alternative reinforcements because of their advantages, including low cost, low density, environmental friendliness, biodegradability, and abundance in nature, as well as their ability to enhance the composite materials' mechanical properties [49–51]. Despite the many significant advantages of natural fillers, they also have drawbacks, such as low thermal stability and high moisture absorption. These weaknesses can lead to poor interfacial adhesion with the matrix, rendering the resulting composites less appealing for certain applications. Consequently, the purpose of this comprehensive chapter is to examine the development of composites from NR and bio-based fillers, with a few examples of NR composite biodegradability studies and reinforcement potential as product applications. Because both the matrix and fillers are derived from natural sources, the resulting composites are biodegradable, environmentally friendly, and sustainable, rendering them particularly noteworthy.

NR composites consist of an NR material that consists of NR matrix and various fillers or reinforcements such as carbon black, silica, clay, cellulose, and starch. These fillers can lower the cost of NR and improve its mechanical, thermal, and electrical properties. However, the presence of fillers can also affect the biodegradability of NR composites depending on their nature, amount, and interaction with the rubber matrix. A biodegradable NR composite is formed by blending NR with other natural or biodegradable materials, resulting in an environmentally friendly composite that can naturally decay or disintegrate over time by microorganisms such as bacteria and fungi [8]. The aim of developing a biodegradable NR composite is to fabricate a material that preserves the favorable characteristics of NR (such as elasticity and strength) while integrating additional biodegradable materials called fillers or additives. Biodegradable fillers can include natural fibers, such as cellulose, jute, hemp, or sisal, as well as other biodegradable materials, such as starch-based components, chitosan, or other plant-derived materials. The biodegradation of NR composites is an important process that can contribute to sustainable and eco-friendly rubber product development and is a potential solution to reduce the environmental impact of the disposal of rubber waste.

By understanding the factors and mechanisms that affect the biodegradability of NR composites, researchers can design and optimize new formulations and processing methods to enhance or control the biodegradation rate and extent of NR composites. The biodegradation process is influenced by many factors, such as the type and concentration of enzymes and microorganisms, temperature and humidity of the environment, oxygen availability, pH level, and presence of nutrients and inhibitors [52]. Depending on relative humidity, microbe reproduction and growth require substantial quantities of water. High levels of humidity enhance hydrolysis by increasing the occurrence of chain-breaking processes [53]. Thus, the polymer degradation rate increased with adequate moisture. Moreover, pH can affect the rate of hydrolysis reactions, and therefore, affect the decomposition rate and growth of microbes. In addition, the reduction temperature of the polymer has a substantial impact on the rate of enzymatic degradation [53]. Enzymes possess distinct active sites that can eliminate a variety of polymers. Biodegradation of NR composites is affected by the characteristics of the composite itself, such as the composition and structure of the NR matrix and the fillers or reinforcements, the degree of crosslinking or vulcanization of the rubber matrix, and the surface area and morphology of the NR composite. In addition, the biodegradation of NR composites involves various mechanisms and pathways (Figure 9.4), with some of the possible mechanisms being hydrolysis, oxidation, depolymerization, assimilation, and mineralization. Moreover, some of the possible pathways include aerobic biodegradation (involving oxygen), anaerobic biodegradation (involving no oxygen), abiotic degradation (involving non-biological factors), or a combination of these pathways [9].

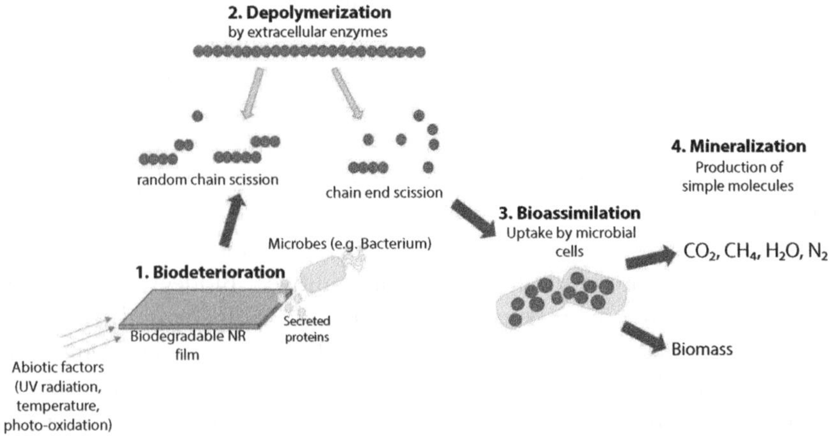

Figure 9.4 Schematic mechanism and pathways of the NR biodegradation.

Several techniques for the biodegradation of NR composites include the following:

(1) Composting: The process of organic matter decomposition in a controlled environment where factors such as temperature, moisture, and oxygen levels are regulated. Composting can be employed to assess the biodegradability of NR composites [54] by quantifying the weight reduction, emission of carbon dioxide, or microbial activity of the samples [55].
(2) Soil burial: A method of burying NR composites in soil and subjecting them to natural environmental conditions including humidity, temperature, and microbes. Soil burial can be employed to assess the biodegradability of NR composites by quantifying the reduction in weight, tensile strength, or alterations in the morphology of the samples [56–60].
(3) Microcosm study: A microcosm study is a useful method for examining the biodegradation mechanisms, kinetics, and routes of NR composites [8]. This is achieved by assessing the chemical, physical, and biological changes that occur in the samples and microorganisms involved. Microcosm investigation is a more precise and regulated approach than soil burial because it enables the manipulation of different environmental factors and microbial communities.

Overall, these methods can measure changes in the physical, chemical, and biological properties of NR composites after exposure to different environmental conditions and microbial communities. Some of the parameters that can be measured are the weight loss, tensile strength, molecular weight, viscosity, carbon dioxide production, and microbial population.

In addition, rubber compounding is a necessary step in the standard rubber production process that makes rubber practically usable for a variety of uses. The vulcanization process is the most important phase of rubber compounding because it produces crosslinking between rubber chains, which improves the physical and mechanical properties. During this phase, fillers and other additives are frequently introduced, such as reinforcing fillers to improve the characteristics, and dyes to change the appearance of the product. In recent years, many reports have surfaced regarding the use of natural fillers as additives or blending components for NR, and the incorporation of such materials has resulted in improved biodegradability of the products by supplying nutrients for microbes during the biodegradation process.

9.3.1 Biodegradation of NR/Cellulose Composite

Cellulose is a widely recognized biodegradable material and several experiments conducted in accordance with ISO 17556 or ASTM D5988 standard tests have yielded cellulose biodegradation results ranging from 80% to 85% [61]. Cellulose, a natural polymer, is increasingly being used in bio-based products owing to its biodegradability, renewability, nontoxicity, economic value, and mechanical properties. Moreover, its low density, thermal conductivity, corrosion resistance, and ease of fabrication make it a popular choice for composite films [62]. Owing to its crystalline structure, cellulose is not soluble in typical organic solvents; however, its mechanical strength and water resistance make it a useful thickening agent in industrial applications. Cellulose can be employed as a natural reinforcing filler to enhance the mechanical and biodegradable properties of various composites. Since cellulose raw materials can be extracted from a wide range of agricultural wastes, which are a plentiful source of renewable biomass, NR composites with cellulosic fiber reinforcements are currently considered an important area of study worldwide.

Recently, biodegradable composite sponges of NR and cellulose fibers (NR-C) were produced using the Dunlop process has been reported [63]. Microcellulose fibers from eucalyptus pulp were used as reinforcement fillers, while sodium alginate (SA) was used as a dispersing agent to enhance the compatibility between cellulose and NR. The NR-C sponges exhibited a

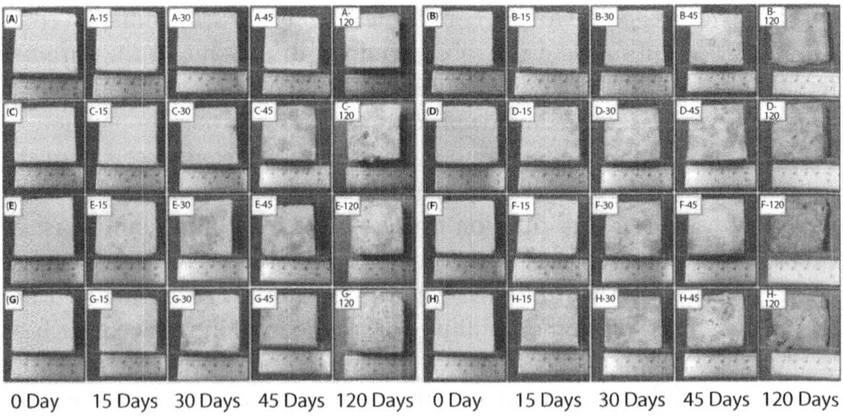

Figure 9.5 Degradation behavior of composite sponges at 0, 15, 30, 45, and 120 days in soil: (A) NR; (B) NR+SA1, (C) NR-C15, (D) NR-C15+SA1, (E) NR-C30, (F) NR-C30+SA1, (G) NR-C45, and (H) NR-C45+SA1. The figure is reprinted from [63]. Copyright (2023), with permission from Elsevier.

better compressive stress, modulus of elasticity, and efficient water absorption capacity. Furthermore, NR-C with the highest cellulose content (45 per hundred rubbers (phr)) and sodium alginate could be biodegraded faster in soil by approximately 21–27% in 120 days, compared to NR sponges (Figure 9.5). Owing to their beneficial properties, sponges are potentially employed as adsorbents or supporting materials in a range of applications, such as agriculture uses like hydroponic planting materials or seed germination sponges, as well as biomedical applications such as wound dressing.

9.3.2 Biodegradation of NR/Starch Composite

Starch is a natural fiber polymer classified as a carbohydrate and is composed of repeating glucose units. Starch is abundant in nearly all plants because of photosynthesis and serves as the primary energy storage compound. Its abundant supply, renewable nature, affordability, and capacity for decomposition make it a valuable resource to produce bioplastics. Starch contains an integral protein that resides in its granular structure. This protein synthesizes a complex carbohydrate chain consisting of amylose and amylopectin, which is consumed by soil bacteria to generate an enzyme that breaks down the matrix, similar to the NR latex backbone chain. For example, NR latex composite films, one with no sago starch (control) and the other with 10 phr sago starch loading, were placed on compost soil for 4 weeks [64]. Biodegradation studies were conducted using films to measure water vapor transmission (WVT) and mass loss. The results demonstrated that the addition of sago starch enhanced the growth and spread of microorganism colonies on NR latex films during biodegradation. Moreover, the results further validated the consumption of sago starch granules as an exclusive energy source for microbial development. The use of sago starch as a substrate for the formation of microbial colonies during the biodegradation of NR latex films yielded favorable outcomes [64].

The utilization of nutrients in NR-filled sago starch latex films was evaluated by examining the microorganism's hyphal structure, which allows for the exploration of nutrients and the growth of mycelium. This process regulates biomass production and promotes active branching. As resources diminish, the mycelium experiences optimal growth and regulates the branching of hyphae to explore NRL film substrates for additional nutrient-rich locations. Mycelium-forming actinomycetes on the surfaces of SS/NRL films after 7 weeks of incubation (Figure 9.6) showed that mycelial hyphal structures envelop sago starch granules and utilize them as a source of nutrition for their growth. The absorption of nutrients occurs through

Figure 9.6 Mycelium-forming actinomycetes on the surfaces of SS/NRL films (10 phr of SS) with hyphae surrounding the SS granules (after 7 weeks of incubation). The figure has been reprinted from [64] and is the author's copyright.

the secretion of an enzyme that breaks down complex amylose and amylopectin chains into smaller compounds. The presence of a mycelial structure on SS/NRL films suggests that these microorganisms can metabolize natural compounds as their energy source. Figure 9.7(a) depicts the densities of the microorganism colonies on NRL surfaces. Observations revealed the formation of an extensive network of hyphae and mycelial structures when the NRL substrates were depleted, resulting in the formation of a hole, as indicated in Figures 9.7(b,c). The formation of an extensive network of hyphae and mycelial structures when NRL substrates were depleted resulted in the formation of a hole in areas where nutrients were reduced. The mycelium functions as a regulator to distribute freshly

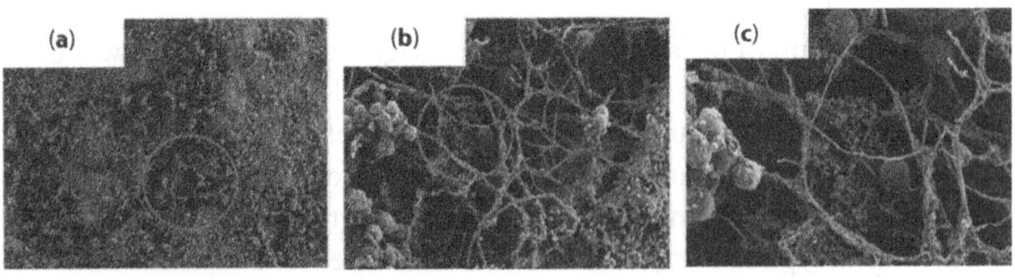

Figure 9.7 Micrograph images of (a) mycelium densities in the hollow region on the surface of SS/NRL film, (b) enlarged focus area ×500 of (a), and (c) enlarged focus area ×1,000 of the red circle from (b). The figure has been reprinted from [64], and is the author's copyright.

produced biomass, supporting the growth of hyphae and development of new branches.

Another NR/starch-based biodegradation study was reported on biopolymer sheets composed of a combination of maleated epoxidized natural rubber (MENR) and natural rubber-g-cassava starch (NR-g-CSt) [65]. In Figure 9.8, it is noticeable that the NR-g-CSt/MENR 10:100 and 20:100 blends underwent total biodegradation within a period of 3 months, resulting in the absence of any detectable specimens. However, samples containing larger amounts of MENR (50, 70, and 100) in the blend exhibited a prolonged degradation period. The samples exhibited a decrease in water absorption owing to the presence of MENR. The reduction in water absorption results in a diminished rate of starch biodegradation due to the indispensability of water for fungal development and the impact of physical variables on enzyme activity.

In another biodegradation study, NR/starch composite work involved blending sago starch fillers with NR latex using two different conditions:

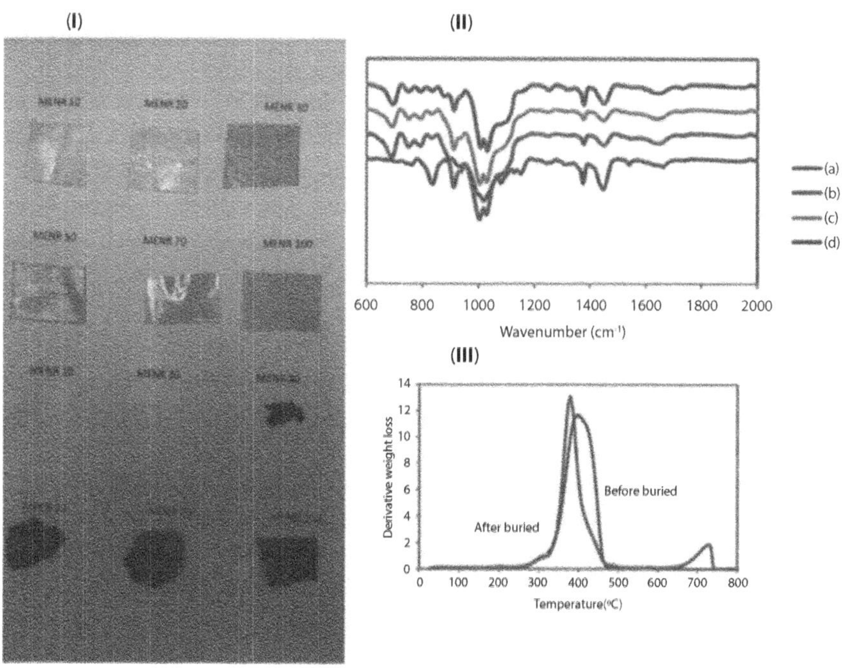

Figure 9.8 The photogram image depicts the NR-g-CSt mix at different ratios of MENR before and after a three-month burial period in natural soil. The figure has been reprinted from [65]. Copyright (2015), with permission from Elsevier.

prior pre-vulcanization and after pre-vulcanization [66]. The composites were buried and evaluated weekly for their mechanical and morphological properties. The optimal starch filler loading was attained at 10 phr, with good tensile properties. Sago starch fillers added before pre-vulcanization improved the filler–rubber interaction, which also demonstrated the practicability of NR latex composite biodegradation.

9.3.3 Biodegradation of NR with other Natural Fibers

There has been a worldwide increase in the manufacturing of high-performance materials derived from affordable and easily accessible natural resources, such as natural fibers. The rising environmental consciousness in society and the more stringent legal obligations concerning environmental protection, together with the implementation of sustainable development principles, demand the ongoing pursuit and advancement of eco-friendly solutions, especially biodegradation of polymer materials. There is an abundance of research on biodegradable NR rubber utilizing natural fibers [67]; for example, a sisal/oil palm hybrid fiber that also reinforced NR composites and their biodegradability and aging was studied by Jacob *et al.* [68]. It was found that the fiber content and chemical treatment caused degradation of the composites. However, chemically modified composites were shown to be less susceptible to biodegradation owing to improved fiber/matrix adhesion and the coupling agents acting as a protective barrier against microorganism action. In terms of preventing degradation, alkali treatment was shown to be more successful than silane coupling agents. The degree of degradation after 12 months of soil exposure was examined using SEM, which showed that the chemically treated rubber composites were more resistant to deterioration. Additionally, the results of the tensile test were also apparent in the SEM images. It was observed that among the fiber composites treated with silane coupling agents, the vinyl silane-treated composite experienced the most significant reduction in tensile strength (approximately 19%), which is consistent with the SEM observations.

NR can also be incorporated into rattan (*Calamus manan*) powder. Rattan is a natural fiber made of cellulose and lignin, which is susceptible to decomposition by microorganisms. NR/rattan composites were added to a silane coupling agent to form better interfacial adhesion between the filler and matrix [69]. However, the silane coupling agent also protected the rattan from degradation.

Another example of a modified natural fiber as filler for NR composites is the sisal fiber (SF) filler coated with styrene butyl acrylate (SBA) [70]. SBA-coated sisal fiber (CF) was reported as a reinforcement biofilter for the NR matrix by enhancing its interfacial adhesion and tensile strength. The coating process also improved the water resistance, particularly at high loading levels. However, when compared to uncoated SF (UF), the biodegradation of the CF/NR composites was retarded because of the coating of SF with hydrophobic SBA, which also reduced the water absorption of the CF/NR composites. This may be due to the coating that protected SF from microorganisms and reduced moisture absorption, which is primarily responsible for the degradation process and essential for the biodegradation process.

Reinforced NR composites with black rice husk ash (BRHA) and white rice husk ash (WRHA) were produced and biodegradability analysis was performed [71]. This process involved the use of latex aqueous microdispersion. The NR matrix may accommodate up to 100 phr of rice husk ashes (RHAs) without experiencing phase separation. The mechanical properties of the composite films were significantly better than those of the pure NR films. The mechanical properties of the composite films reinforced with WRHA were superior to those of films reinforced with BRHA. However, the composites filled with BRHA exhibited a larger elongation at break. The addition of calcium chloride ($CaCl_2$) enhanced the tensile strength of the film, while reducing its flexibility. After 3 months, the films exhibited biodegradability in soil (Figure 9.9), with a

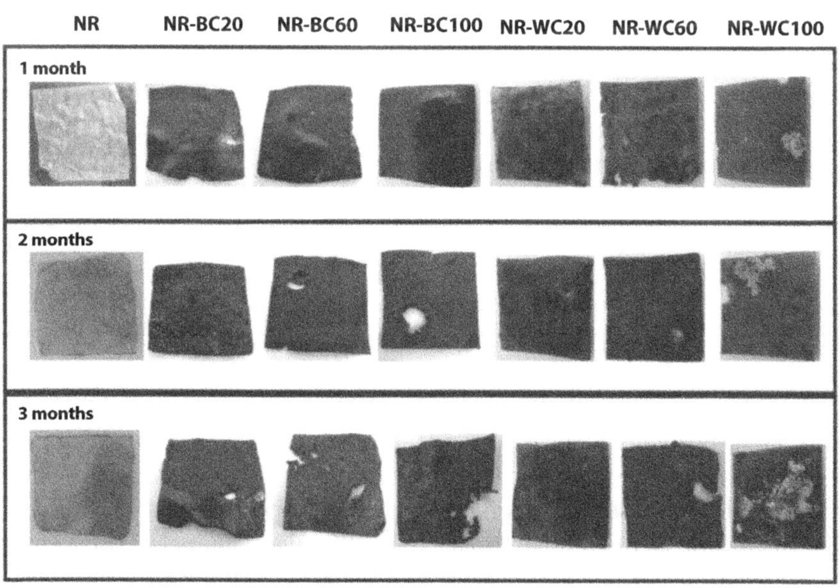

Figure 9.9 An examination of the degradation of NR-BC and NR-WC in soil over a period of 3 months using visual analysis. The figure has been reprinted from [71]. Copyright (2021), by publisher MDPI, Basel, Switzerland.

weight reduction ranging from 7.6% to 18.3% of the original dry weight. The addition of RHAs increased the dielectric constant and dielectric loss factor of the composite films. Based on the acquired features, these composites show promise for extended advancement as flexible conductive substrates or semiconducting polymer films for electronic products.

9.4 Biodegradation of NR and their Nanocomposites

The development of advanced nanotechnology in NR composites has been strengthened by nanosized renewable fillers. The interface between the NR matrix and nanofillers has a substantial impact on the mechanical, thermal, and barrier properties of the NR nanocomposites. Research on NR has shifted towards using bionanomaterials instead of synthetic materials, creating environmentally friendly nanocomposites using natural fillers. These biomaterials aim to reduce the use of hazardous synthetic materials and demonstrate exceptional qualities that surpass those of traditional composites. A nanocomposite is a solid material with many phases, one of which has dimensions in the nanoscale range. Compared with traditional composites, the addition of a small quantity (between 5 and 10 wt.%) of fillers in the nanometric range will result in highly potent reinforcing effects [72]. This was attributed to the significant increase in the surface-to-volume ratio, resulting in a substantial surface area available for interaction with the matrix [73]. Nanocomposite materials have attracted the attention of researchers and industrialists. Compared to natural nanomaterials, synthetic nanomaterials lack facile processability, biocompatibility, and biodegradability [74]. Moreover, biopolymers and nanosized reinforcements can be used to produce biodegradable nanocomposites with improved thermal, barrier, and mechanical properties.

Researchers have been working on the processing of nanocomposites to enhance their mechanical properties. These materials use a matrix with nanosized reinforcement elements, which are considered nanoparticles when at least one dimension is <100 nm. The growing interest in bio-based products reduces dependence on fossil fuels and offers unique properties that are not found in conventional composites. Therefore, biodegradable bio-based nanocomposites are being considered as materials for the future. Depending on their biological origin, nanoelements can be prepared from a variety of renewable resources, including chitin, starch, and cellulose. Bio-based nanocomposites are highly intriguing in this age of technological advancement and where a sustainable economy is at the forefront. Given the widespread recognition of bio-based nanocomposites as biodegradable nanomaterials, there is a need for publications that specifically address the biodegradation investigations of NR-based nanocomposites. Nonetheless, the following biodegradability analyses of NR nanocomposites are reviewed.

9.4.1 Biodegradation of NR/Nanocellulose Composite

Cellulose, especially in nanoform (nanocellulose) can meet the reinforcing filler requirements of elastomers, which are a key type of polymeric material with a unique mix of strength and flexibility. Because nanocellulose has a high degree of tailoring ability, the characteristics of elastomer nanocomposites can be modified to meet specific performance requirements [8]. Nanocellulose can be used to create green NR composites with good thermal and mechanical properties [75] and is commonly regarded as an environmentally friendly additive for enhancing the strength of elastomers [76].

Cellulose whiskers, derived from bleached sugarcane bagasse kraft pulp, were employed as reinforcing fillers in NR by the casting method to produce biodegradable nanocomposites [77]. Nanoparticles that resemble long crystalline rods are called "whiskers" for short, often produced by acid hydrolysis after the digestion of amorphous cellulosic domains. Incorporating cellulose whiskers into a hydrophobic polymer matrix, such as rubber, reduces the water vapor permeability while enhancing the thermomechanical properties and tensile strength of the nanocomposites. Additionally, cellulose whiskers considerably boost the biodegradation of rubber in the soil. For example, neat rubber lost 19% of its weight after 4 weeks, whereas NR with 7.5% and 12.5% cellulose whiskers lost 62% and 71%, respectively (Figure 9.10). This is because of the faster disintegration of cellulose, which microorganisms consume faster than rubber. This process leads to increased porosity and void formation, weakening the rubber film and causing it to fragment into smaller pieces. As a result, nanocomposite films with cellulose whiskers degraded more quickly than those without cellulose. The accelerated degradation of cellulose is due to the microbial disruption of bonds within cellulose macromolecules [77].

A comparative study conducted by Abraham *et al.* (2012) examined the mechanical properties and biodegradation of NR composites with nanocellulose using the vermicomposting method, focusing on the crosslinking of the matrix [78]. *Eudrilus eugeniae*, an epigeic vermicompost worm, was used to decompose the matrix. The addition of nanocellulose increased the tensile strength and modulus of both the crosslinked and non-crosslinked composites. The crosslinked composites exhibited the most favorable

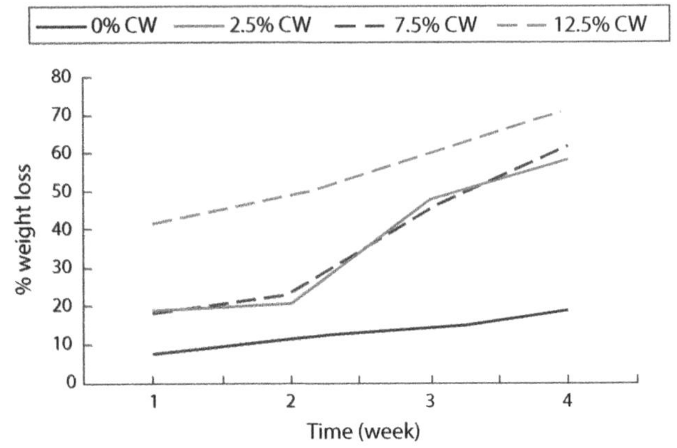

Figure 9.10 Bio-disintegration of rubber/cellulose whiskers nanocomposites in soil (up to 4 weeks). The figure has been reprinted from [77]. Copyright (2010) by Elsevier.

characteristics. However, the degradation rate of the non-crosslinked samples was higher than that of the crosslinked ones. This study observed that the biodegradation rate using vermicomposting was relatively greater in non-cross-connected composites than in crosslinked composites. In addition, this study demonstrated the efficacy of *E. eugeniae* in NR/cellulose composites, regardless of whether they were crosslinked or non-crosslinked. Figure 9.11 shows the composite sample photos, both crosslinked and non-crosslinked, after the biodegradation test.

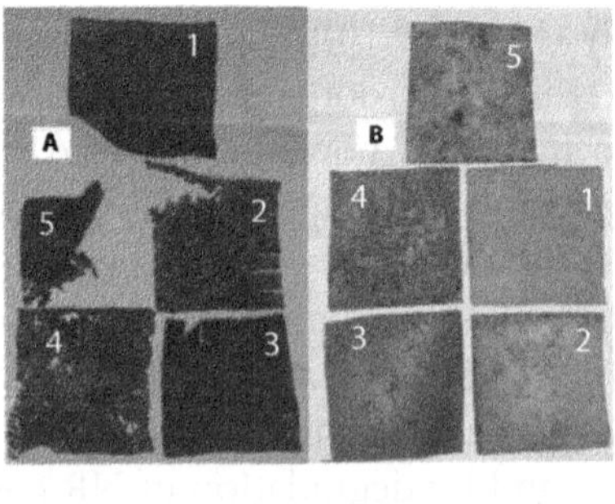

Figure 9.11 The post-biodegradation pictures of the NR composite samples.
(A) Composite without crosslinking and (B) Composite with crosslinking. The figure has been reprinted from [78] with copyright (2012) from Elsevier.

9.4.2 Biodegradation of NR/Starch Nanocrystal Composite

Recently, there has been growing interest in starch nanocrystal (SNC)—as a nanoscale material because of its abundant, renewable, and biodegradable nature. SNCs are platelet-like particles produced when the semicrystalline structure of starch granules is disrupted by the acid hydrolysis of amorphous components [79]. SNCs are highly transparent and have

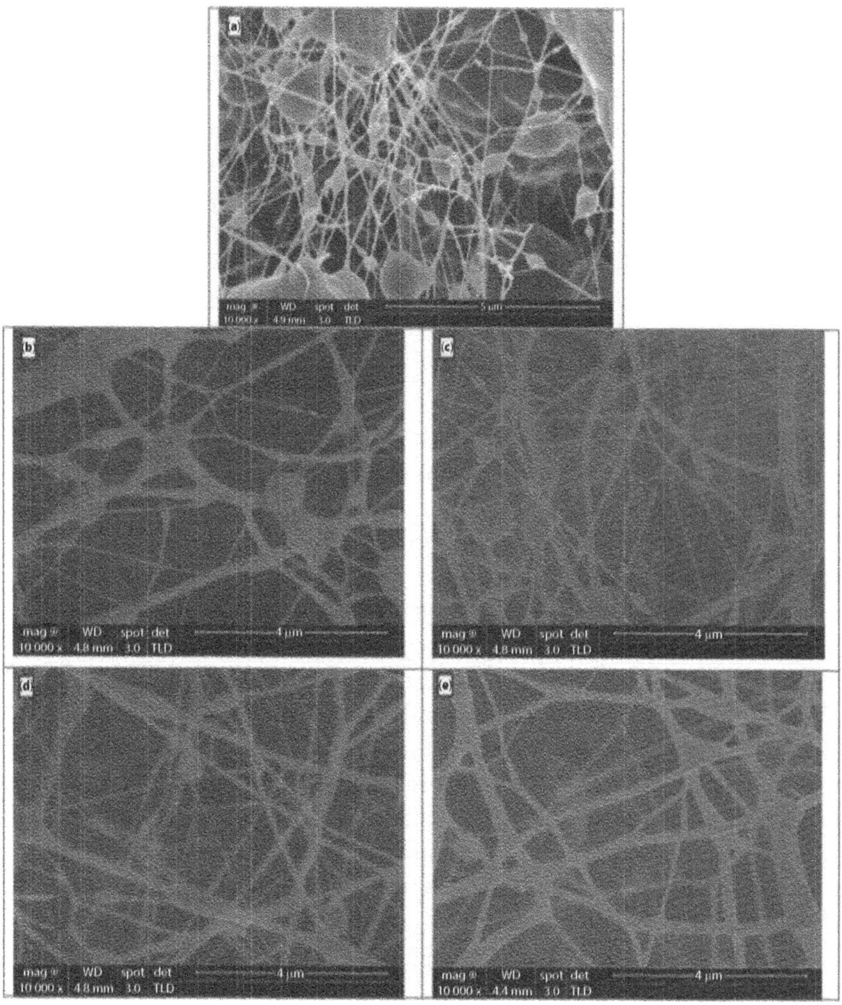

Figure 9.12 SEM images of (a) CNR/PVA, (b) CNR/PVA/K-2.5, (c) CNR/PVA/K-5, (d) CNR/PVA/S-2.5, and (e) CNR/PVA/S-5. The figure has been reprinted from [80]. Copyright (2019) by Elsevier.

the potential to enhance the mechanical, barrier, and electrical characteristics of the polymer matrices. Consequently, many polymers have been combined with SNCs as nanofillers for the preparation of polymer nanocomposites. A modified NR, chloroacetated NR with Polyvynil alcohol (CNR/PVA), was produced as a film and encapsulated kaolin and maize starch at a fixed weight ratio [80]. The aim was to create non-woven nanofiber films for biomedical applications using biodegradable materials with added fillers. Encapsulation of starch and kaolin enhanced the mechanical properties and increased the rate of degradation. Biodegradation of kaolin nanofiber films was observed in vials containing phosphate-buffered saline (PBS) with alpha-amylase. The biodegradation rate of encapsulated kaolin nanofiber films was comparatively lower than that of starch encapsulation, both in the absence and presence of enzymes. The most significant enzymatic degradation was observed at a starch concentration of 2.5 wt.%, but, when starch concentration was increased to 5%, the degradation rate decreased slightly, possibly due to a denser scaffold structure as agreed as SEM micrograph shown in Figure 9.12.

9.5 Conclusions

Recently, there has been a surge in the utilization of biodegradable materials for a wide variety of applications owing to their renewability, sustainability, and eco-friendliness. In addition, advantageous qualities, including biocompatibility, biodegradability, nontoxicity, and low density, serve as catalysts for the substitution of synthetic materials. The significance of natural fillers such as cellulose, chitin, and starch in the development of biodegradable NR composites and nanocomposites is outlined in this chapter. The utilization of these fillers is crucial for the development of reinforced NR composites. In addition, the natural nanofillers demonstrated enhanced strength owing to augmentation of the specific area. Although numerous researchers have described various mechanisms for achieving enhanced properties, no definitive conclusions have been reached regarding the most effective mechanisms. By implementing recently developed characterization techniques and a theoretical approach, the efficacy of the reinforcement mechanisms can be optimized. In addition, the biodegradability of the composites is affected by the biodegradation environment and conditions. This implies that comparing the biodegradability of polymers based solely on the extent of weight loss or biodegradation reported in various studies is unreasonable. Nevertheless, research on the biodegradation of polymers under similar conditions is limited.

Additionally, the present review relied on assessing the impact of natural fibers, especially cellulose and starch, as additives on the biodegradability of the resultant NR composites. Despite the advantageous properties exhibited by biodegradable materials, certain limitations significantly impede their potential for broader applications. For example, cellulose, glucose, and chitin are incompatible with NR; thus, surface modification of the fillers or NR matrix is required to circumvent this drawback. As a result, a considerable number of researchers have focused on improving the interfacial adhesion and interaction between natural additives and NR matrices. Additionally, they emphasized the potential of these composite materials for a wider range of applications. Despite the widespread use of biodegradable NR composites, additional research is required to identify the following significant drawbacks: (i) inadequate durability, (ii) susceptibility to moisture absorption, and (iii) inability for outdoor and high load applications. This review concludes that the potential of biodegradable NR materials with enhanced efficiency could contribute to the prevention of environmental damage.

References

1. Thomas, S.K., Parameswaranpillai, J., Krishnasamy, S., Begum, P.M.S., Nandi, D., Siengchin, S., George, J.J., Hameed, N., Salim, N.V., Sienkiewicz, N., A comprehensive review on cellulose, chitin, and starch as fillers in natural rubber biocomposites. *Carbohydr. Polym. Technol. Appl.*, 2, 100095, 2021.
2. Xu, A., Zhang, X., Wu, S., Xu, N., Huang, Y., Yan, X., Zhou, J., Cui, Z., Dong, W., Pollutant Degrading Enzyme: Catalytic Mechanisms and Their Expanded Applications. *Molecules*, 26, 16, 475, 2021.
3. Liu, D., Maguire, R.J., Lau, Y.L., Pacepavicius, G.J., Okamura, H., Aoyama, I., Factors affecting chemical biodegradation. *Environ. Toxicol.*, 15, 5, 476–483, 2000.
4. Nhung, N.T.H., Nguyen, X.T., Long, V.D., Wei, Y., Fujita, T., A Review of Soil Contaminated with Dioxins and Biodegradation Technologies: Current Status and Future Prospects. *Toxics*, 10, 6, 278, 2022.
5. Lipczynska-Kochany, E., Effect of climate change on humic substances and associated impacts on the quality of surface water and groundwater: A review. *Sci. Total Environ.*, 640–641, 1548–1565, 2018.
6. Nguyen, L.H., Nguyen, H.D., Tran, P.T., Nghiem, T.T., Nguyen, T.T., Dao, V.L., Phan, T.N., To, A.K., Hatamoto, M., Yamaguchi, T., Kasai, D., Fukuda, M., Biodegradation of natural rubber and deproteinized natural rubber by enrichment bacterial consortia. *Biodegradation*, 31, 4, 303–317, 2020.

7. Ahmad, N., Abnisa, F., Daud, W.M.A.W., Potential use of natural rubber to produce liquid fuels using hydrous pyrolysis – a review. *RSC Adv.*, 6, 73, 68906–68921, 2016.
8. Bosco, F. and Mollea, C., Biodegradation of Natural Rubber: Microcosm Study. *Water Air Soil Pollut.*, 232, 6, 227, 2021.
9. Ali Shah, A., Hasan, F., Shah, Z., Kanwal, N., Zeb, S., Biodegradation of natural and synthetic rubbers: A review. *Int. Biodeterior. Biodegrad.*, 83, 145–157, 2013.
10. Mollea, C. and Bosco, F., Natural rubber biodegradation by Alternaria alternata and Penicillium chrysogenum isolates. *Biorem. J.*, 24, 2–3, 112–128, 2020.
11. Basik, A.A., Sanglier, J.J., Yeo, C.T., Sudesh, K., Microbial Degradation of Rubber: Actinobacteria. *Polymers (Basel)*, 13, 12, 1989, 2021.
12. Baumeister, D., Berekaa, M., Kroppenstedt, R., Linos, A., Taxonomic characterization of two rubber degrading bacteria belonging to the species Gordonia polyisoprenivorans and analysis of hyper variable regions of 16S rDNA sequences. *FEMS Microbiol. Lett.*, 205, 2, 277–282, 2001.
13. Arenskötter, M., Bröker, D., Steinbüchel, A., Biology of the metabolically diverse genus Gordonia. *Appl. Environ. Microbiol.*, 70, 6, 3195–3204, 2004.
14. Linos, A., Berekaa, M.M., Steinbüchel, A., Kim, K.K., Sproer, C., Kroppenstedt, R.M., Gordonia westfalica sp. nov., a novel rubber-degrading actinomycete. *Int. J. Syst. Evol. Microbiol.*, 52, 4, 1133–1139, 2002.
15. Yahya, Y.S.R., Azura, A.R., Ahmad, Z., Effect of curing systems on thermal degradation behaviour of natural rubber (SMR CV 60). *J. Phys. Sci.*, 22, 2, 1–14, 2011.
16. Fainleib, A., Pires, R.V., Lucas, E.F., Soares, B.G., Degradation of non-vulcanized natural rubber-renewable resource for fine chemicals used in polymer synthesis. *Polímeros*, 23, 441–450, 2013.
17. Mastalygina, E., Varyan, I., Kolesnikova, N., Gonzalez, M.I.C., Popov, A., Effect of natural rubber in polyethylene composites on morphology, mechanical properties and biodegradability. *Polymers (Basel)*, 12, 2, 437, 2020.
18. Francesca, B. and Chiara, M., Mycoremediation in Soil, in: *Environmental Chemistry and Recent Pollution Control Approaches*, S.-N. Hugo, M.-T. Mario Alfonso, F. Robina, D. Rajendra, R. Sara, (Eds.), IntechOpen, Rijeka, 2019.
19. Muniandy, K., Othman, N., Ismail, H., Characterization and Properties of Rattan Fibre/Natural Rubber Biocomposites, in: *Green Biocomposites: Manufacturing and Properties*, M. Jawaid, S.M. Sapuan, O.Y. Alothman, (Eds.), Springer International Publishing, Cham, 2017.
20. Sarkar, B. and Mandal, S., Microbial Degradation of Natural and Synthetic Rubbers, in: *Microbial Bioremediation & Biodegradation*, M.P. Shah, (Ed.), Springer Singapore, Singapore, 2020.
21. Kasai, D., Poly(cis-1,4-isoprene)-cleavage enzymes from natural rubber-utilizing bacteria. *Biosci. Biotechnol. Biochem.*, 84, 6, 1089–1097, 2020.

22. Yikmis, M. and Steinbüchel, A., Historical and recent achievements in the field of microbial degradation of natural and synthetic rubber. *Appl. Environ. Microbiol.*, 78, 13, 4543–4551, 2012.
23. De Vries, O., Zersetzung von Kautschuk-Kohlenwasserstoff durch Pilze. *Zentbl. Bakteriol. Parasitenkd. Infektionskr.*, 74, 22–24, 1928.
24. Nayanashree, G. and Thippeswamy, B., Natural rubber degradation by Aspergillus niger and Penicillium sp. *Int. J. Recent Sci. Res.*, 4, 9, 1337–1341, 2013.
25. Bosco, F., Antonioli, D., Casale, A., Gianotti, V., Mollea, C., Laus, M., Malucelli, G., Biodegradation of unvulcanized natural rubber by microorganisms isolated from soil and rubber surface: A preliminary study. *Biorem. J.*, 22, 1–2, 43–52, 2018.
26. Birke, J., Röther, W., Jendrossek, D., RoxB Is a Novel Type of Rubber Oxygenase That Combines Properties of Rubber Oxygenase RoxA and Latex Clearing Protein (Lcp). *Appl. Environ. Microbiol.*, 83, 14, e00721-17, 2017.
27. Seidel, J., Schmitt, G., Hoffmann, M., Jendrossek, D., Einsle, O., Structure of the processive rubber oxygenase RoxA from Xanthomonas sp. *Proc. Natl. Acad. Sci.*, 110, 34, 13833–13838, 2013.
28. Sharma, V., Siedenburg, G., Birke, J., Mobeen, F., Jendrossek, D., Prakash, T., Correction: Metabolic and taxonomic insights into the Gram-negative natural rubber degrading bacterium Steroidobacter cummioxidans sp. nov., strain 35Y. *PLoS One*, 13, 7, e0200399, 2018.
29. Tsuchii, A. and Takeda, K.J.A., Rubber-degrading enzyme from a bacterial culture. *Appl. Environ. Microbiol.*, 56, 1, 269–274, 1990.
30. Kasai, D., Imai, S., Asano, S., Tabata, M., Iijima, S., Kamimura, N., Masai, E., Fukuda, M., Identification of natural rubber degradation gene in Rhizobacter gummiphilus NS21. *Biosci. Biotechnol., Biochem.*, 81, 3, 614–620, 2017.
31. Linos, A., Reichelt, R., Keller, U., Steinbüchel, A., A gram-negative bacterium, identified as Pseudomonas aeruginosa AL98, is a potent degrader of natural rubber and synthetic cis-1, 4-polyisoprene. *FEMS Microbiol. Lett.*, 182, 1, 155–161, 2000.
32. Bode, H.B., Kerkhoff, K., Jendrossek, D., Bacterial degradation of natural and synthetic rubber. *Biomacromolecules*, 2, 1, 295–303, 2001.
33. Imai, S., Ichikawa, K., Muramatsu, Y., Kasai, D., Masai, E., Fukuda, M., Isolation and characterization of Streptomyces, Actinoplanes, and Methylibium strains that are involved in degradation of natural rubber and synthetic poly(cis-1,4-isoprene). *Enzyme Microb. Technol.*, 49, 6, 526–531, 2011.
34. Rose, K., Tenberge, K.B., Steinbüchel, A., Identification and Characterization of Genes from Streptomyces sp. Strain K30 Responsible for Clear Zone Formation on Natural Rubber Latex and Poly(cis-1,4-isoprene) Rubber Degradation. *Biomacromolecules*, 6, 1, 180–188, 2005.

35. Nanthini, J. and Sudesh, K., Biodegradation of Natural Rubber and Natural Rubber Products by Streptomyces sp. Strain CFMR 7. *J. Polym. Environ.*, 25, 3, 606–616, 2017.
36. Watcharakul, S., Röther, W., Birke, J., Umsakul, K., Hodgson, B., Jendrossek, D., Biochemical and spectroscopic characterization of purified Latex Clearing Protein (Lcp) from newly isolated rubber degrading Rhodococcus rhodochrous strain RPK1 reveals novel properties of Lcp. *BMC Microbiol.*, 16, 1–13, 2016.
37. Berekaa, M.M., Linos, A., Reichelt, R., Keller, U., Steinbüchel, A., Effect of pretreatment of rubber material on its biodegradability by various rubber degrading bacteria. *FEMS Microbiol. Lett.*, 184, 2, 199–206, 2000.
38. Luo, Q., Hiessl, S., Poehlein, A., Steinbüchel, A., Microbial gutta-percha degradation shares common steps with rubber degradation by Nocardia nova SH22a. *Appl. Environ. Microbiol.*, 79, 4, 1140–1149, 2013.
39. Sarkar, B., Gupta, A.M., Shah, M.P., Mandal, S., Poly-cis-isoprene Degradation by Nocardia sp. BSTN01 Isolated from Industrial Waste. *Appl. Biochem. Biotechnol.*, 194, 8, 3333–3350, 2022.
40. Ibrahim, E.M.A., Arenskötter, M., Luftmann, H., Steinbüchel, A., Identification of poly (cis-1, 4-isoprene) degradation intermediates during growth of moderately thermophilic actinomycetes on rubber and cloning of a functional lcp homologue from Nocardia farcinica strain E1. *Appl. Environ. Microbiol.*, 72, 5, 3375–3382, 2006.
41. Linh, D.V., Huong, N.L., Tabata, M., Imai, S., Iijima, S., Kasai, D., Anh, T.K., Fukuda, M., Characterization and functional expression of a rubber degradation gene of a Nocardia degrader from a rubber-processing factory. *J. Biosci. Bioeng.*, 123, 4, 412–418, 2017.
42. Sato, S., Honda, Y., Kuwahara, M., Kishimoto, H., Yagi, N., Muraoka, K., Watanabe, T., Microbial scission of sulfide linkages in vulcanized natural rubber by a white rot basidiomycete, ceriporiopsis s ubvermispora. *Biomacromolecules*, 5, 2, 511–515, 2004.
43. Borel, M., Kergomard, A., Renard, M.F., Degradation of natural rubber by Fungi imperfecti. *Agric. Biol. Chem.*, 46, 4, 877–881, 1982.
44. de Lima, D.R., Vieira, I.R.S., Dutra da Rocha, E.B., Furtado de Sousa, A.M., Augusto da Costa, A.C., Furtado, C.R.G., Biodegradation of natural rubber latex films by highlighting the crosslinked bond. *Ind. Crops Prod.*, 204, 117290, 2023.
45. Vilanova, C., Marín, M., Baixeras, J., Latorre, A., Porcar, M., Selecting Microbial Strains from Pine Tree Resin: Biotechnological Applications from a Terpene World. *PLoS One*, 9, 6, e100740, 2014.
46. Shah, A.A., Hasan, F., Shah, Z., Mutiullah, Hameed, A., Degradation of polyisoprene rubber by newly isolated Bacillus sp. AF-666 from soil. *Appl. Biochem. Microbiol.*, 48, 37–42, 2012.
47. Linos, A., Berekaa, M.M., Reichelt, R., Keller, U., Schmitt, J., Flemming, H.-C., Kroppenstedt, R.M., Steinbüchel, A., Biodegradation of cis-1, 4-polyisoprene

rubbers by distinct actinomycetes: microbial strategies and detailed surface analysis. *Appl. Environ. Microbiol.*, 66, 4, 1639–1645, 2000.
48. Strotmann, U., Reuschenbach, P., Schwarz, H., Pagga, U., Development and evaluation of an online CO(2) evolution test and a multicomponent biodegradation test system. *Appl. Environ. Microbiol.*, 70, 8, 4621–4628, 2004.
49. Zhang, C., Dan, Y., Peng, J., Turng, L.-S., Sabo, R., Clemons, C., Thermal and Mechanical Properties of Natural Rubber Composites Reinforced with Cellulose Nanocrystals from Southern Pine. *Adv. Polym. Technol.*, 33, S1, 21448, E1–E7, 2014.
50. Kumar C M, P., B, A.R., Kumar, M., R., C.P., Natural nano-fillers materials for the Bio-composites: A review. *J. Indian Chem. Soc.*, 99, 10, 100715, 2022.
51. Sienkiewicz, N., Dominic, M., Parameswaranpillai, J., Natural Fillers as Potential Modifying Agents for Epoxy Composition: A Review. *Polymers*, 14, 2, 265, 2022.
52. Haider, T.P., Völker, C., Kramm, J., Landfester, K., Wurm, F.R., Plastics of the Future? The Impact of Biodegradable Polymers on the Environment and on Society. *Angew. Chem. Int. Ed.*, 58, 1, 50–62, 2019.
53. Varyan, I., Kolesnikova, N., Xu, H., Tyubaeva, P., Popov, A., Biodegradability of Polyolefin-Based Compositions: Effect of Natural Rubber. *Polymers (Basel)*, 14, 230, 2022, DOI: 10.3390/polym14030530.
54. Clark, T., *Enhancing the Biodegradation of Waste Rubber*. International Latex Conference, Fairlawn Ohio, 2015.
55. Zhao, X., Ji, K., Kurt, K., Cornish, K., Vodovotz, Y., Optimal mechanical properties of biodegradable natural rubber-toughened PHBV bioplastics intended for food packaging applications. *Food Packag. Shelf Life*, 21, 100348, 2019.
56. Yahya, S.R., Rashid, A.A., Azahari, B., Soil Burial Studies for Biodegradation of Natural Rubber Latex Films. *Adv. Mater. Res.*, 844, 406–409, 2014.
57. Sareena, C., Sreejith, M.P., Ramesan, M.T., Purushothaman, E., Biodegradation behaviour of natural rubber composites reinforced with natural resource fillers–monitoring by soil burial test. *J. Reinf. Plast. Compos.*, 33, 5, 412–429, 2014.
58. Ramasamy, S., Ismail, H., Munusamy, Y., Soil burial, tensile properties, morphology, and biodegradability of (rice husk powder)-filled natural rubber latex foam. *J. Vinyl Add. Tech.*, 21, 2, 128–133, 2015.
59. de Lima, D.R., Vieira, I.R.S., da Rocha, E.B.D., de Sousa, A.M.F., da Costa, A.C.A., Furtado, C.R.G., Biodegradation of natural rubber latex films by highlighting the crosslinked bond. *Ind. Crops Prod.*, 204, 117290, 2023.
60. Daud, S., Ismail, H., Bakar, A.A., Soil burial study of palm kernel shell-filled natural rubber composites: The effect of filler loading and presence of silane coupling agent. *BioResources*, 11, 4, 8686–8702, 2016.
61. Mistriotis, A., Papardaki, N.-G., Provata, A., Biodegradation of Cellulose in Laboratory-Scale Bioreactors: Experimental and Numerical Studies. *J. Polym. Environ.*, 27, 12, 2793–2803, 2019.

62. Chen, Z., Aziz, T., Sun, H., Ullah, A., Ali, A., Cheng, L., Ullah, R., Khan, F.U., Advances and Applications of Cellulose Bio-Composites in Biodegradable Materials. *J. Polym. Environ.*, 31, 6, 2273–2284, 2023.
63. Suppanucroa, N., Nimpaiboon, A., Boonchuay, K., Khamkeaw, A., Phisalaphong, M., Green composite sponge of natural rubber reinforced with cellulose filer using alginate as a dispersing agent. *J. Mater. Res. Technol.*, 27, 3119–3130, 2023.
64. Misman, M. and Azura, A., Utilization of Starch to Accelerate the Growth of Degrading Microorganisms on the Surface of Natural Rubber Latex Films. *J. Chem. Chem. Eng.*, 7, 137–144, 2013.
65. Riyajan, S.-A., Robust and biodegradable polymer of cassava starch and modified natural rubber. *Carbohydr. Polym.*, 134, 267–277, 2015.
66. Izmar, M.H., Afiq, M.M., Azura, A.R., Effects of different additions of sago starch filler on physical and biodegradation properties of pre-vulcanized NR latex composites. *Composites, Part B*, 43, 7, 2746–2750, 2012.
67. Shinoj, S., Visvanathan, R., Panigrahi, S., Kochubabu, M., Oil palm fiber (OPF) and its composites: A review. *Ind. Crops Prod.*, 33, 1, 7–22, 2011.
68. Jacob, M., Thomas, S., Varughese, K.T., Biodegradability and aging studies of hybrid biofiber reinforced natural rubber biocomposites. *J. Biobased Mater. Bioenergy*, 1, 1, 118–126, 2007.
69. Muniandy, K., Ismail, H., Othman, N.J.B., Biodegradation, morpholgical, and ftir study of rattan powder-filled natural rubber composites as a function of filler loading and a silane coupling agent. *BioResources*, 7, 1, 0957–0971, 2012.
70. Abdel-Hakim, A., El-Wakil, A.E.-A.A., El-Mogy, S., Halim, S., Effect of fiber coating on the mechanical performance, water absorption and biodegradability of sisal fiber/natural rubber composite. *Polym. Int.*, 70, 9, 1356–1366, 2021.
71. Sintharm, P. and Phisalaphong, M., Green Natural Rubber Composites Reinforced with Black/White Rice Husk Ashes: Effects of Reinforcing Agent on Film's Mechanical and Dielectric Properties. *Polymers (Basel)*, 13, 6, 882, 2021.
72. Bouthegourd, E., Rajisha, K.R., Kalarical, N., Saiter, J.M., Thomas, S., Natural rubber latex/potato starch nanocrystal nanocomposites: Correlation morphology/electrical properties. *Mater. Lett.*, 65, 23, 3615–3617, 2011.
73. Hambir, S., Bulakh, N., Jog, J.P., Polypropylene/clay nanocomposites: Effect of compatibilizer on the thermal, crystallization and dynamic mechanical behavior. *Polym. Eng. Sci.*, 42, 9, 1800–1807, 2002.
74. Rajisha, K.R., Maria, H.J., Pothan, L.A., Ahmad, Z., Thomas, S., Preparation and characterization of potato starch nanocrystal reinforced natural rubber nanocomposites. *Int. J. Biol. Macromol.*, 67, 147–153, 2014.
75. Gupta, A., Pal, A.K., Patwa, R., Dhar, P., Katiyar, V., Green composites with excellent barrier properties, in: *Advanced Green Composites*, p. 321, Wiley, J., Sons, I.H., NJ, USA, 2018.

76. Jiang, W., Shen, P., Yi, J., Li, L., Wu, C., Gu, J., Surface modification of nanocrystalline cellulose and its application in natural rubber composites. *J. Appl. Polym. Sci.*, 137, 39, 49163, 2020.
77. Bras, J., Hassan, M.L., Bruzesse, C., Hassan, E.A., El-Wakil, N.A., Dufresne, A., Mechanical, barrier, and biodegradability properties of bagasse cellulose whiskers reinforced natural rubber nanocomposites. *Ind. Crops Prod.*, 32, 3, 627–633, 2010.
78. Abraham, E., Elbi, P.A., Deepa, B., Jyotishkumar, P., Pothen, L.A., Narine, S.S., Thomas, S., X-ray diffraction and biodegradation analysis of green composites of natural rubber/nanocellulose. *Polym. Degrad. Stab.*, 97, 11, 2378–2387, 2012.
79. Xu, C., Chen, C., Wu, D., The starch nanocrystal filled biodegradable poly(ε-caprolactone) composite membrane with highly improved properties. *Carbohydr. Polym.*, 182, 115–122, 2018.
80. Azarian, M.H., Boochathum, P., Kongsema, M., Biocompatibility and biodegradability of filler encapsulated chloroacetated natural rubber/polyvinyl alcohol nanofiber for wound dressing. *Mater. Sci. Eng., C*, 103, 109829, 2019.

Index

Printed and bound by CPI Group (UK) Ltd, Croydon, CR0 4YY
17/09/2026
14964280-0001